T0211330

PARTIAL DIFFERENTIAL EQUATIONS
IN FLUID DYNAMICS

This book is concerned with partial differential equations applied to fluids problems in science and engineering. This work is designed for two potential audiences. First, this book can function as a text for a course in mathematical methods in fluid mechanics in nonmathematics departments or in mathematics service courses. The authors have taught both. Second, this book is designed to help provide serious readers of journals (professionals, researchers, and graduate students) in analytical science and engineering with tools to explore and extend the missing steps in an analysis. The topics chosen for the book are those that the authors have found to be of considerable use in their own research careers. These topics are applicable in many areas, such as aeronautics and astronautics; biomechanics; chemical, civil, and mechanical engineering; fluid mechanics; and geophysical flows. Continuum ideas arise in other contexts, and the techniques included have applications there as well.

Isom H. Herron is a Professor of Mathematical Sciences at Rensselaer Polytechnic Institute. After completing his Ph.D. at Johns Hopkins and a postdoctoral position at California Institute of Technology, he was in the Mathematics Department at Howard University for many years, and he has held visiting appointments at Northwestern University, University of Maryland, MIT, and Los Alamos National Laboratory. Professor Herron's research is in one of the richest areas of applied mathematics: the theory of the stability of fluid flows. Common applications are to phenomena in the atmosphere and the oceans, to problems of the motion of ships and aircraft, and to internal machinery. Modern approaches involve new techniques in operator theory, energy methods, and dynamical systems. His current research interests are in stability of rotating magneto-hydrodynamic flows and more complicated geophysical flows, such as groundwater, for which mathematical models are still being developed.

Michael R. Foster is Adjunct Professor of Mathematical Sciences at Rensselaer Polytechnic Institute and Emeritus Professor of Aeronautical Engineering at The Ohio State University, where he has been a faculty member since 1970, following the completion of his Ph.D. at California Institute of Technology and a postdoctoral appointment at MIT. Professor Foster has held visiting appointments at Lehigh University, University College London, the University of Dundee, and the University of Manchester. His specialty is theoretical fluid dynamics, generally using a combination of asymptotic methods and computation. Professor Foster focused for many years in geophysical fluid dynamics, working with colleagues at Arizona State University and the University of Dundee; more recently, his collaborations have been at the University of Dundee and at the University of Manchester and with colleagues in mathematics at Ohio State. His more recent research has been in two areas: directional solidification problems, including Bridgman devices and dendritic crystal growth, and boundary layers in dilute suspensions, especially the singularities of standard models.

Partial Differential Equations in Fluid Dynamics

Isom H. Herron

Rensselaer Polytechnic Institute

Michael R. Foster

The Ohio State University

and

Rensselaer Polytechnic Institute

CAMBRIDGE
UNIVERSITY PRESS

CAMBRIDGE
UNIVERSITY PRESS

32 Avenue of the Americas, New York NY 10013-2473, USA

Cambridge University Press is part of the University of Cambridge.

It furthers the University's mission by disseminating knowledge in the pursuit of
education, learning and research at the highest international levels of excellence.

www.cambridge.org
Information on this title: www.cambridge.org/9781107427211

First published 2008
First paperback edition 2014

A catalogue record for this publication is available from the British Library

Library of Congress Cataloguing in Publication data

Herron, Isom H., 1946–
Partial differential equations in fluid dynamics / Isom H. Herron, Michael R. Foster.
   p.   cm.
Includes bibliographical references and index.
ISBN 978-0-521-88824-0 (hardcover)
1. Fluid dynamics –Mathematics – Textbooks.   2. Differential equations, Partial – Textbooks.
I. Foster, Michael R., 1943–   II. Title.
TA357.H43   2008
532'.0501515353–dc22   2008014569

ISBN 978-0-521-88824-0 Hardback
ISBN 978-1-107-42721-1 Paperback

Both of us would like to thank our wives, Myra Herron and Jeanne Foster,
without whose love and understanding this project would not
have been possible.

Contents

Preface

In this book, it is our intent to equip graduate students in applied mathematics and engineering with a range of classical analytical methods for the solution of partial differential equations. In our research specialties, numerical methods, on the one hand, and perturbation and variational methods, on the other, constitute contemporary tools that are explicitly not covered in this book, since there are significant books that are devoted to those topics specifically.

This book grew not from the authors' desire to write a textbook but rather from many years for each of us in compiling notes to be distributed to our graduate students in courses devoted to the solution of partial differential equations. One of us taught mostly engineers (MRF at The Ohio State University), and the other (IH at Howard University and later at Rensselaer Polytechnic Institute) mostly mathematicians. It surprised us to learn, on becoming re-acquainted at RPI, how similar are our perspectives about this material, and in particular the level of rigor with which it ought to be presented. Further, both of us wanted to create a book that would include many of the techniques that we have learned one way or another but are quite simply not in books.

The topics chosen for the book are those that we have found to be of considerable use in our own research careers. These are topics that are applicable in many areas, such as aeronautics and astronautics; biomechanics; chemical, civil, and mechanical engineering fluid mechanics; and geophysical flows. Continuum ideas arise in other contexts, and the techniques we expound have applications there as well. The first chapter is a review of complex variables. Our view is that this topic is basic to applied mathematics and deserves to be solidified early on. An immediate application of these ideas occurs in Chapter 2, on special functions. Chapter 3, "Eigenvalue Problems," is based on attempting to understand a particular problem, the stability/instability of flow between rotating cylinders. It leads to a development of Sturm-Liouville theory as a means of understanding Synge's (1933) proof of Rayleigh's criterion. Chapter 4 introduces and discusses the use of Green's functions for solving non-homogeneous linear boundary-value problems. There is an appendix with background material on linear differential equations. Chapters 5 and 6 develop the techniques of solving partial differential equations by Laplace transforms and Fourier transforms, respectively. Applications of these techniques are made to several different problems in Chapter 7, the types of problems that are normally only found in the research literature. Chapter 8, asymptotic methods for definite integrals, could logically come earlier in the book. However, it appears later so that techniques for differential equations could be available as examples. Each chapter features exercises that extend both the theory and applications involved, often exposing the reader to areas outside the text.

Acknowledgments

As usual when authors embark on such a project, there are many persons who have been directly or indirectly responsible for the shape and (hopefully!) usefulness of this book. We list here a few, without much comment, though such might be warranted. MRF: my list is headed by Philip Saffman, my Ph.D. advisor, who started me thinking about the world of fluid dynamics in mathematical terms; my admiration for him as a researcher, a teacher, and a person has not dimmed. Others are, in no particular order, Keith Stewartson, Odus Burggraf, Richard Bodonyi, David Walker, Peter Davies, Frank Smith, Greg Baker, Saleh Tanveer, Peter Duck, and Rich Hewitt. IH: as my teachers, I owe a debt of gratitude to F. B. Hildebrand, L. N. Howard, and S. H. Davis. We would both like to thank our many graduate students over the years for suggestions, criticisms, typo corrections, and the like.

Review of Analytic Function Theory

1.1 Preamble

Complex analysis is the foundation for everything in this book. Special functions, integral transforms, Green's functions, orthogonal function expansions, and classical asymptotic techniques like steepest descent cannot be properly understood or used without a thorough understanding of analytic function theory. We provide here only a review; the student for whom this is a first exposure to the subject ought to consult other texts that treat these topics exclusively. There are a vast number of such books – many of them are very helpful. Among them, we cite a very complete book for applied mathematicians, engineers, and scientists by Carrier, Krook, and Pearson and the exhaustive treatise by Markushevich.

Because the subject was given birth by the need to solve problems in fluid dynamics and electromagnetism, there is also a significant library of books on those topics that make intensive use of complex-variable methods. In the area with which the authors are most familiar, the classic book on hydrodynamics by Milne-Thomson is a great resource. In particular, there is much attention paid there to conformal mappings – a topic not discussed here because it is not directly helpful in most solution methods for partial differential equations – with some obvious exceptions.

1.2 Fundamentals of Complex Numbers

A complex number, c, is defined by

$$c = a + ib, \quad i \equiv \sqrt{-1}, \tag{1.1}$$

where a and b are real numbers. The quantities a and b are called, respectively, the "real" and "imaginary" parts of the complex number, c. Thus, we will write $a = \Re\text{eal}(c)$ and $b = \Im\text{mag}(c)$. It is often convenient to work with the *complex conjugate* of the number, c, which we will denote always in this book by the notation $\bar{c}$. (In some books, the notation c^* is used.) The complex conjugate of c is obtained by

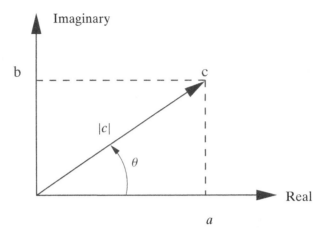

Figure 1.1. Argand diagram for a complex number.

everywhere replacing i by $(-i)$. Hence, $\bar{c} = a - ib$. The *modulus* of the complex number is a norm, defined by $|c| = \sqrt{a^2 + b^2}$. Note, then, that

$$|c|^2 = a^2 + b^2 = c\bar{c} = \bar{c}c.$$

Complex numbers may also then be written in *polar* form, so that

$$c = |c|\cos\theta + i|c|\sin\theta = |c|e^{i\theta},$$

with the angle as indicated in Figure 1.1. This angle θ will be called the "argument" of z. Hence, $\theta = \arg(z)$.

Any complex number can be displayed in a plane of numbers by specifying either (a, b) or $(|c|, \theta)$. The horizontal axis is the axis of *real parts*, and the vertical axis is for the *imaginary parts*. The conjugate of any complex number is then located at its reflection in the horizontal axis.

For doing arithmetic with these numbers, clearly addition and multiplication result in

$$c_1 + c_2 = (a_1 + a_2) + i(b_1 + b_2),$$
$$c_1 \times c_2 = (a_1 a_2 - b_1 b_2) + i(a_1 b_2 + a_2 b_1).$$

If a complex number is written as the sum of a real part and i times its imaginary part, then division can be done by multiplying numerator and denominator by the complex conjugate of the denominator. Consider the following example:

$$\frac{1 + 2i}{2 - i} = \frac{1 + 2i}{2 - i}\frac{2 + i}{2 + i} = \frac{5i}{2^2 + 1^2} = i.$$

1.2.1 Complex Roots and Logarithms

Before turning to more advanced questions of differentiation, integration, and so on, we note something quite useful about roots that is mysterious with real quantities but makes sense in the complex plane. Suppose we wish to find the cube root of 8. The real root, the one the calculator gives you, or you have memorized, is 2. However, according to the fundamental theorem of algebra, there are *three* cube roots of 8. We can find them as follows:

$$z = 8^{\frac{1}{3}} = \left(8e^{2n\pi i}\right)^{\frac{1}{3}}. \tag{1.2}$$

The quantity n is an integer. Since

$$e^{i\theta} = \cos\theta + i\sin\theta,$$

we see that $\exp(2n\pi i) = \sin(2n\pi) + i\sin(2n\pi) \equiv 1$. Continuing with what we wrote in (1.2), we have

$$z = 8^{\frac{1}{3}} = \left(8e^{2n\pi i}\right)^{\frac{1}{3}} = 2e^{2n\pi i/3}.$$

Therefore, there are *three* roots, spaced about the origin 120 degrees apart.

Consider, as a second example, the roots of the polynomial

$$\lambda^4 + 16 = 0. \tag{1.3}$$

Rearranging,

$$\lambda = (-16)^{\frac{1}{4}} = \left(16e^{(2n\pi+\pi)i}\right)^{\frac{1}{4}} = 2e^{i\pi/4+n\pi i/2}.$$

So, the four roots of the quartic equation all have modulus 2 and are spaced around the origin 90 degrees apart, with the $n = 0$ one located at $z = \sqrt{2}(1 + i)$.

The exponential function e^z is extended to complex numbers as

$$e^z = e^{x+iy} = e^x(\cos y + i\sin y),$$

so that Euler's identity is still valid. Notice that $e^z = 1$, if $z = 2n\pi i$, and n is any integer. Thus, e^z is periodic with period $2\pi i$,

$$e^{z+2\pi i} = e^z.$$

The logarithm is the inverse of the exponential function, defined as the "multivalued function":

$$\log z = \log|z| + i\arg z + 2\pi ni, \quad n = 0, \pm 1, \pm 2, \ldots.$$

Many authors use $\ln|z|$ as an equivalent name for $\log|z|$. In any case, the *principal branch of the logarithm* is defined by taking $n = 0$ and $-\pi < \arg z < \pi$. (Sometimes, it may be more suitable to take the principal branch to be given by $0 < \arg z < 2\pi$.) Thus, for example,

$$\log i = \log|i| + i\arg i = \log 1 + i\pi/2 = \pi i/2,$$

with either choice of principal branch, while

$$\log(-1) = \log 1 + i\arg(-1) = \pi i,$$

with the second choice.

Fractional powers are defined through the logarithm

$$z^\alpha = e^{\alpha \log z}, \quad z \neq 0.$$

So z^α is a multivalued function when α is not an integer. For example,

$$i^i = e^{i \log i} = e^{i(\log 1 + i \arg i + 2n\pi i)}, \quad n = 0, 1, 2, \ldots$$

For its principal value, take $n = 0$ and $i^i = e^{-\pi/2}$. Much of this type of analysis extends to the inverse trigonometric and inverse hyperbolic functions. For example, since

$$\sin w = \frac{e^{iw} - e^{-iw}}{2i} = z,$$

say, then

$$w = \sin^{-1} z = \frac{1}{i} \log \left[iz + (1 - z^2)^{1/2} \right].$$

If, $z \neq \pm 1$, $(1 - z^2)^{1/2}$ has two possible values. In particular, to evaluate $\sin^{-1}(2)$, we have

$$\sin^{-1}(2) = \frac{1}{i} \log \left[2i \pm i\sqrt{3} \right]$$

$$= \begin{cases} \frac{1}{i} \left(\log(2 - \sqrt{3}) + i \left(\frac{\pi}{2} + 2n\pi \right) \right) \\ \frac{1}{i} \left(\log(2 + \sqrt{3}) + i \left(\frac{\pi}{2} + 2n\pi \right) \right) \end{cases}, \quad n = 0, \pm 1, \pm 2, \ldots$$

$$= \frac{\pi}{2} + 2n\pi - i \log(2 - \sqrt{3}), \ \frac{\pi}{2} + 2n\pi + i \log(2 + \sqrt{3}), \ n = 0, \pm 1, \pm 2, \ldots.$$

Now, observe that since $(2 + \sqrt{3})(2 - \sqrt{3}) = 1$ and $\log(2 + \sqrt{3}) = -\log(2 - \sqrt{3})$, therefore

$$\sin^{-1}(2) = \frac{\pi}{2} + 2n\pi \pm i \log(2 + \sqrt{3}), \quad n = 0, \pm 1, \pm 2, \ldots.$$

1.3 Analytic Functions

Just as the function $y = f(x)$ assigns one real number, y, to another real number, x, so we can define functions of a complex variable, say, $z = x + iy$. Then, one analogy that might be drawn is to a vector function of two variables (x, y). Hence, for example, the function $f(z) = z^2$ takes the value 1 for $z = 1$, the value -4 for $z = 2i$, and the value $(-3 + 4i)$ for $z = 1 + 2i$. The independent variable is defined over a plane like that shown in Figure 1.1. As in the case of any complex number, this function can be split into real and imaginary parts, so that we could write the function in terms of two real functions of real variables; that is,

$$f = U(x, y) + i V(x, y), \quad z = x + iy. \tag{1.4}$$

1.3.1 Limits and Continuity

The theory of complex functions is so closely tied to that of functions of two real variables, and functionally similar it turns out, to that of a single real variable, that many of the underlying processes are often taken for granted. It is important to note that a sequence of points $z_n \to z_0$, if and only if, $(x_n + i y_n) \to (x_0 + i y_0)$. For a function $f(z)$, the process $\lim_{z \to z_0} f(z)$ takes place as

$$\lim_{(x,y) \to (x_0, y_0)} [U(x, y) + i V(x, y)] = \lim_{(x,y) \to (x_0, y_0)} U(x, y) + i \lim_{(x,y) \to (x_0, y_0)} V(x, y).$$

Thus, the definition of *continuity*,

$$\lim_{z \to z_0} f(z) = f(z_0),$$

means that

$$\lim_{(x,y) \to (x_0, y_0)} [U(x, y) + i V(x, y)] = U(x_0, y_0) + i V(x_0, y_0),$$

as well. In particular, this means that the indicated limits must be independent of the direction in which they are taken.

So, for elementary functions, such as polynomials, the limits are as to be expected:

$$\lim_{z \to 3i} (z^2 + 9) = 0.$$

Even for quotients, one may conclude that

$$\lim_{z \to 3i} \frac{z^2 + 9}{z - 3i} = \lim_{z \to 3i} \frac{(z + 3i)(z - 3i)}{z - 3i}$$

$$= \lim_{z \to 3i} (z + 3i) = 6i.$$

Continuous functions are defined by their limits so $f(z) = |z|$ is everywhere continuous. However, $f(z) = \arg z$ is discontinuous at each point on the nonpositive real axis, with a cut on the non-positive real axis.

1.3.2 Differentiation

Let us consider the derivative of this function of the complex variable, z, by analogy with real variables. Then,

$$\frac{df}{dz} = \lim_{\Delta z \to 0} \left[\frac{f(z + \Delta z) - f(z)}{\Delta z} \right] \tag{1.5}$$

$$= \lim_{\Delta z \to 0} \left[\frac{U(x + \Delta x, y + \Delta y) + i V(x + \Delta x, y + \Delta y) - U(x, y) - i V(x, y)}{\Delta x + i \Delta y} \right]$$

$$= \lim_{\Delta z \to 0} \left[\frac{U(x + \Delta x, y + \Delta y) - U(x, y) + i (V(x + \Delta x, y + \Delta y) - V(x, y))}{\Delta x + i \Delta y} \right],$$

after collecting the real and imaginary parts. The partial derivatives U_x and V_x are just those when y is held constant; $\Delta y = 0$. Likewise, U_y and V_y arise when x is held constant; $\Delta x = 0$. The result is that the derivative df/dz may be written in two ways:

$$\frac{df}{dz} = \begin{cases} U_x + iV_x, & \text{if } \Delta y = 0 \\ V_y - iU_y, & \text{if } \Delta x = 0 \end{cases}. \tag{1.6}$$

Using Taylor's theorem for a function of two variables, the expression (1.5) can be simplified. That is, the limit may be taken along some line in the complex plane along which both $\Delta x \to 0$ and $\Delta y \to 0$. The difference quotient then becomes

$$\frac{df}{dz} = \lim_{\Delta z \to 0} \left[\frac{(U_x + iV_x)\Delta x + i(V_y - iU_y)\Delta y + \text{higher-order terms}}{\Delta x + i\Delta y} \right],$$

which may be simplified and the limit performed. The result is

$$\frac{df}{dz} = \frac{(U_x + iV_x) + iS(V_y - iU_y)}{1 + iS}, \tag{1.7}$$

where $S \equiv \Delta y/\Delta x$ is the slope of the line in the complex z-plane along which the limit is performed. The cases where $S = 0$ and $S = \infty$ were obtained in (1.6). Taking a cue from several real variables, it would be sensible if the derivative of a complex function at any location z were independent of the direction of approach to that point – a kind of generalization of the idea of continuity for real functions. Thus, we give the following definition:

Definition 1.1. *A function f, of a complex variable, z, is "analytic" in a region $\mathcal{R}$ of the complex plane if its derivative exists and is single-valued in that region.*

That means, from the above results, that f' should be independent of the constant S, and inspection shows that such independence can be achieved if and only if $U_x + iV_x = V_y - iU_y$. On equating real and imaginary parts, we obtain the Cauchy–Riemann equations, and the associated theorem is as follows in Theorem 1.1.

Theorem 1.1. *A necessary and sufficient condition for a function, $f = U + iV$, of a complex variable z to be analytic in a region $\mathcal{R}$ of the complex plane is that*

$$\frac{\partial U}{\partial x} = \frac{\partial V}{\partial y},$$

$$\frac{\partial V}{\partial x} = -\frac{\partial U}{\partial y}. \tag{1.8}$$

everywhere in $\mathcal{R}$, and U_x, U_y, V_x and V_y exist in $\mathcal{R}$.

Equations (1.8) are known as the Cauchy–Riemann equations.

Going back to (1.7), and inserting the Cauchy–Riemann equations, we regain the formulas for the derivative – when it exists, namely,

$$\frac{df}{dz} = \frac{\partial U}{\partial x} + i\frac{\partial V}{\partial x} = \frac{\partial V}{\partial y} - i\frac{\partial U}{\partial y}.$$

Note that since $z = x + iy$ and $\bar{z} = x - iy$, these equations may be inverted, so that $x = (z + \bar{z})/2$ and $y = (z - \bar{z})/(2i)$. Then, from (1.4), in general, f is a function of both z and $\bar{z}$. However, it can easily be shown, that the derivative cannot be independent of S unless f is independent of $\bar{z}$. Thus, we have the result (see Exercise 1.3)

Corollary 1.1. *A necessary condition for a function of a complex variable to be analytic anywhere in the complex plane is that $\partial f / \partial \bar{z} \equiv 0$ in the plane.*

Let's put it a different way: Suppose we choose any two functions U and V of x and y. Then, we can create, from those functions, a function of a complex variable by writing $f = U + iV$ as in (1.4). Then, as noted above, this function will in general be a function of both z and $\bar{z}$. What Corollary 1.1 says is that if $\bar{z}$ appears in the expression for f, it *cannot* be an analytic function; it is impossible. (A necessary condition!) If f depends on z alone, then the function *may* be analytic in some portion of the complex z-plane.

Now, for some examples. Consider the function $f = x^2 - y^2 + 2ixy$. The reader can easily verify that the Cauchy–Riemann equations are indeed satisfied. The partial derivatives all exist in the *finite* z-plane, so we conclude that this function is analytic in the finite z-plane. Such a function is said to be an "entire function." Note, in terms of the corollary noted above, that this function can also be written as $f = z^2$. As a second example, consider $f = x^2 - y^2 - 2ixy$. The Cauchy–Riemann equations (1.8) are NOT satisfied for any z. Hence, the function is not analytic anywhere in the plane. (In fact, this function is $f = \bar{z}^2$ and hence cannot be analytic!) As a final example, consider the function $f = 1/z$. The real and imaginary parts are given by $U = x/(x^2 + y^2)$ and $V = -y/(x^2 + y^2)$. Again, the Cauchy–Riemann equations are satisfied in the plane, but note that the partial derivatives are unbounded at the origin. Hence, this function $f = 1/z$ is analytic everywhere except at $z = 0$. The origin, in this case, is a *singularity* of the function.

1.3.3 Harmonic Functions

For an analytic function $f = U + iV$, note that

$$\frac{\partial^2 U}{\partial x^2} = \frac{\partial^2 V}{\partial x \partial y} = -\frac{\partial^2 U}{\partial y^2} \implies \nabla^2 U = 0.$$

Both Cauchy–Riemann equations (1.8) were used in this derivation. In the same way, V can also be shown to be a solution of Laplace's equation. In this derivation, it was tacitly assumed that if f is analytic, so is f'. This is true and will be demonstrated later in Section 1.4.

At a point $z_0 = x_0 + iy_0$, where $f'(z_0) = 0$, $\nabla U = 0$ and $\nabla V = 0$, so the point (x_0, y_0) is a critical point for both of the conjugate functions. Furthermore, since U and V are harmonic, the critical point will be a saddlepoint as long as f is not a constant.

The real and imaginary parts of polynomials are thereby harmonic functions. In a similar manner, with restrictions on the domains, other harmonic functions may be constructed. For example,

$$\log z = \log \sqrt{x^2 + y^2} + i \arctan\left(\frac{y}{x}\right),$$

has harmonic components on the whole plane, except at the origin.

1.3.4 Note on Fluid Dynamics

In incompressible fluid flow, we know that both the velocity potential, ϕ, and the stream function, ψ, obey Laplace's equation. So, we can build an analytic function $F = \phi + i\psi$ to describe the fluid flow. Note that Equations (1.8) are, with these functions,

$$u = \frac{\partial \phi}{\partial x} = \frac{\partial \psi}{\partial y},$$

$$v = \frac{\partial \phi}{\partial y} = -\frac{\partial \psi}{\partial x}.$$

Therefore, instead of working with real functions ϕ and ψ, we can deal with a *complex potential F*. From this derivative formula, the fluid velocity components are simply related to the derivative of F, that is,

$$F' = \phi_x + i\psi_x = u - iv.$$

So, for example, stagnation points in a flow are found by putting $F' = 0$.

A thorough development of these ideas can be found in many places; a good graduate-level reference is a book by Milne-Thomson.

1.4 Integration and Cauchy's Theorem

Consider a complex function, $f(z)$. Along some curve C from point a to point b in the plane, choose any set of $(N + 1)$ points $\{z_k\}$, separated by intervals $\{\Delta z_k\}$, where $\Delta z_k = z_{k+1} - z_k$, along a segment of the curve from $a = z_0$ to $z_N = b$, such that

$$\sup_N \sum_{k=0}^{N} |\Delta z_k|$$

is finite. The curve, is called a "rectifiable curve." Consider the sum

$$S_N \equiv \sum_{k=0}^{N} f(z_k)\Delta z_k. \tag{1.9}$$

In order for the terms in this sum to have meaning, we assume f to be continuous along the curve C. If we take the limit as $N \to \infty$, define the limit of (1.9) as an integral (Markushevich, Vol. I),

$$S_\infty = \int_a^b f(z)dz. \tag{1.10}$$

This definition of the integral is similar to that of a line integral in the plane. The ideas of line integration are often used in the evaluation of complex integrals. There is an additional potential tool in the Cauchy–Riemann equations. However, to employ the Cauchy–Riemann equations on a region $\mathcal{R}$ containing C, f should be analytic. In particular, it may be shown that this integral is independent of the path chosen between a and b, if f obeys the Cauchy–Riemann equations along C, that is, if the function is analytic along C. The region $\mathcal{R}$, whether bounded or unbounded must be *simply connected*, that is, whenever $\mathcal{R}$ contains a simple, rectifiable, closed curve C (*Jordan curve*), it also contains the interior of C.

It may also be demonstrated that the familiar antiderivative forms for standard functions of real variables remain valid for antiderivatives of analytic functions, so that the antiderivative of $\exp(kz)$, for example, is $\exp(kz)/k$.

It follows that *Cauchy's theorem* is crucial to being able to do integration in the complex plane. The theorem is stated as follows in Theorem 1.2.

Theorem 1.2. *Let $\mathcal{R}$ be a simply connected region. If a function f, is single-valued and analytic on and inside a closed, rectifiable path C in $\mathcal{R}$, then $\oint_C f(z)dz = 0$.*

A simple proof involving only Green's theorem in the plane is as follows:

Write the analytic function f as

$$f(z) = u(x, y) + iv(x, y).$$

Call the region inside C as D. Then using the ideas of line integration underlying Equation (1.10), we take $dz = dx + idy$ so that

$$\oint_C f(z)dz = \oint_C (udx - vdy) + i\oint_C (vdx + udy)$$

$$= -\iint_D \left(\frac{\partial v}{\partial x} + \frac{\partial u}{\partial y}\right)dxdy + i\iint_D \left(\frac{\partial u}{\partial x} - \frac{\partial v}{\partial y}\right)dxdy.$$

By application of Green's theorem and the Cauchy–Riemann equations, each of the last two integrals is zero and the result has been proved. The proof may be found in standard complex-variable texts (Markushevich, Vol. I).

Consider the integral of the function z^n, with n an integer, carried out around a circle of radius r, centered at the origin, and denoted here by C. Hence,

$$\oint_C z^n dz = ir^{n+1}\int_0^{2\pi} e^{i(n+1)\theta}d\theta. \tag{1.11}$$

The Cauchy theorem states only that this integral must be zero for $n \geq 0$. Clearly, on doing the integration, we see that the integral is exactly zero for all values of n different from -1. For $n = -1$, however, the value of the integral is $2\pi i$. Therefore, we can write

$$\oint_C z^n dz = \begin{cases} 0, & n \neq -1 \\ 2\pi i, & n = -1 \end{cases}. \tag{1.12}$$

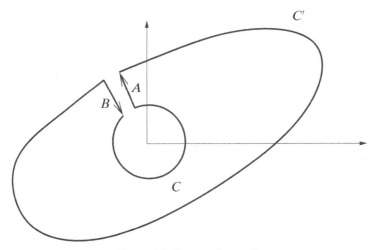

Figure 1.2. Integration path.

This result is actually more general than it appears. Consider another curve, say C', which is of arbitrary shape, which also encircles the origin and wholly contains C – f is analytic everywhere on C' and in the region between C and C'. As shown in Figure 1.2, consider now a new closed curve that is constructed by using C and C' except at a small segment of each, where the two curves are connected with two line segments A and B. Around a new closed curve formed going around C to A, along A to C', around C' in the clockwise direction, then back to C along B, the integral is zero, according to the Cauchy theorem. Since f is analytic between C and C', as we let A approach B, the integrals along those two paths just cancel, leaving the integrals along C and C' to add to zero. If we change the direction of integration along C' to the usual, positive (counterclockwise) direction, then we find that the integral around C' is exactly equal to the integral along C. Thus, the result in Equation (1.12) is independent of the shape of the path.

This discussion shows how Cauchy's theorem may be extended to regions with "holes," which are *multiply connected* regions.

The question naturally arises, does the condition $\oint_C f(z)dz = 0$ imply that f is analytic? The answer to this is known as **Morera's Theorem** and is simply

Theorem 1.3. *Let $\mathcal{R}$ be a open connected region. If a function f is continuous in $\mathcal{R}$ and $\oint_C f(z)dz = 0$ for all closed curves C in $\mathcal{R}$, then $f(z)$ is analytic in $\mathcal{R}$.*

The proof will not be presented here, but does depend on showing that such an $f(z)$ is the derivative of analytic function $F(z)$ and hence is analytic.

1.4.1 Cauchy's Integral Formula

The expression given by (1.12) is indicative of the importance of a more general line integral for analytic functions. This result, called "Cauchy's integral formula," is stated as

Theorem 1.4. *Let $\mathcal{R}$ be a simply connected region. If a function f, is single-valued and analytic on and inside a closed, rectifiable path C in $\mathcal{R}$, and if z_0 is a point within $\mathcal{R}$, then*

$$\frac{1}{2\pi i} \oint_C \frac{f(z)}{z - z_0} dz = f(z_0). \tag{1.13}$$

As a demonstration of the plausibility of this result, by the previous discussion deform the contour C into a circle C_0 of radius ε, about $z = z_0$:

$$\frac{1}{2\pi i} \oint_{C_0} \frac{f(z)}{z - z_0} dz = \frac{f(z_0)}{2\pi i} \oint_{C_0} \frac{dz}{z - z_0} + \frac{1}{2\pi i} \oint_{C_0} \frac{f(z) - f(z_0)}{z - z_0} dz,$$

so that on C_0, $z(t) = \varepsilon e^{it}$. The first term on the right has the value we desire so we need to conclude that the second term is 0. Since $f(z)$ is analytic within C_0, including at $z = z_0$, $\left| f(z) - f(z_0) \right| \leq M_\varepsilon$ on C_0. By this reasoning, the second term on the right is bounded by M_ε, which goes to 0 as $\varepsilon \to 0$.

Cauchy's integral formula may be extended under the same hypotheses to derivatives of f, that is,

$$f^{(m-1)}(z_0) = \frac{(m-1)!}{2\pi i} \oint_C \frac{f(z)}{(z - z_0)^m} dz, \tag{1.14}$$

$m = 1, 2, \ldots$. This formally looks as if one can differentiate under the integral sign, and it this result that permits the determination of the Taylor series for an analytic function that will be discussed in Section 1.4.2.

Remark 1.1. *Mean value property; critical points.*

A useful immediate application of Equation (1.13) is to derive the mean value property for analytic functions. Suppose C is a circle given by $z(t) = z_0 + re^{it}$, then

$$f(z_0) = \frac{1}{2\pi i} \int_0^{2\pi} \frac{f(z(t))}{z(t) - z_0} z'(t) dt.$$

Since $z'(t) = ire^{it} = i(z(t) - z_0)$,

$$f(z_0) = \frac{1}{2\pi} \int_0^{2\pi} f(z_0 + re^{it}) dt. \tag{1.15}$$

The value of f at the center of the circle is the average of the values over the circle. Of course these values are complex, but one real quantity associated with f is its modulus $|f|$. From the mean-value theorem follows the *maximum modulus property*, which is found by taking moduli of both sides of (1.15)

$$\left| f(z_0) \right| \leq \frac{1}{2\pi} \int_0^{2\pi} \left| f(z_0 + re^{it}) \right| dt \leq M, \tag{1.16}$$

where M is the maximum of $|f|$ on the circle. Moreover, if f is not constant, its maximum modulus is attained on the boundary of any disk surrounding it. Otherwise, if $\left| f(z_0) \right|$ is a maximum, then $\left| f(z_0 + re^{it}) \right| \leq \left| f(z_0) \right|$, and, by a continuity argument if strict inequality holds for any t, it must hold over some finite t-interval, and thus be shown to contradict (1.16). More generally, on a bounded region, the maximum modulus of a nonconstant analytic function will be attained on the boundary. There is also a *minimum modulus property*, that is, if f is analytic and $f \neq 0$ on the region, then $|f|$ has its minimum on the boundary. This follows by considering $1/f$ and applying the maximum modulus property.

Besides inherent interest in these results, these properties have important implications for harmonic functions. Thus, it is possible to show, from the maximum and minimum modulus properties, that if

$$f(z) = e^{g(z)},$$

where $g(z) = u + iv$ is analytic and nonconstant on a bounded region, then $|f| = e^{u(x,y)}$ must have its maximum and minimum values on the boundary since $f \neq 0$. Likewise, $u(x, y)$ must have its maximum and minimum values on the boundary. All of this allows us to conclude that, as a function of two real variables, *the critical points of the harmonic function u, where $\nabla u = \mathbf{0}$ must be saddlepoints*. Now, the critical points of v are precisely those of u, where $g' = 0$. Thus, we are able to conclude that *the critical points of an analytic function are saddlepoints for its real and imaginary parts*.

For example, if we examine the function $g(z) = z^2$, it has its real and imaginary parts:

$$u(x, y) = x^2 - y^2, \quad v(x, y) = 2xy.$$

Both of these quadric surfaces have a saddlepoint at $(x, y) = (0, 0)$, which corresponds to the point where $g' = 0$.

The critical point characterizations are important in the theory of asymptotic expansions of integrals, which is discussed in Chapter 8.

1.4.2 Singularities and Series

Singularities

The only function that is analytic over all of the complex plane, including infinity, is a constant. (This is a statement of the Liouville theorem – see (Carrier, Krook, and Pearson).) Thus, any other analytic function will have locations or regions in which it fails to be analytic. A location, z_o, where the function fails to be analytic is a *singularity* of the function. So, for example, $f = z^2$ has a singularity at infinity, and $f = 1/(z + i)$ has a singularity at $z = -i$.

Singularities are of two general types: isolated and nonisolated.

Definition 1.2. *A function has an isolated singularity at $z = z_o$ if the function is analytic and single-valued for values of z arbitrarily close to z_o.*

A singularity that does not have that property is not isolated. Nonisolated singularities are, for the most part, what we will call "branch points." Some examples follow below.

Isolated Singularity Examples.

1. A function $f(z)$ may be bounded near a singularity, say z_0. If this is so and the function's limit $z \to z_0$ exists, and that by assigning this limiting value to a function produces an analytic function, then the singularity is said to be *removable*. For example, the function

$$f(z) = \frac{\sin z}{z}$$

has a removable singularity at $z = 0$, because it has the limit 1 as $z \to 0$, and the resulting function is analytic at $z = 0$.

2. The function $f = 1/(z+i)$, which is unbounded at $z = -i$, has the derivative $f' = -1/(z+i)^2$, which also fails at $z = -i$, but is finite, single-valued, and arbitrarily close, say on $0 < |z+i| < \varepsilon$. Thus, the singularity is *isolated*.

3. The function $f = \exp(1/z)$ has derivative $f' = -\exp(1/z)/z^2$. Again, the derivative is unbounded at $z = 0$, but arbitrarily close, on $0 < |z| < \varepsilon$, it is finite and single-valued.

Nonisolated Singularity Example. Consider the function $f = z^{1/3}$. Its derivative is $f' = 1/(3z^{2/3})$, which is unbounded at the origin, and, hence, there is a singularity at the origin. If we use a polar representation for the variable $z = r \exp(i\theta)$, then the function is is given by

$$f = r^{1/3} e^{i\theta/3}.$$

At the location $z = 8 + i0$, $\theta = 0$, and therefore $f = 2$. If we encircle the origin, and return to the same point, $r = 8$ still, but now $\theta = 2\pi$, and therefore $f = 2 \exp(2\pi i/3) = i\sqrt{3} - 1$. What has happened is that there are two different values of the function at a location *not* close to the singularity at the origin. Hence, the singularity's effects are *not* isolated. Generally, fractional powers and logarithms are examples of such branch-point singularities – that is, the function (or its derivative) has multiple values, or "branches."

Isolated Singularities Come in Two Varieties, poles and essential singularities. Consider a function, $f(z)$, that has an isolated singularity at $z = a$.

Definition 1.3. *Define a new function $g \equiv (z - a)^m f(z)$. If this new function is analytic (by virtue of the removable singularity) at $z = a$ for some positive integer m, then f is said to have a pole of order n at $z = a$, where n is the smallest value of m that makes g analytic.*

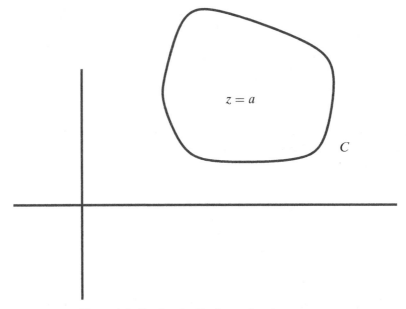

Figure 1.3. Region for Taylor series about $z = a$.

Isolated singularities that are not poles are essential singularities. In the previous examples, the function $f = 1/(z + i)$ has a pole of order 1 at $z = -i$ because $(z + i) f$ is analytic there. However, the function $f = \exp(1/z)$ has an essential singularity at the origin because $z^m f$ is not analytic there for any choice of m.

It is important to observe that L'Hospital's rule holds for analytic functions. That is, if f and g are both analytic at $z = z_0$, when $f(z_0) = g(z_0) = 0$, but when $g'(z_0) \neq 0$, then

$$\lim_{z \to z_0} \frac{f(z)}{g(z)} = \frac{f'(z_0)}{g'(z_0)}.$$

This is true because

$$\lim_{z \to z_0} \frac{f(z)}{g(z)} = \lim_{z \to z_0} \frac{f(z) - 0}{z - z_0} \cdot \frac{z - z_0}{g(z) - 0}$$

$$= \lim_{z \to z_0} \frac{\frac{f(z) - f(z_0)}{z - z_0}}{\frac{g(z) - g(z_0)}{z - z_0}} = \frac{f'(z_0)}{g'(z_0)}.$$

Taylor and Laurent Series

Theorem 1.5. *Suppose that a point $z = a$ lies inside a region $\mathcal{R}$, of the complex plane, bounded by some curve C, in which a function, f, is analytic. (See Figure 1.3.) Then, it is possible to write that function in a Taylor series:*

$$f = f(a) + f'(a)(z - a) + \frac{1}{2!} f''(a)(z - a)^2 + \frac{1}{3!} f'''(a)(z - a)^3 + \cdots . \qquad (1.17)$$

Analyticity in the neighborhood of the point z = a is necessary and sufficient for the existence of such a series.

For example, the function $\sin z$ has the series

$$\sin z = z - \frac{z^3}{3!} + \frac{z^5}{5!} + \cdots .$$

We can use the *ratio test for series convergence*, which states that a series $\sum q_n$ is absolutely convergent if $\lim_{n \to \infty} |q_{n+1}|/|q_n| < 1$. In this case, $|q_{n+1}|/|q_n| = |z|^2/[(2n+3)(2n+2)]$, which goes to zero for any finite $|z|$, hence the series has an infinite radius of convergence – which makes sense because $\sin z$ has no singularities in the finite plane.

As a second example, consider the function $f = 1/(z+i)$. Using the standard formula for the sum of a geometric series, we can deduce its Taylor series about the origin,

$$\frac{1}{z+i} = \frac{1}{i} \frac{1}{1-iz} = -i + z + iz^2 + \cdots . \tag{1.18}$$

The ratio test gives $|q_{n+1}|/|q_n| = |z|$, so that the radius of convergence is 1 – exactly the distance from the origin to the singularity at $z = -i$.

The derivation of the Taylor series (1.17) for analytic functions may be carried out as follows. Change the notation in (1.13) to read

$$f(z) = \frac{1}{2\pi i} \oint_C \frac{f(\xi)}{\xi - z} d\xi.$$

Consider point z such that $|z - a| < |\xi - a|$ for some a inside C. Then,

$$f(z) = \frac{1}{2\pi i} \oint_C \frac{f(\xi)}{\xi - a} \left(\frac{\xi - a}{(\xi - a) - (z - a)} \right) d\xi$$

$$= \frac{1}{2\pi i} \oint_C \frac{f(\xi)}{\xi - a} \frac{d\xi}{\left[1 - \frac{z-a}{\xi-a} \right]}.$$

Now

$$\left| \frac{z - a}{\xi - a} \right| < 1,$$

so

$$\left[1 - \frac{z - a}{\xi - a} \right]^{-1} = \sum_{n=0}^{\infty} \left(\frac{z - a}{\xi - a} \right)^n,$$

and

$$f(z) = \frac{1}{2\pi i} \oint_C \frac{f(\xi)}{\xi - a} \sum_{n=0}^{\infty} \left(\frac{z - a}{\xi - a} \right)^n d\xi$$

$$= \sum_{n=0}^{\infty} c_n (z - a)^n,$$

for some constants c_n. Comparing those terms with (1.14) we see that

$$c_n = \frac{1}{2\pi i} \oint_C \frac{f(\xi)}{(\xi - a)^{n+1}} d\xi = \frac{f^{(n)}(a)}{n!}, \quad n = 0, 1, \ldots. \tag{1.19}$$

This expression for the Taylor series (1.17) is identical to the series for real functions, which means, of course, that use of the integral representation (1.19) is not usually needed in practice. Thus, for the function

$$f(z) = \frac{\log(1 + z)}{z},$$

which has a removable singularity at $z = 0$ and branch-point singularity at $z = -1$, one may compute, valid on $0 < |z| < 1$,

$$\frac{\log(1 + z)}{z} = \frac{z - \frac{z^2}{2} + \frac{z^3}{3} + \cdots + \frac{(-1)^n z^n}{n} + \cdots}{z}$$

$$= 1 - \frac{z}{2} + \frac{z^2}{3} + \cdots + \frac{(-1)^n z^{n-1}}{n} + \cdots.$$

Consequently $\lim_{z \to 0} \frac{\log(1+z)}{z} = 1$.

A Taylor series has a generalization in the theory of complex variables that does not arise in real variable theory. Consider a function f that has an isolated singularity at $z = a$. Let C_i be a closed curve completely surrounding $z = a$ and arbitrarily close to it. Suppose that f is analytic exterior to C_i but is inside of another closed curve C_o. So, the curve C_i lies wholly inside C_o. Then we may write, using Cauchy's integral formula on an annular region

$$2\pi i f(z) = \oint_{C_0} \frac{f(\xi)}{\xi - z} d\xi - \oint_{C_i} \frac{f(\xi)}{\xi - z} d\xi,$$

where z is a point lying between the two curves and the origin is at $z = a$. Then we make a similar expansion for Taylor series by writing

$$2\pi i f(z) = \oint_{C_0} \frac{f(\xi)}{\xi\left(1 - \frac{z}{\xi}\right)} d\xi - \oint_{C_i} \frac{f(\xi)}{z\left(\frac{\xi}{z} - 1\right)} d\xi$$

$$= \sum_{n=0}^{\infty} z^n \oint_{C_0} \frac{f(\xi)}{\xi^{n+1}} d\xi + \sum_{n=1}^{\infty} z^{-n} \oint_{C_i} f(\xi)\xi^{n-1} d\xi,$$

since $|\xi| > |z|$ with respect to C_0, while $|\xi| < |z|$, with respect to C_i. We identify

$$a_n = \oint_{C_0} \frac{f(\xi)}{\xi^{n+1}} d\xi \text{ and } a_{-n} = \oint_{C_i} f(\xi)\xi^{n-1} d\xi.$$

Replacing z with $z - a$, we have sketched the following theorem.

Theorem 1.6. *(Markushevich, Vol. II). If f is analytic within an annular region between C_i and C_o, then it can be written in the form of a convergent Laurent series, as*

$$f(z) = \sum_{-\infty}^{\infty} a_n(z - a)^n. \tag{1.20}$$

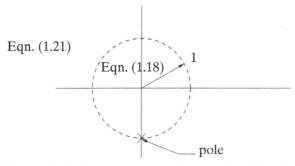

Figure 1.4. Series convergence regions for $f = 1/(z+i)$.

It follows that if the series truncates at a finite number of negative n terms, say $a_{-m} \neq 0$ for some positive integer m, but $a_j = 0$ for all $j < -m$, then $z = a$ is a pole of order m. For example, the function we have examined, $f = 1/(z+i)$, can be written in the form

$$\frac{1}{z+i} = \frac{1}{z}\frac{1}{1+i/z} = \frac{1}{z}\left(1 - \frac{i}{z} - \frac{1}{z^2} + \cdots\right). \tag{1.21}$$

This is the Taylor series for the function about $z = \infty$, with a singularity at $z = -i$. The ratio test shows that the series converges for $|z| > 1$, there are an infinite number of negative-power terms, which give it the appearance of a Laurent series. For another example, the function, $g = 1/z(z+i)$, can be written in the form

$$\frac{1}{z(z+i)} = \frac{1}{z}\left(-i + z + iz^2 + \cdots\right) = -\frac{i}{z} + 1 + iz + \cdots, \tag{1.22}$$

also based on the expansion we did in (1.18). This is a Laurent series about $z = 0$, for a function with a singularity at $z = -i$. It has one term with a negative power because it has a simple pole at $z = 0$. The function g may also be expressed as

$$\frac{1}{z(z+i)} = \frac{i}{(z+i)\left[1 - \frac{z+i}{i}\right]}$$

$$= \sum_{n=0}^{\infty}(-i)^{n+1}(z+i)^{n-1},$$

which converges for $0 < |z+i| < 1$, as shown in Figure 1.4.

On the other hand, consider the function $f = \exp(1/z)$. Its series is

$$e^{1/z} = 1 + \frac{1}{z} + \frac{1}{2!z^2} + \frac{1}{3!z^3} + \cdots.$$

Note the infinite number of negative n terms, which indicate the essential singularity at $z = 0$, because this series converges for all z but not zero.

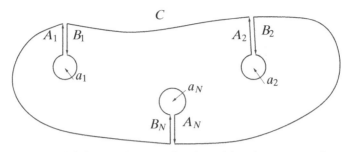

Figure 1.5. Integration path for the residue theorem proof.

1.4.3 The Residue Theorem

We know, based on our discussion of Laurent series in Section 1.4.2, that we can write

$$f(z) = \sum_{-\infty}^{\infty} a_n z^n, \tag{1.23}$$

inside some annulus $\mathcal{R}$ about $z = 0$, in which f is analytic. Consider the integral of this function f along a closed curve, C_a, that encircles the origin and lies wholly in $\mathcal{R}$. Provided that the series is convergent, as it must be inside the annulus, the integral of f is the sum of the integrals of the terms in the series on the right-hand side of (1.23). Of all of those integrals, according to (1.12), the only one that is nonzero is the one for which $n = -1$. Therefore, we conclude that

$$\oint_{C_a} f(z)dz = 2\pi i a_{-1}. \tag{1.24}$$

The quantity a_{-1} is the "residue" of f at $z = 0$. Now suppose we consider the integral of a function f along a closed curve C, inside of which at $\{a_k\}$ the function f has N isolated singular points. That path can be modified to a new path, as shown in Figure 1.5, with each path C_k about z_k lying inside an annulus about a_k, in which the function is analytic and around which there is a Laurent series.

As in the argument above, because f is analytic in the regions occupied by the connecting lines $\{A_k, B_k\}$, as these lines approach each other, the sum of the integrals along each pair is zero. Therefore, we find that

$$\oint_C f(z)dz = \sum_{k=1}^{N} \oint_{C_k} f(z)dz. \tag{1.25}$$

For each integral in the sum in (1.25), the result (1.24) may be applied, and hence we have the result, known as the "residue theorem." The statement is as follows:

Theorem 1.7. *(Markushevich, Vol. II). If $f(z)$ is analytic inside the curve C, except for a finite number of isolated singular points at $\{a_k\}$, then*

$$\oint_C f(z)dz = 2\pi i \sum_{k=1}^{N} \text{Res}(f(z))\Big|_{z=a_k}. \tag{1.26}$$

It is important to emphasize that this result is valid *only* for functions that have isolated singular points inside of C.

Residues may be evaluated directly from the Laurent series, but that can be cumbersome. Working from (1.13) and L'Hospital's rule, it can be easily verified that, for a first-order pole at $z = a$, and with the function f written in the form $f = N(z)/D(z)$, the residue is $N(a)/D'(a)$. For example,

$$f(z) = \frac{z}{(\sin kz)^2}$$

has a simple pole at $z = 0$. Consequently, its residue is

$$\varrho = \lim_{z \to 0} zf(z) = \lim_{z \to 0} \frac{z^2}{(\sin kz)^2} = \frac{1}{k^2},$$

but the residue is also seen to be

$$\varrho = \lim_{z \to 0} \frac{z/\sin kz}{\frac{d}{dz} \sin kz} = \lim_{z \to 0} \frac{z/\sin kz}{k \cos kz} = \frac{1}{k^2}.$$

For a pole of higher order, say n, it can be shown from (1.14) that the residue is given by

$$\operatorname{Res}(f)\Big|_a = \frac{1}{(n-1)!} \frac{d^{n-1}}{dz^{n-1}} \left((z-a)^n f(z)\right)\Big|_{z=a}. \tag{1.27}$$

1.5 Application to Real Integrals

In this section, we use the residue theorem to work out a variety of real definite integrals, of the kinds that might arise in a variety of applied settings, and particularly in problems solved by integral transforms. The following, then, is simply a set of examples.

Example 1.1. Evaluate

$$I \equiv \int_0^\infty \frac{dx}{1+x^2}. \tag{1.28}$$

Notice that I is half of the integral from $-\infty$ to ∞. Consider a closed, contour integral,

$$\oint_C \frac{dz}{1+z^2},$$

along a path shown in Figure 1.6.

The integral is two integrals, one along the real axis, say I_1, and one along the semicircle, say I_R. Let $R \to \infty$, $I_1 \to 2I$. What of I_R? On the semicircle, we write $z = R\exp(i\theta)$. Then,

$$I_R = i \int_0^\pi \frac{Re^{i\theta}\,d\theta}{1+R^2 e^{2i\theta}}.$$

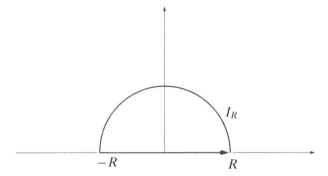

Figure 1.6. Integration path for Example 1.1.

Using the Schwarz inequality, this quantity is easily bounded as

$$|I_R| \leq \frac{R}{R^2 - 1} \int_0^\pi d\theta = \frac{\pi R}{R^2 - 1}.$$

Clearly, as $R \to \infty$, $|I_R|$ has an upper bound tending to zero, so $I_R \to 0$. Thus, we conclude that the contour integral is exactly $2I$. The contour integral may be evaluated by residues. The only enclosed singularity is at $z = i$, and the residue there is obviously $1/2i$. Therefore, $2I = 2\pi i \times (1/2i)$, so the real integral required is $I = \pi/2$.

Example 1.2. The value of I in Example 1.1 is alternatively worked out by elementary methods. However, if we now consider

$$I \equiv \int_0^\infty \frac{dx}{(1 + x^2)(a^2 + x^2)^2}, \quad a \neq 1, \tag{1.29}$$

there are no simple alternative methods for evaluation. The procedure is identical to Example 1.1, with the same choice of contour. The integral along the real axis, as $R \to \infty$, is $2I$, and the semicircle integral can be shown to vanish in the limit. Therefore, we come finally to

$$I = \pi i \sum \text{Res inside } C. \tag{1.30}$$

The residues are at i, as before, and at ia. The latter pole is a second-order pole, so we need to use the more general residue formula, (1.27). Thus,

$$\text{Res}(f)|_i = \frac{1}{2i(a^2 - 1)^2},$$

and

$$\text{Res}(f)|_{ia} = \left[\frac{d}{dz} \left(\frac{1}{(z^2 + 1)(z + ia)^2} \right) \right]_{ia} = \frac{1 - 3a^2}{4ia^3(a^2 - 1)^2}.$$

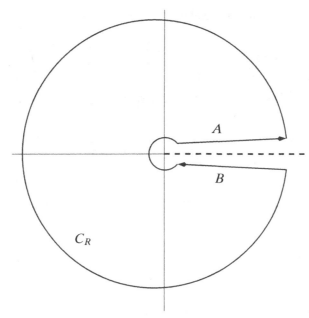

Figure 1.7. Integration path for Example 1.3. Dashed line represents the branch cut.

Combining this formula with (1.30) leads to the final result,

$$\int_0^\infty \frac{dx}{(1+x^2)(a^2+x^2)^2} = \frac{\pi(1+2a)}{2a^3(a+1)^2}. \tag{1.31}$$

Notice the result is valid when $a = 1$.

The first two examples were both done easily because the integrands are even, and so the integral along the real axis is twice the desired result. In the next example, we use a standard trick to evaluate the integral (see (Carrier, Krook, and Pearson)).

Example 1.3. We want to evaluate the integral,

$$I \equiv \int_0^\infty \frac{x\,dx}{1+x^3}. \tag{1.32}$$

Clearly, the integral from $-\infty$ to zero of the same integrand has a quite different value than I, and may be infinite. Hence, we need an alternative. Consider the integral,

$$\oint_C \frac{z\log z\,dz}{1+z^3}.$$

We cut the plane along $\Re\text{eal}(z) \geq 0$, to make the integrand analytic, requiring that for $z = r\exp(i\theta)$, $0 \leq \arg z < 2\pi$. Consider the integral around the path C shown in Figure 1.7.

So, the closed-path integral is composed of integrals along four segments of
C, namely,

$$\oint \frac{z \log z}{1 + z^3} dz = \int_A f dz + \int_{C_R} f dz + \int_B f dz + \int_{C_\epsilon} f dz.$$

If the radius of the circle C_R is R, where $z = R e^{i\theta}$, $R > 1$. Then that integral
can be bounded as follows:

$$\left| \int_{C_R} \frac{z \log z}{1 + z^3} dz \right| \leq \int_0^{2\pi} \frac{R^2 |\log R + i\theta|}{|1 + R^3 \exp(3i\theta)|} d\theta \leq \int_0^{2\pi} \frac{R^2 \sqrt{(\log R)^2 + \theta^2}}{R^3 - 1} d\theta$$

$$\leq \int_0^{2\pi} \frac{R^2 \sqrt{(\log R)^2 + 4\pi^2}}{R^3 - 1} d\theta \leq \frac{R^2 \sqrt{(\log R)^2 + 4\pi^2}}{R^3 - 1} 2\pi.$$

Clearly, as $R \to \infty$, this integral goes to zero. A similar thing can be done with
the integral around C_ϵ, where $z = \epsilon e^{i\theta}$, $\epsilon < 1$,

$$\left| \int_{C_\epsilon} \frac{z \log z}{1 + z^3} dz \right| \leq \int_0^{2\pi} \frac{\epsilon^2 |\log \epsilon + i\theta|}{|1 + \epsilon^3 \exp(3i\theta)|} d\theta \leq \frac{\epsilon^2 \sqrt{(\log \epsilon)^2 + 4\pi^2}}{1 - \epsilon^3} 2\pi.$$

On finding the limit as the circle radius, $\epsilon \to 0$, this integral also goes to zero.
Thus, we are left with integrals along A and B only, with their limits as $0 \leq r \leq$
∞. Evaluating those integrals along the paths A and B, we have

$$\oint_C \frac{z \log z dz}{1 + z^3} = \int_0^\infty \frac{r \log r dr}{1 + r^3} - \int_0^\infty \frac{r(\log r + 2\pi i) dr}{1 + r^3}.$$

Because the two integrals involving the $\log r$ cancel, we are left simply with
$I = -\sum \text{Res}$, from (1.26). That summing of residues must be done with some
care. The general residue formula is simply $\text{Res} = \log z/(3z)$. The poles are located at the three roots of (-1), which are at $\{\exp(i\pi/3) \exp(i\pi), \exp(5i\pi/3)\}$.
Therefore,

$$I = -\sum \text{Res} = -\frac{i\pi/3}{3e^{i\pi/3}} - \frac{i\pi}{3e^{i\pi}} - \frac{i5\pi/3}{3e^{i5\pi/3}} = \frac{2\pi \sqrt{3}}{9}.$$

Example 1.4. Evaluate

$$I \equiv \int_0^{2\pi} \frac{d\theta}{a + b \cos \theta}.$$

This integral is quite different. It suggests that we consider some function integrated around a unit circle. Hence, we write $z = \exp(i\theta)$, on the unit circle. In
that case, we find that $\cos \theta = (z + z^{-1})/2$ and $d\theta = -i dz/z$. With these substitutions, and taking a and b to be positive for now, for convenience, we have

$$I = \frac{2}{i} \oint_C \frac{dz}{b(z^2 + 1) + 2az},$$

and C is the unit circle. The first-order poles of the integrand are located at

$$z = -\frac{a}{b} \pm \sqrt{\frac{a^2}{b^2} - 1}.$$

In the case that $a < b$, both poles lie on the unit circle. (Actually, from the integral, we note that the integrand of the real integral, I, is actually singular at two locations.) Thus, I must be rethought in that case, and a Cauchy principal-value integral must be invoked – but more of that later (see Example 1.9). For now, we impose the restriction, $a > b$, for which case the poles are on the real axis and only one pole is inside of the circle. Thus,

$$I = 4\pi \operatorname{Res}|_{-a/b+\sqrt{a^2/b^2-1}} = \frac{2\pi}{\sqrt{a^2 - b^2}}.$$

Example 1.5. Evaluate

$$I \equiv \int_0^{2\pi} e^{a \sin \theta} d\theta. \tag{1.33}$$

Using the same trick as in Example 1.4, we can rewrite this integral as being around the unit circle C,

$$I = \oint_C e^{az/(2i)-a/(2iz)} \frac{dz}{iz}.$$

The integrand has an essential singular point at the origin, so we cannot use (1.27) to evaluate the residue. However, using Taylor's theorem for each of the exponential terms in the integrand, we find that the integrand of the contour integral for I is

$$\frac{1}{iz} \left[\sum_{k=0}^{\infty} \frac{1}{k!} \left(\frac{az}{2i} \right)^k \right] \left[\sum_{n=0}^{\infty} \frac{1}{n!} \left(-\frac{a}{2iz} \right)^n \right].$$

These two series, when multiplied together, form the Laurent series for the integrand. In that case, the coefficient of $1/z$ in the series is the residue. That occurs whenever the indices n and k are identical in the multiplied series. Hence, using (1.26) again,

$$I = 2\pi \sum_{n=0}^{\infty} \frac{1}{(n!)^2} \left(\frac{a}{2} \right)^{2n} = 2\pi I_0(a),$$

where we have noted that this integral is essentially the modified Bessel function of the first kind! (See Section 2.3.2.)

Example 1.6. Evaluate

$$F_n \equiv \int_0^{\infty} \frac{x^n dx}{(x^2 + 1)^2}. \tag{1.34}$$

Standard techniques reveal that this integral exists only for n in the open range $(-1, 3)$. Consider the closed contour integral around the path we used for Example 1.3, this time with the function $z^n/(1 + z^2)^2$. As before, the integral around the large circle vanishes as $R \to \infty$, so we have, on finding the integrals above and below the branch cut,

$$\oint_C \frac{z^n dz}{(1+z^2)^2} = \left[1 - e^{2n\pi i} \right] F_n. \tag{1.35}$$

The residues are for the second-order poles located at $\pm i$, so we have

$$\sum \text{Res} = \frac{e^{i\pi(n-1)/2}}{(2i)^3}(2n-2)i + \frac{e^{i3\pi(n-1)/2}}{(-2i)^3}(-2n+2)i = -\frac{1}{2}(1-n)e^{in\pi} \sin\left(\frac{n\pi}{2}\right).$$

Combining with (1.35), we have

$$F_n = \frac{(1-n)\pi}{4 \cos\left(\frac{n\pi}{2}\right)}.$$

This procedure used the fact that the origin is a branch point for the integrand function, so it is interesting that, for the case $n = 1$, for which the function does *not* have a branch point, this formula – using the L'Hospital Rule – gives the correct answer for $n = 1$, namely, $F_1 = 1/2$, and also the correct answer for $n = 0$, $F_0 = \pi/4$.

We will find in Chapter 6 that we will have occasion to have to evaluate integrals with cosines and sines in them.

Example 1.7. Consider the following example

$$I = \int_{-\infty}^{\infty} \frac{\cos(kx)}{x^2 + a^2} dx.$$

In this equation, a and k are real quantities. As we will see, it is convenient to write this integral as the real part of a complex integral. Thus,

$$I = \Re\text{eal} \left\{ \int_{-\infty}^{\infty} \frac{e^{ikx}}{x^2 + a^2} dx \right\}. \tag{1.36}$$

Here, we use the contour shown in Figure 1.6. Therefore,

$$\oint_C \frac{e^{ikz} dz}{z^2 + a^2} = \int_{-R}^{R} \frac{e^{ikx}}{x^2 + a^2} dx + \int_{C_R} \frac{e^{ikz} dz}{z^2 + a^2}. \tag{1.37}$$

The second of these integrals, the one around C_R, can be bounded in the usual way, as

$$\left| \int_{C_R} \frac{e^{ikz} dz}{z^2 + a^2} \right| \leq \int_0^{\pi} \frac{\left| e^{ik(R\cos\theta + i\sin\theta)} \right| R d\theta}{\left| (Re^{i\theta})^2 + a^2 \right|} \leq \int_0^{\pi} \frac{e^{-kR\sin\theta}}{R^2 - 1} R d\theta.$$

Note that for $k > 0$, the exponential in the integral is less than or equal to one, so we have

$$\left| \int_{C_R} \frac{e^{ikz} dz}{z^2 + a^2} \right| \leq \frac{R}{R^2 - 1} \pi.$$

Clearly, this integral vanishes for $R \to \infty$, so then (1.37), on taking that limit, becomes

$$\oint_C \frac{e^{ikz}dz}{z^2 + a^2} = \int_{-\infty}^{\infty} \frac{e^{ikx}}{x^2 + a^2} dx.$$

We know that the contour integral above is equal to 2π times the residue at ia. Since this is a simple pole,

$$\text{Res}\big|_{z=ia} = \frac{e^{-ka}}{2ia}. \tag{1.38}$$

Thus, combining (1.38) with (1.36), we have the result that

$$I \equiv \int_{-\infty}^{\infty} \frac{\cos(kx)dx}{x^2 + a^2} = \Re\text{eal}\left\{2\pi i \frac{e^{-ka}}{2ia}\right\} = \frac{\pi}{a}e^{-ka}, \ k > 0.$$

Now, if, instead, $k < 0$, we must choose a contour closed by a semicircle in the *lower* half-plane rather than the upper-half-plane. In this case, doing this calculation again, we obtain the result $\pi e^{ka}/a$. Thus, in general, we have the result

$$I = \frac{\pi}{a}e^{-|k|a}.$$

1.5.1 Principal Values; Jordan's Lemma

Example 1.8. In real integration, there are two standard types of improper integrals. First, for integrals defined on $(-\infty, \infty)$, we may define

$$\int_{-\infty}^{\infty} f(x)dx = \lim_{R \to \infty} \int_{-R}^{R} f(x)dx, \tag{1.39}$$

when this limit exists. It may be that (1.39) exists, but the related integral

$$\lim_{\substack{a \to -\infty \\ b \to \infty}} \int_a^b f(x)dx$$

does not. In that case, (1.39) is a "Cauchy principal value."

There is a second type of Cauchy principal value integral in which the interval of integration may be finite. This is discussed in Example 1.9 later in this section.

Jordan's Lemma

This result is sometimes used in conjunction with the principal-value approach. Let

$$I(R) = \int_{C(R)} e^{ikz}g(z)dz,$$

where $C(R)$ is the arc $z = Re^{i\theta}$, $0 \le \theta \le \pi$, and $a > 0$. If $\left|g(Re^{i\theta})\right| \le G(R) \to 0$ as $R \to \infty$, then $\lim_{R \to \infty} I(R) = 0$.

We see how this goes as follows:

$$I = \int_{C(R)} e^{ikz}g(z)dz = \int_0^{\pi} ie^{ikR\cos\theta}e^{-kR\sin\theta}g(Re^{i\theta})Re^{i\theta}d\theta.$$

Hence,

$$|I| \leq G(R)R \int_0^\pi e^{-kr\sin\theta} d\theta = 2G(R)R \int_0^{\pi/2} e^{-kr\sin\theta} d\theta,$$

after applying the given estimates. However, a more refined estimate is needed. Notice that for

$$0 \leq \theta \leq \frac{\pi}{2}, \quad \sin\theta \geq \frac{2\theta}{\pi}.$$

Thus,

$$|I| \leq 2G(R) \int_0^{\pi/2} e^{-2kR\theta/\pi} d\theta$$

$$= \frac{\pi}{k} G(R)(1 - e^{-kR}) \to 0,$$

as $R \to \infty$.

Example 1.8

As an illustration of this result, we evaluate, as a Cauchy principal value,

$$\int_{-\infty}^{\infty} \frac{xe^{ikx}}{1+x^2} dx.$$

Convert this integral to a contour over the curve C given in Figure 1.6. This integral is

$$\oint_C \frac{ze^{ikz}}{1+z^2} dz = \text{sgn}(k)2\pi i \text{Res}(f(z))|_{z=\pm i}.$$

The poles occur at $z = \pm i$. We make the estimate

$$|g(Re^{i\theta})| = \left| \frac{Re^{i\theta}}{1+(Re^{i\theta})^2} \right|$$

$$\leq \frac{R}{R^2-1} = G(R) \to 0,$$

as $R \to \infty$. This is the same type of estimate as in the example in Section 1.5. Moreover, Jordan's lemma applies. We have the result that

$$\int_{-\infty}^{\infty} \frac{xe^{ikx}dx}{x^2+a^2} = \begin{cases} 2\pi i \frac{e^{-k}}{2}, k > 0 \\ -2\pi i \frac{e^k}{2}, k < 0 \end{cases} = \text{sgn}(k)\pi i e^{-|k|}.$$

Example 1.9. Evaluate

$$R \equiv \int_{-\infty}^{\infty} \frac{I_0\left(\sqrt{k^2+1}a\right)\sin(kx)dk}{kI_0\left(\sqrt{k^2+1}\right)}. \tag{1.40}$$

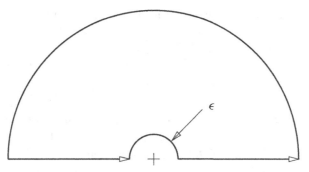

Figure 1.8. Integration path for Equation (1.41).

In this expression, I_0 is the modified Bessel function of first kind (Section 2.3.2), order zero. Though it appears that there are branch points at $\pm i$, a careful examination, say, *via* series, reveals that the function is analytic at $\pm i$. We evaluate this integral by writing it as the imaginary part of an expression that contains a related integral,

$$R = \Im\mathrm{mag} \left[\fint_{-\infty}^{\infty} \frac{I_0\left(\sqrt{k^2+1}\,a\right) e^{ikx}dk}{k\,I_0\left(\sqrt{k^2+1}\right)} - i\pi \frac{I_0(a)}{I_0(1)} \right] + \pi \frac{I_0(a)}{I_0(1)}. \qquad (1.41)$$

We have introduced a new notation here – indicated by the dash across the integral sign. It denotes a second type of "Cauchy principal-value integral," which is a real integral for a function with a simple zero in the denominator, and is defined by

$$\fint_a^b \frac{f(x)dx}{x-x_o} \equiv \lim_{\epsilon \to 0} \left[\int_a^{x_o-\epsilon} \frac{f(x)dx}{x-x_o} + \int_{x_o+\epsilon}^b \frac{f(x)dx}{x-x_o} \right], \quad \text{for } a < x_o < b.$$

This definition is necessary because the integrand of the quantity introduced in (1.41) has a zero denominator at $k = 0$. The subtracted quantity removes the contribution near the origin pole, as we will see. As a contour integral this may thought of as integration around a pole through a polar angle of π, and then taking the limit. On occasion, it may necessary to integrate around a pole through some other fraction of 2π (a complete circuit), and then take the limit. The contribution to the principal value thereby depends on this angle.

The Bessel function, $I_0(z)$, behaves like $z^{-1/2}\exp(z)$, for $z \to \infty$, so clearly this integral exists only for $0 < a < 1$. In the analysis following, we also take the variable $x > 0$ throughout. Here we use a path like that used for Example 1.1 except that there is a cutout around the origin because of the pole there. It appears as shown in Figure 1.8.

The integral from $-\infty$ to $-\epsilon$ and then from ϵ to ∞ is, as $\epsilon \to 0$, is the Cauchy principal-value integral of (1.41). The integral around the circle of radius ϵ can be shown as $\epsilon \to 0$, $-i\pi\,\mathrm{Res}(0)$, which is equal to $-i\pi I_0(a)/I_0(1)$. So, the terms in the square brackets in (1.41) together constitute the closed contour integral of that k-function, because it is easy to show that the integral on the large-radius semicircle vanishes as the radius goes to infinity. Therefore, we have

$$R = \Im\text{mag}\left[\oint_C \frac{I_0\left(\sqrt{z^2+1}\,a\right)e^{izx}dz}{zI_0\left(\sqrt{z^2+1}\right)}\right] + \pi\frac{I_0(a)}{I_0(1)}. \tag{1.42}$$

There is a somewhat tricky point here, because the integrand clearly has a pole at the origin. However, it is easily seen that the residue there is purely imaginary, so it contributes nothing to R, once the imaginary part of the integral is taken. The modified Bessel function $I_0(z)$ is an entire function in z, but since $I_0(z) = J_0(iz)$, where J_0 is the Bessel function of the first kind, which has an infinity of zeroes, located, we will say, at $ij_{0,n}$, $n = 1, 2, 3, \ldots$. (It also has negative zeroes, but those are not enclosed in C.) Therefore, there will be poles at the zeroes of the denominator, that is, at

$$\sqrt{z^2+1} = ij_{0,n}.$$

Solving, the upper-half-plane zeroes are at $z_n = i\sqrt{1 + j_{0,n}^2}$. Using the fact that $I_0' = I_1$, where I_1 is the modified Bessel function of order one, we can write down the formula for the residue

$$\text{Res} = \left[\frac{I_0(\sqrt{z^2+1}\,a)\sqrt{z^2+1}}{I_1(\sqrt{z^2+1})z^2}e^{izx}\right]_{z_n} = \frac{J_0(j_{0,n}a)j_{0,n}}{J_1(j_{0,n})(j_{0,n}^2+1)}e^{-x\sqrt{1+j_{0,n}^2}}.$$

Therefore, on summing over all the residues and taking the imaginary part, we have the result

$$R = 2\pi\sum_{n=1}^{\infty}\frac{J_0(j_{0,n}a)j_{0,n}}{J_1(j_{0,n})(j_{0,n}^2+1)}e^{-|x|\sqrt{1+j_{0,n}^2}} + \pi\frac{I_0(a)}{I_0(1)},$$

where we have used the identities, $I_1(ix) = iJ_1(x)$ and $I_0(ix) = J_0(x)$. Furthermore, we have appended the fact that if the problem is redone for $x < 0$, the answer is the same except that x is replaced by $-x$. (For $x < 0$, the contour is closed in the lower-half-plane instead, and so we use the conjugate pole locations, $\bar{z}_n$. Everything else is essentially the same.)

Example 1.10. In this example, we seek to evaluate the integral

$$I = \int_0^{\pi}\frac{\theta\sin\theta d\theta}{\alpha^2+1-2\alpha\cos\theta}. \tag{1.43}$$

We notice first that the integrand is even, so that the integral is half of the integral from $-\pi$ to π. So the integral from $-\pi$ to π can be viewed as an integration around a circular path in a complex plane, and therefore we write $z = \exp(i\theta)$; as in a Example 1.4, the integral, I, then becomes

$$I = \frac{i}{4}\int_{C_o}\frac{(z^2-1)\log z dz}{z[(\alpha^2+1)z - \alpha(z^2+1)]}, \tag{1.44}$$

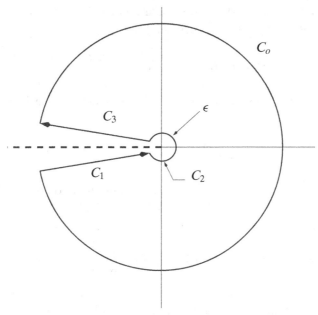

Figure 1.9. Integration paths for Example 1.10. (The dashed line denotes the branch cut.)

where C_o denotes the unit circle from $-\pi$ to π. Because of the presence of the logarithm in the integrand, we cut the plane along the negative real axis. Let the various segments be denoted by C_1, C_2, C_3 as well as C_o as shown in the sketch in Figure 1.9.

Denoting for the moment the integrand in (1.44) by $H(z)$,

$$\oint H dz = \int_{C_1} H dz + \int_{C_2} H dz + \int_{C_3} H dz - 4i\,I. \tag{1.45}$$

Working out the integral around C_2 first, we write $z = \epsilon \exp(i\theta)$ and substitute in the integrand. Noting that we will take the limit $\epsilon \to 0$ in the result, we have

$$\int_{C_2} H dz = \int_{\pi}^{-\pi} \frac{1}{\alpha}(\log \epsilon + i\theta)d\theta + \mathcal{O}(\epsilon) = -\frac{2\pi i}{\alpha}\log \epsilon + \mathcal{O}(\epsilon). \tag{1.46}$$

Consider next the integrals along the cut, on C_1 and C_3. We have

$$\int_{C_1} H dz + \int_{C_3} H dz = \int_1^{\epsilon} \frac{(r^2 - 1)(\log r + i\pi)dr}{r[-(\alpha^2 + 1)r - \alpha(r^2 + 1)]}$$

$$+ \int_{\epsilon}^1 \frac{(r^2 - 1)(\log r - i\pi)dr}{r[-(\alpha^2 + 1)r - \alpha(r^2 + 1)]}$$

$$= \frac{2\pi i}{\alpha} \int_{\epsilon}^1 \frac{(r^2 - 1)dr}{r(r + \alpha)(r + 1/\alpha)},$$

and this integral can be done by standard methods, using, say, partial fractions. The completed integration process leads to

$$\int_{C_1} H\,dz + \int_{C_3} H\,dz = \frac{2\pi i}{\alpha} \log \epsilon + \frac{2\pi i}{\alpha} \log \left(\frac{(1+\alpha)^2}{\alpha} \right).$$

Combining this result with that of (1.46) and substituting it into (1.45), then letting $\epsilon \to 0$, we obtain

$$\oint H\,dz = \frac{2\pi i}{\alpha} \log \left(\frac{(1+\alpha)^2}{\alpha} \right) - 4i\,I. \tag{1.47}$$

It remains, then, to evaluate residues. The denominator of (1.44) is easily factored, and that shows that there are poles at α and $1/\alpha$. If we take $\alpha < 1$ for the moment, then only the former pole lies inside the contour, so that

$$\text{Res}|_{z=\alpha} = -\frac{\log \alpha}{\alpha}.$$

Since the closed path integral, which is the left-hand side of (1.47), is equal to $2\pi i$ times this residue, substitution into (1.47) allows us to solve for I. Hence,

$$I = \frac{\pi}{\alpha} \log(1 + \alpha), \quad \text{for } 0 < \alpha < 1.$$

The $\alpha > 1$ result is easily obtained by noting that, from (1.43), $I(1/\alpha) = \alpha^2 I(\alpha)$. It then follows that

$$I = \frac{\pi}{\alpha} \log \left(\frac{1+\alpha}{\alpha} \right), \quad \text{for } \alpha > 1.$$

REFERENCES

George F. Carrier, Max Krook, and Carl E. Pearson, *Functions of a Complex Variable: Theory and Technique*, SIAM Publications, Philadelphia, 2005.

E. A. Coddington and R. Carlson, *Linear Ordinary Differential Equations*, SIAM Publications, Philadelphia, 1997.

A. I. Markushevich, *Theory of Functions of a Complex Variable*, Vols. I, II, III, New York : Chelsea Publishing Co., 1977.

L. M. Milne-Thomson, *Theoretical Hydrodynamics*, 4th edition, Macmillian, New York, 1960.

EXERCISES

1.1 If t is a real parameter, the point $z = z(t) = Re^{it}$ travels on a circle of radius R. What is the locus of the points

(a) $$z = ae^{it} + be^{-it} \quad (a, b \text{ real})?$$

(b) $$z = \cos 2t + i \cos t \quad (0 \le t \le \pi)?$$

1.2 (a) Given that

$$\cosh z = \frac{e^z + e^{-z}}{2},$$

show that

$$\cosh^{-1} z = \log \left[z + (z^2 - 1)^{1/2} \right].$$

(b) Find all of the solutions to $\cosh z + 2 = 0$.

1.3 (a) Show that

$$\tan^{-1} z = \frac{1}{2i} \log \left(\frac{i - z}{i + z} \right).$$

(b) Given that formally from calculus,

$$\int \frac{dx}{x^2 + a^2} = \int \frac{1}{2ai} \left(\frac{1}{x - ia} - \frac{1}{x + ia} \right) dx$$

$$= \frac{1}{2ai} \log \left(\frac{x - ia}{x + ia} \right) + C,$$

reconcile the two results.

1.4 Verify Corollary 1.1: *A necessary condition for a function of a complex variable to be analytic anywhere in the complex plane is that $\partial f / \partial \bar{z} \equiv 0$ in the plane.*
Hint: Show, by a method similar to the argument on page 4, that the derivative cannot be independent of S unless f is independent of $\bar{z}$.

1.5 Show that the function

$$\frac{1}{z} = \frac{x - iy}{x^2 + y^2},$$

has harmonic components on the whole plane, except at $z = 0$.

1.6 Show that if ϕ and ψ are harmonic, then

$$u = \phi_x \phi_y + \psi_x \psi_y \text{ and}$$

$$v = \frac{1}{2} (\phi_x^2 + \psi_x^2 - \phi_y^2 - \psi_y^2)$$

satisfy the Cauchy–Riemann equations.

1.7 Use of complex variables can be extremely helpful in understanding the properties of the solutions of ordinary differential equations. Consider the equation for $u(z)$,

$$(z^2 + 4)u'' + 3zu' + u = 0.$$

Use of the substitution

$$u = \sum_{n=0}^{\infty} a_n z^n$$

gives two analytic solutions, and that the radius of convergence of each series is finite. Locate the singular points of the differential equation. Explain on the basis of a known theorem (see e.g., (Coddington and Carlson)), why this behavior is to be expected.

1.8 Use Cauchy's theorem, Cauchy's integral formula and its generalization to evaluate

$$\oint_C \frac{\log(1-z)}{z^n} dz,$$

where C is the positively oriented circle $|z| = \frac{1}{2}$, for the three cases $n = 0, n = 1$, and $n = 2$.

1.9 Answer the questions below for the function, f, of a complex variable, z, given by

$$f(z) = \frac{\sin z}{z^2(\log z + 1)}.$$

(a) Indicate the location of all singularities, and classify them.

(b) If the plane is cut along the negative real axis from the origin to negative infinity, evaluate $f(i)$.

(c) With the plane cut as in part (b), what is the value of the integral $\oint_C f dz$, using the path shown in Figure 1.10?

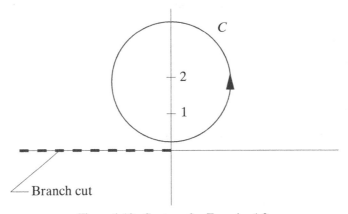

Figure 1.10. Contour for Exercise 1.9.

1.10 (a) Locate all of the singularities of the following functions, and classify each singularity as a pole, essential singularity, or branch point. If branch points occur, make a choice for branch cut(s), and indicate in a sketch.

$$A: \quad f(z) = \frac{\sin z}{z+1},$$

$$B: \quad f(z) = \frac{e^{-\sqrt{z}}}{z^2 - iz - 1},$$

$$C: \quad f(z) = \frac{\sin(z^{1/2})}{z^{3/2} + z^{7/2}},$$

$$D: \quad f(z) = \sin(z^{-1/4}).$$

(b) Evaluate each of the above functions – in planes suitably cut in the case of branch points – at locations $z = 1$ and $z = -1 + i$.

(c) Give the first few terms in either a Taylor or Laurent series near the origin, for each of the functions in (a), where possible.

(d) In the cases for which no such series can be constructed, note the reason. When a series is possible, give the annulus of validity, for example, $|z| < 1$ for A.

1.11 (a) Locate all singularities of the following functions, and classify each singularity as a pole, essential singularity, or branch point. If branch points occur, make a choice for branch cut(s), and indicate the branch cuts and singularities in a sketch.

$$\text{A:} \quad f(z) = \frac{e^z}{z + 1 + i},$$

$$\text{B:} \quad f(z) = \frac{\cos \sqrt{z}}{z^2 + 2iz + 3},$$

$$\text{C:} \quad f(z) = \frac{\sin(z^{3/2})}{z^{5/2} + z^{7/2}},$$

$$\text{D:} \quad f(z) = \cos(z^{-1/3}).$$

(b) Evaluate each of the above functions – in planes suitably cut in the case of branch points – at locations $z = 1$ and $z = -1 + i$.

(c) Give the first few terms in either a Taylor or Laurent series near the origin, for each of the functions in (a) B–D, where possible.

(d) In the cases for which no such series can be constructed, note the reason. When a series is possible, give the annulus of validity. For example, for A, the Taylor series around the origin has the form, $f \sim \frac{1}{1+i} + \frac{1}{2}z + \mathcal{O}(z^3) + \cdots$, for $|z| < \sqrt{2}$.

1.12 (a) Locate all singularities of the following functions, and classify each singularity as a pole, essential singularity, or branch point. If branch points occur, make a choice for branch cut(s), and indicate the branch cuts and singularities in a sketch.

$$\text{A:} \quad f(z) = \frac{\cos(z^3)}{(z + 1)^2},$$

$$\text{B:} \quad f(z) = \frac{z^{1/4}}{z^2 + 2z + 1 - i},$$

$$\text{C:} \quad f(z) = \frac{z^{1/4} \sin(z^{1/4})}{z^{1/2} + \frac{1}{3}z^{5/2}},$$

$$\text{D:} \quad f(z) = \cosh(z^{-1/2}).$$

(b) Evaluate each of the above functions – in planes suitably cut in the case of branch points – at locations $z = -2$ and $z = 1 - i$. All answers should be numerical, in the form $a + ib$.

(c) Give the first few terms in either a Taylor or Laurent series near the origin, for each of the functions in (a), where possible. In the cases for which no such series can be constructed, note the reason. When a series is possible, give the annulus of validity, for example, $|z| < 1$ for A.

1.13 Consider the following function, with branch points at $z = \pm 1$,

$$f(z) = \left(\frac{z+1}{z-1}\right)^{\frac{1}{4}}.$$

Suppose that the plane is cut to the right from $z = 1$ and cut to the left from $z = -1$. Determine the value of the function at $z = i$, $z = 0$, and $z = 1 + i2/\sqrt{3}$. (You will need to use two sets of polar coordinates, defined from *each* of the singularities.)

1.14 The *circulation* of a fluid flow with velocity $\mathbf{u}$ around a closed contour C is defined as

$$K(C) = \oint_C \mathbf{u} \cdot \mathbf{ds}.$$

Consider a two-dimensional incompressible flow with no singularities on an open region $\mathcal{R}$. It is observed that there is zero circulation around every closed contour interior to $\mathcal{R}$. Use Morera's theorem to show that $\mathbf{u} = (u, v)$ has a complex potential and thus is a potential flow. *Hint*: Let $f(z) = u - iv$, and note that because the flow is incompressible, $u_x + v_y = 0$, and thus there is a stream function $\psi(x, y)$, such that

$$u = \frac{\partial \psi}{\partial y} \text{ and } v = -\frac{\partial \psi}{\partial x}.$$

1.15 It can be shown (See (Milne-Thomson, p. 173)) that the aerodynamic force on a closed two-dimensional object can be written as a contour integral,

$$X - iY = \frac{1}{2}i\rho \oint_C \left(\frac{dF}{dz}\right)^2 dz,$$

where C is any contour that encloses the object, and the force vector in the (x, y) plane is $X\mathbf{i} + Y\mathbf{j}$; ρ is the fluid density; $F(z)$ is the "complex potential" of Section 1.3.4. In general, we can write the complex potential for flow at a speed U past an object as

$$F(z) = Uz + iK \log z + \sum_{n=1}^{\infty} \frac{a_n}{z^n}.$$

Show using the residue theorem that the aerodynamic force is given by

$$X + iY = 2\pi\rho U K i.$$

Note that if K is real, then the force is normal to the flow direction, and that if $K \equiv 0$ – no circulation – then there is no force.

1.16 Using the contour in Figure 1.7, show that

$$\int_0^\infty \frac{dx}{x^{1/2}(1+x^2)} = \frac{\pi}{\sqrt{2}}.$$

1.17 Using an integration path like that shown in Figure 1.7, show that

$$\int_0^\infty \frac{x^{\frac{1}{n}}}{1+x^2} dx = \frac{\pi}{2\cos\left(\frac{\pi}{2n}\right)}, \quad n > 1.$$

1.18 Using the integration contour shown in Figure 1.7, work out the real integral

$$\int_0^\infty \frac{x^n}{1+x^2} dx, \quad -1 < n < 1.$$

1.19 Use the integration path from Figure 1.7 to show that

$$I \equiv \int_0^\infty \frac{x^p dx}{(x+a)(x+b)} = \frac{b^p - a^p}{b-a}\pi\csc(p\pi),$$

if a and b are positive constants and $0 < p < 1$. (You don't need to add a logarithm in this case; simply do the integral $\oint z^p dz/((z+a)(z+b))$.)

1.20 An infinitely long cylinder of radius d is rotating at angular speed Ω above a plane; the gap between the two is very narrow and filled with fluid, so a lubrication approximation is possible. The nearest separation of the cylinder and plane is h_o. It can be shown that the torque per unit length acting on the rotating cylinder is given by

$$T = 3\mu\Omega\sqrt{\frac{d^5}{h_o}}\int_{-\infty}^\infty \frac{u^2 du}{(u^2+1)^2}.$$

Work out the integral to give a formula for the torque. (Here, μ is the fluid viscosity.) *Method:* Use the path shown in Figure 1.6 to evaluate the integral. You may confirm your answer by a trigonometric substitution.

1.21 Using the integration contour shown in Figure 1.7, show that

$$I = \int_0^\infty \frac{(\log x)dx}{(x+a)(x+b)} = \frac{\log(a/b)\log(ab)}{2(a-b)},$$

by starting with the integral

$$\oint \frac{(\log z)^2 dz}{(z+a)(z+b)},$$

where we require that

$$a > 0, \quad b > 0.$$

1.22 Use the contour of Figure 1.7 to show that

$$\int_0^\infty \frac{u^{s-1}du}{1+u} = \frac{\pi}{\sin(\pi s)}, \quad 0 < \Re\text{eal}(s) < 1.$$

1.23 Following the procedures for a similar integral in Example 1.4, show that

$$I \equiv \int_0^{2\pi} \frac{\sin^2\theta\,d\theta}{A + \cos\theta} = 2\pi A \left[1 - \frac{\sqrt{A^2 - 1}}{A} \right],$$

where $|A| \geq 1$.

1.24 Using the path shown, work out the value of the principal-value integral shown.

$$\fint_{-\infty}^{\infty} \frac{\cos(bx)\,dx}{x^2 - a^2} = -\frac{\pi}{b}\sin(ab).$$

Hint: Make the integral the real part of an integral involving e^{ibx}.

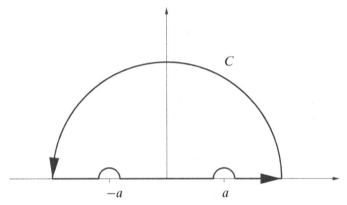

Figure 1.11. Contour for Exercise 1.24.

1.25 Evaluate the integral

$$\int_0^{2\pi} \frac{\cos(k\theta)\,d\theta}{a^2 + \sin^2\theta},$$

if k has an integer value.

If k is not an integer, discuss the difficulties of doing the evaluation, but do not attempt to carry it out.

2

Special Functions

2.1 Preamble

The term "special function" has come to be applied to any of a set of functions that have arisen in a wide variety of applications in engineering and physics. What these functions have in common is that they may not be written down in terms of elementary functions and standard transcendental functions, and most often they must be computed approximately, either by summing an infinite series or doing a quadrature. There is a vast literature on this subject, and we do not intend to repeat all of the elegant details about these functions that may be found in the classic works by Whittaker and Watson, Watson, and Jeffreys and Jeffreys, to name a few. Readers who require more details are referred to those canonical works and others cited in this chapter. For instance, in recent years, the handbook by Abramowitz and Stegun has proved to be an invaluable resource, and it is supplemented by an electronic update (Lozier et al.). We turn to an example in which a special function arises in solving a problem in fluid dynamics.

2.1.1 Batchelor's Trailing Vortex

In a model of the flow in a trailing-line vortex (Batchelor) described in cylindrical coordinates (x, r, θ), the Navier–Stokes equations reduce to the axisymmetric form

$$U\frac{\partial u}{\partial x} = -\frac{1}{\rho}\frac{\partial p}{\partial x} + \nu\left(\frac{\partial^2 u}{\partial r^2} + \frac{1}{r}\frac{\partial u}{\partial r}\right), \tag{2.1}$$

$$-\frac{w^2}{r} = -\frac{1}{\rho}\frac{\partial p}{\partial r}, \tag{2.2}$$

$$U\frac{\partial w}{\partial x} = \nu\left(\frac{\partial^2 w}{\partial r^2} + \frac{1}{r}\frac{\partial w}{\partial r} - \frac{w}{r^2}\right), \tag{2.3}$$

for the velocity components (u, v, w) and the pressure p, where the density ρ is constant with kinematic viscosity ν. The region of interest is far downstream where a boundary-layer approximation has been made, so that $v \ll u$, while by the wake approximation $|u - U| \ll U$, the *free-stream speed*. The appropriate boundary conditions are $u \to U$, $w \to 0$ as $x \to \infty$, $v \to \infty$.

Since Equation (2.3) is decoupled from the rest, it may be solved first. A change of dependent variable $\chi = rw$ simplifies it to

$$U \frac{\partial \chi}{\partial x} = v \left(\frac{\partial^2 \chi}{\partial r^2} - \frac{1}{r} \frac{\partial \chi}{\partial r} \right). \tag{2.4}$$

The variable χ is proportional to the circulation around the x-axis. A similarity solution is sought in terms of a nondimensional variable η given as

$$\eta = \frac{U r^2}{4 v x}. \tag{2.5}$$

We expect that the edge of the vortex should correspond to a "potential vortex," so χ should take a constant value there. Hence, we write $\chi(r, x) = F(\eta)$ and (2.4) becomes

$$F''(\eta) + F'(\eta) = 0.$$

The boundary conditions are $F \to 0$ as $\eta \to 0$ and $F \to F_0$ as $\eta \to \infty$. This constant coefficient equation is easy to solve. The solution is

$$F = F_0(1 - e^{-\eta}).$$

More challenging is the pressure distribution that this implies. From (2.2), if $p \to p_0$ as $r \to \infty$, then

$$p - p_0 = -\rho \int_r^\infty \frac{w^2}{r} dr = -\rho \int_r^\infty \frac{\chi^2}{r^3} dr.$$

Allowing for the change of variable (2.5), we obtain

$$p - p_0 = -\frac{\rho F_0^2 U}{8 v x} \int_\eta^\infty \frac{(1 - e^{-y})^2}{y^2} dy.$$

The function,

$$P(\eta) = \int_\eta^\infty \frac{(1 - e^{-y})^2}{y^2} dy,$$

is a special function but it has not been widely studied. It was employed and evaluated by Batchelor. It is related to more commonly used functions. For instance,

$$P(\eta) = \int_\eta^\infty (1 - e^{-y})^2 d \left(-\frac{1}{y} \right)$$

$$= \frac{(1 - e^{-\eta})^2}{\eta} + \int_\eta^\infty \frac{2(1 - e^{-y}) e^{-y}}{y} dy$$

$$= \frac{(1 - e^{-\eta})^2}{\eta} + 2 \int_\eta^\infty \frac{e^{-y}}{y} dy - \int_\eta^\infty \frac{2 e^{-2y}}{y} dy. \tag{2.6}$$

Hence, we introduce the standard *exponential integral function*, ei,

$$\text{ei}(\eta) \equiv \int_{\eta}^{\infty} \frac{e^{-y}}{y} dy. \tag{2.7}$$

The last integral in (2.6) is expressible in terms of ei by the change of variable $2y = \xi$ so that

$$\int_{\eta}^{\infty} \frac{2e^{-2y}}{y} dy = \int_{2\eta}^{\infty} \frac{2e^{-\xi}}{\xi} d\xi,$$

and finally we obtain

$$P(\eta) = \frac{(1 - e^{-\eta})^2}{\eta} + 2\text{ei}(\eta) - 2\text{ei}(2\eta).$$

Should the need arise to express ei as a complex-valued function, it may defined for all η except the origin and the negative real axis, which corresponds to a branch cut.

Note that Equation (2.1) may then, in principle, be solved for u, but that also involves other special functions, and so will not be discussed further. The interested reader may refer to the article by (Batchelor).

2.2 The Gamma Function

It turns out that, in dealing with many of the special functions in this chapter, understanding of the gamma function is essential. For real p, the gamma function may be represented by the integral

$$\Gamma(p) = \int_0^{\infty} t^{p-1} e^{-t} dt, \ p > 0. \tag{2.8}$$

The theory of the gamma function is intimately connected with that of the Laplace transform.

From the expression (2.8) using integration by parts,

$$\Gamma(p + 1) = p\Gamma(p). \tag{2.9}$$

If p is an integer $\Gamma(p + 1) = p!$.

The value of $\Gamma\left(\frac{1}{2}\right)$ is of particular interest:

$$\Gamma\left(\frac{1}{2}\right) = \int_0^{\infty} t^{-1/2} e^{-t} dt.$$

With the change of variable $t = x^2$,

$$\Gamma\left(\frac{1}{2}\right) = 2 \int_0^{\infty} e^{-x^2} dx.$$

Hence,

$$\left(\Gamma\left(\frac{1}{2}\right)\right)^2 = 4 \left(\int_0^\infty e^{-x^2} dx\right)\left(\int_0^\infty e^{-y^2} dy\right)$$

$$= 4 \int_0^\infty \int_0^\infty e^{-x^2-y^2} dx dy$$

$$= 4 \int_0^{\pi/2} \int_0^\infty e^{-r^2} r\, dr\, d\theta$$

$$= 2 \int_0^{\pi/2} d\theta = \pi,$$

after making a change to the polar coordinates. Thus, $\Gamma(1/2) = \sqrt{\pi}$.

The gamma function is also defined for complex arguments, so

$$\Gamma(z) = \int_0^\infty t^{z-1} e^{-t} dt, \quad \Re\mathrm{eal}(z) > 0. \tag{2.10}$$

The constraint on $\arg(z)$ is required to assure convergence of the integral at $t = 0$. The extension process (2.9) for real arguments still applies. The gamma function is then defined for all $\Re\mathrm{eal}(z) > 0$, but we must work a bit harder to handle the function in the left half-plane. It turns out that the gamma function is actually a *meromorphic function*, i.e., single-valued and analytic except for poles, in this case at the nonpositive integers. This can be shown from the representation (2.10) by writing

$$\int_0^\infty t^{z-1} e^{-t} dt = \int_0^1 t^{z-1} e^{-t} dt + \int_1^\infty t^{z-1} e^{-t} dt.$$

In the first integral on the right, replacing e^{-t} with its power series and integrating term-by-term,

$$\int_0^1 t^{z-1} e^{-t} dt = \sum_{n=0}^\infty \frac{(-1)^n}{n!} \int_0^1 t^{n+z-1} dt = \sum_{n=0}^\infty \frac{(-1)^n}{n!(z+n)}.$$

This last series converges as long as $z + n \neq 0$. So the definition (2.10), though not valid for all z, may be extended (by *analytic continuation*) to a form that is convenient for $\Re\mathrm{eal}(z) \leq 0$, $z \neq 0, -1, \ldots$, as

$$\Gamma(z) = \sum_{n=0}^\infty \frac{(-1)^n}{n!(z+n)} + \int_1^\infty t^{z-1} e^{-t} dt.$$

The pole locations are immediately evident from the first term, and the second term – the integral – can be easily shown to be an entire function.

Remark 2.1. Euler's Reflection Formula. *One important identity in the use of the gamma function is*

$$\Gamma(z)\Gamma(1-z) = \frac{\pi}{\sin(\pi z)}.$$

This is "Euler's reflection formula." It can be demonstrated as follows: From (2.8), we have

$$\Gamma(x)\Gamma(y) = \int_0^\infty \int_0^\infty e^{-(s+t)}t^{x-1}s^{y-1}dt\,ds.$$

Setting $s + t = v$ leads to

$$\Gamma(x)\Gamma(y) = \int_0^\infty \int_0^v e^{-v}t^{x-1}(v-t)^{y-1}dt\,dv.$$

The substitution $t = wv$ gives

$$\int_0^\infty \int_0^v e^{-v}t^{x-1}(v-t)^{y-1}dt\,dv = \int_0^\infty e^{-v}v^{x+y-1}dv \int_0^1 w^{x-1}(1-w)^{y-1}dw$$

$$= \Gamma(x+y)\int_0^1 w^{x-1}(1-w)^{y-1}dw.$$

Consequently, we have shown that

$$\frac{\Gamma(x)\Gamma(y)}{\Gamma(x+y)} = \int_0^1 w^{x-1}(1-w)^{y-1}dw. \tag{2.11}$$

One last transformation is to set

$$w = \frac{u}{1+u}.$$

Then,

$$\frac{\Gamma(x)\Gamma(y)}{\Gamma(x+y)} = \int_0^\infty \frac{u^{x-1}du}{(1+u)^{x+y}}.$$

In the last integral set $x = z$, $y = 1 - z$. This shows

$$\Gamma(z)\Gamma(1-z) = \int_0^\infty \frac{u^{z-1}du}{1+u}.$$

We saw in Chapter 1 (Exercise 1.22) that this integral has the value $\pi/\sin(\pi z)$, $0 < \Re\mathrm{eal}(z) < 1$. The product can be extended to all z except the integers.

The function defined in (2.11) is the *beta function* denoted by $B(x, y)$; either side of that equation is, at times, given as the defining expression.

2.2.1 Different Forms for $\Gamma(z)$

There is a related integral representation of $\Gamma(z)$ that is valid at all the integers. It is most easily expressed by

$$\frac{1}{\Gamma(z)} = \frac{i}{2\pi}\int_{C_\Gamma}(-t)^{-z}e^{-t}dt,$$

where C_Γ is a path that begins a ∞e^{i0}, encircles the origin in a counterclockwise direction, and ends at $\infty e^{2\pi i}$. The derivation of this result is based on *Hankel's definition* of the gamma function

$$\Gamma(z) = \frac{-1}{2i \sin \pi z} \int_{C_\Gamma} (-t)^{z-1} e^{-t} dt = \frac{e^{-\pi i z}}{2i \sin \pi z} \int_{C_\Gamma} t^{z-1} e^{-t} dt. \qquad (2.12)$$

for the same contour. Then one makes use of the reflection identity $\Gamma(z)\Gamma(1 - z) = \pi / \sin \pi z$.

As with other examples, the contour is on different sides of the branch cut. That is, above the axis

$$t^z = e^{z \log t} = e^{z \log s} = s^z, \quad s > 0, \quad \text{is real},$$

and below the axis

$$t^z = e^{z(\log t + 2\pi i)} = s^z e^{2\pi i z}, \quad s > 0, \quad \text{is real}.$$

Thus, the integral on the right in (2.12) can be written as

$$\int_{C_\Gamma} t^{z-1} e^{-t} dt = \lim_{R \to \infty} \left[\int_R^\varepsilon s^{z-1} e^{-s} ds + \int_{C_\varepsilon} t^{z-1} e^{-t} dt + e^{2\pi i z} \int_\epsilon^R s^{z-1} e^{-s} ds \right]. \qquad (2.13)$$

Notice that as $\varepsilon \to 0$, the integral along C_ε will vanish. Let $t = \varepsilon e^{i\theta}$. Then,

$$\int_{C_\varepsilon} t^{z-1} e^{-t} dt = i \int_0^{2\pi} (\varepsilon e^{i\theta})^{z-1} e^{-\varepsilon(\cos \theta + i \sin \theta)} d\theta.$$

If we let $z = x + iy$, then,

$$i\theta z = i\theta(x + iy) = i\theta x - \theta y,$$

and

$$\varepsilon^z = e^{z \log \varepsilon} = e^{x \log \varepsilon} e^{iy \log \varepsilon} = \varepsilon^x e^{iy \log \varepsilon}.$$

Thus,

$$\left| \int_{C_\varepsilon} t^{z-1} e^{-t} dt \right| \le \varepsilon^x \int_0^{2\pi} e^{-y\theta} e^{-\varepsilon \cos \theta} d\theta. \qquad (2.14)$$

Since $x = \Re\text{eal}(z) > 0$, and the integral on the left of (2.14) is finite, then $|\int_{C_\varepsilon} t^{z-1} e^{-t} dt| \to 0$, as $\varepsilon \to 0$. Equations (2.12)–(2.13) then become

$$\Gamma(z) = \frac{e^{-\pi i z}}{2i \sin \pi z} \int_{C_\Gamma} t^{z-1} e^{-t} dt$$

$$= \lim_{R \to \infty} \left[\frac{e^{-\pi i z}(e^{2\pi i z} - 1)}{2i \sin \pi z} \int_0^R s^{z-1} e^{-s} ds \right]$$

$$= \int_0^\infty s^{z-1} e^{-s} ds, \quad \Re\text{eal}(z) > 0.$$

This is the standard form for the Γ-function.

There are many other identities involving the gamma function, which will be introduced as needed.

2.3 Functions Defined by Differential Equations

A number of the solutions of ordinary differential equations lead to special functions. Sometimes these functions arise simply because of the geometry of the system being analyzed. So, for fluid flow in a rectangular channel, the Cartesian coordinate system leads to trigonometric functions, while for flow through a circular pipe, the cylindrical coordinate system leads to Bessel functions.

Several of the special functions can be obtained as special cases of the equation,

$$x^2(1 + r_m x^m)y'' + x(p_0 + p_m x^m)y' + (q_0 + q_m x^m)y = 0, \tag{2.15}$$

where m is a positive integer (Hildebrand, p. 138). Some, even if their coefficients are not polynomials, may be transformed to be polynomials.

2.3.1 Legendre Functions

Legendre's Equation. If a potential flow solution is sought in space using spherical coordinates (r, θ, ϕ), independent of the circumferential angle θ, there arises a need to solve

$$\sin \phi \Phi''(\phi) + \cos \phi \Phi'(\phi) + n(n+1) \sin \phi \Phi(\phi) = 0, \quad 0 < \phi < \pi,$$

where n is an integer. The substitutions $x = \cos(\phi)$, $y(x) = \Phi(\phi)$, lead to

$$(1 - x^2)y'' - 2xy' + n(n+1)y = 0. \tag{2.16}$$

This is an equation of the form (2.15) with $m = 2$, $r_2 = -1$, $p_0 = 0$, $p_2 = -2$, $q_0 = 0$, and $q_2 = n(n+1)$. Then it is possible to look for a series solution of the form $y(x) = \sum_{k=0}^{\infty} c_k x^k$, and the polynomial solutions that arise, when n is a nonnegative integer are the Legendre polynomials $P_n(x)$. These are $P_0(x) = 1$, $P_1(x) = x$, $P_2(x) = (3x^2 - 1)/2$, and $P_3(x) = (5x^3 - 3x)/2$, ... normalized by the condition $P_n(1) = 1$. A general formula for $P_n(x)$ may be given as a series,

$$P_n(x) = \sum_{k=0}^{\left[\frac{n}{2}\right]} (-1)^k \frac{(2n - 2k)!}{2^n(n-k)!(n-2k)!} x^{n-2k},$$

where $\left[\frac{n}{2}\right]$ is the greatest integer less than or equal to $\frac{n}{2}$. An important companion result is *Rodrigues's formula*

$$P_n(x) = \frac{1}{2^n n!} \frac{d^n}{dx^n} (x^2 - 1)^n.$$

(It should be noted that there is a second, linearly independent solution of (2.16), generally written as Q_n; see Exercise 2.18.)

The Legendre polynomials are an example of a class of *orthogonal polynomials* which are defined as solutions of second-order linear differential equations. We will consider another set, Chebyshev polynomials, in Section 2.3.5. The general theory of orthogonal polynomials is best treated by the methods discribed in Chapter 3. The *associated Legendre equation* (see Exercise 2.8, equation (2.46)) arises when the separation assumption is made in Laplace's equation with asymmetry. There are many other special functions defined by differential equations that are not of the form (2.15), but an understanding of how to treat this equation goes far in providing a necessary analysis in other cases.

Example 2.1. **Mathieu's Equation** arises in studying potential flow in elliptic cylindrical coordinates. It is given by

$$y''(\theta) + (\lambda + 16q_1 \cos 2\theta)y(\theta) = 0, \quad 0 \le \theta \le \pi. \tag{2.17}$$

The substitution $x = \cos^2 \theta = (\cos 2\theta + 1)/2 \Rightarrow \cos 2\theta = 2x - 1$. The differential equation is then

$$4x(1 - x)y''(x) + (2 - 4x)y'(x) + (\lambda - 16q_1 + 32q_1 x)y(x) = 0.$$

Though it is not of the form (2.15), the equation may be solved by using a series and by other means. The solutions of (2.17) have been studied because they are typical of linear equations with periodic coefficients; for example, see (Lozier et al.)

2.3.2 Bessel Functions

These special functions are among the most important and the most studied of the genre. They can be defined by the solutions of a differential equation

$$x^2 y'' + xy' + (x^2 - v^2)y = 0. \tag{2.18}$$

This equation arises naturally when solutions are sought to certain partial differential equations with variable coefficients. It is noted that (2.18) is of the form (2.15) with $m = 2$, $r_2 = 0$, $p_0 = 1$, $p_2 = 0$, $q_0 = 1$, and $q_2 = -v^2$. Equations of this form are easily solvable by power series about $x = 0$ in the form

$$y(x) = \sum_{k=0}^{\infty} c_k x^{mk+s}.$$

Here s is a parameter determined by solutions of the *indicial equation*:

$$s^2 + (p_0 - 1)s + q_0 = 0.$$

So, for the Bessel equation $s = \pm v$. The resulting *recurrence relation* for the choice $s = v \ge 0$ is

$$4(k + 1)(k + 1 + v)c_{k+1} + c_k = 0, \quad k = 0, 1, 2, \ldots.$$

Since $\lim_{k \to \infty} |c_{k+1}/c_k| = 0$, the series has an infinite radius of convergence. This function, represented by the series, may now be understood to be a function of a complex variable, z, and from our discussion in Chapter 1, it is evident that if v is an integer, the solution is analytic, and, in fact, an entire function. On the other hand, for noninteger values of v, clearly this function has a branch-point singularity at the origin. Because the point $x = 0$ is a (regular) singular point of the equation, no other such solution is to be expected. The normalized solution is the *Bessel function of order v* written as

$$J_v(x) = \sum_{k=0}^{\infty} \frac{(-1)^k (\frac{x}{2})^{2k+v}}{\Gamma(k + v + 1)k!}. \tag{2.19}$$

It may be verified that the two solutions, $J_v(x)$ and $J_{-v}(x)$, where

$$J_{-v}(x) = \sum_{k=0}^{\infty} \frac{(-1)^k (\frac{x}{2})^{2k-v}}{\Gamma(k - v + 1)k!},$$

are linearly independent for noninteger values of v. However, if v is an integer, say n, as it is in many applications, then the gamma functions become factorials, but the factorials are not defined for negative arguments. In the last summation, when $v = n$, let $p = k - n$,

$$J_{-n}(x) = \sum_{p=0}^{\infty} \frac{(-1)^{p+n} (\frac{x}{2})^{2p+n}}{p!(p + n)!} = (-1)^n J_n(x). \tag{2.20}$$

The series (2.20) can also be derived by taking $s = -v$ from the indicial equation. In any case, this means that in the case of integer v, the second solution to Bessel's equation must be found by some other means. The method of reduction of order can be used to show that such a solution will exist, and it will have a singularity at $x = 0$. The second solution may be written in a variety of forms but, in most common usage, is $Y_v(x)$, defined as

$$Y_v(x) = \frac{J_v(x) \cos(v\pi) - J_{-v}(x)}{\sin(v\pi)}. \tag{2.21}$$

If v is an integer, this function is defined in the limiting sense, and L'Hospital's rule must be employed to evaluate the result. This is a non-trivial limit because of the ways in which v occurs in the series. Terms involving $\log(x)$ arise so that the solution of problems on domains containing $x = 0$ are not usually described by means of $Y_v(x)$. However, on an interval, $0 < a < x < b$, both $J_v(x)$ and $Y_v(x)$ are likely to be needed.

If v is not an integer, $Y_v(x)$ is still well-defined and plays an important role in many studies; it is not a third linearly independent solution but encompasses just the behavior of the Bessel functions involved; as $x \to \infty$, for instance. The form of the series solution when v is an integer may be found in handbooks, such as Lozier et al., but we will see shortly that by integral representations much of this work can be obviated.

Modified Bessel Functions

An important related equation related to (2.18) is

$$x^2 y'' + x y' - (x^2 + v^2) y = 0. \tag{2.22}$$

Its solutions are called "modified Bessel functions." A canonical pair of solutions $I_v(x)$ and $K_v(x)$ is given by,

$$I_v(x) = \sum_{k=0}^{\infty} \frac{(\frac{x}{2})^{2k+v}}{\Gamma(k+v+1)k!}, \tag{2.23}$$

$$K_v(x) = \frac{\pi}{2} \frac{(I_{-v}(x) - I_v(x))}{\sin(v\pi)}, \quad v \text{ not an integer.} \tag{2.24}$$

When v is an integer, then $K_v(x)$ is defined in a limiting sense, and terms involving $\log(x)$ arise. (See (Lozier et al.) or (Watson)).

2.3.3 Hypergeometric Functions

The *hypergeometric* differential equation is

$$x(1 - x)y'' + (\gamma - (\alpha + \beta + 1)x)y' - \alpha\beta y = 0. \tag{2.25}$$

It too is of the form (2.15) with $m = 1$, $p_0 = \gamma$, $p_1 = \alpha + \beta + 1$, $q_0 = 0$, $q_1 = -\alpha\beta$, $r_1 = -1$. As long as γ is neither zero, nor a negative integer, the power series solution to (2.25) leads to

$$y = 1 + \frac{\alpha\beta}{1 \cdot \gamma} x + \frac{\alpha(\alpha + 1)\beta(\beta + 1)}{2!\gamma(\gamma + 1)} x^2 + \cdots.$$

This solution is defined as

$$_2F_1(\alpha; \beta; \gamma; x) = \frac{\Gamma(\gamma)}{\Gamma(\alpha)\Gamma(\beta)} \sum_{k=0}^{\infty} \frac{\Gamma(\alpha + k)\Gamma(\beta + k)}{k!\Gamma(\gamma + k)} x^k. \tag{2.26}$$

The series converges absolutely for $|x| < 1$. The convergence at either the endpoints of the interval or on the unit circle in the complex plane depends on the coefficients, α, β, and γ. Absolute convergence on $|x| = 1$ is assured if $\Re\text{eal}(\alpha + \beta - \gamma) < 0$.

The hypergeometric series terminates for certain values of α and β. Rather than examine that situation in general, we note that using the transformation $x \rightarrow 1 - 2x$, $\alpha = -n$, $\beta = n + 1$, $\gamma = 1$ in (2.25), results in (2.16), so it is possible to make the connection that

$$_2F_1\left(-n; n + 1; 1; \frac{1}{2}(1 - x)\right) = P_n(x),$$

which are the Legendre polynomials. Other such connections will be observed later in Section 2.3.5.

2.3.4 Confluent Hypergeometric Functions

In the hypergeometric equation, there are three adjustable parameters, which though of great generality leads to a function of four variables including x. Of less generality is the equation obtained when $\beta \to \infty$. This is done by setting $x = x/\beta$ and then taking the infinite limit. There is a price to paid however. The singularity at infinity is a confluence of two regular singularities, rendering it an irregular singularity. The resulting differential equation is

$$xy'' + (\gamma - x)y' - \alpha y = 0.$$

It too is a form of (2.15), with $m = 1$, $r_1 = q_0 = 0$, $p_0 = \gamma$, $p_1 = -1$, and $q_1 = -\alpha$. The series solutions are based on *Kummer's function*:

$$M(\alpha, \gamma; x) = 1 + \frac{\alpha x}{\gamma} + \frac{\alpha(\alpha + 1)}{\gamma(\gamma + 1)2!} + \cdots$$

$$= \frac{\Gamma(\gamma)}{\Gamma(\alpha)} \sum_{k=0}^{\infty} \frac{\Gamma(\alpha + k)}{\Gamma(\gamma + k)} \frac{x^k}{k!}. \tag{2.27}$$

This series converges for all $|x| < \infty$. It terminates if α is a negative integer.

An alternative notation for Kummer's function is $_1F_1(\alpha; \gamma; x)$.

2.3.5 Chebyshev Functions; Worked Example

Chebyshev's Equation. Though not usually arising from the method of separation of variables for partial differential equations as some other special functions do, Chebyshev functions are ideally suited to the numerical solution of many problems. They satisfy

$$(1 - x^2)y'' - xy' + p^2 y = 0, \quad -1 < x < 1. \tag{2.28}$$

This is an equation of the form (2.15) with $m = 2$, $r_2 = -1$, $p_0 = 0$, $p_2 = -1$, $q_0 = 0$, and $q_2 = p^2$. It is possible then to look for a series solution of the form $y(x) = \sum_{k=0}^{\infty} c_k x^k$. The recurrence formula is given by

$$c_{k+2} = \frac{k^2 - p^2}{(k + 2)(k + 1)} c_k, \quad k = 0, 1, 2, \ldots.$$

So, if p is an integer, one of the series terminates. These are called the "Chebyshev polynomials" $T_p(x)$. They are $T_0(x) = 1$, $T_1(x) = x$, $T_2(x) = 2x^2 - 1$, and $T_3(x) = 4x^3 - 3x, \ldots$, normalized by the condition $T_p(1) = 1$. A general formula for $T_p(x)$ may be given as a series,

$$T_p(x) = \frac{p}{2} \sum_{k=0}^{\left[\frac{p}{2}\right]} (-1)^k \frac{(p - k - 1)!}{k!(p - 2k)!} (2x)^{p-2k},$$

where $\left[\frac{p}{2}\right]$ is the greatest integer less than or equal to $\frac{p}{2}$. An important transformation in working with the Chebyshev equation is to set $x = \cos\theta$ in (2.28).

With $y = f(\theta)$, there results $f'' + p^2 f = 0$, $-\pi \le \theta \le \pi$. Hence, $f(\theta) = \cos p\theta$ is the even solution of this equation, and

$$\cos p\theta = T_p(\cos\theta) = T_p(x) \tag{2.29}$$

is an alternate representation for the Chebyshev polynomials.

The Chebyshev polynomials also have a connection to the hypergeometric functions (2.26). In this case, it is

$$T_p(x) = {}_2F_1\left(-p; p; \frac{1}{2}; \frac{1-x}{2}\right).$$

There is a second set of Chebyshev polynomials, of the second kind, related to the solution $f(\theta) = \sin p\theta$ shown above. These are given by

$$U_p(x) = (p+1){}_2F_1\left(-p; p+2; \frac{3}{2}; \frac{1-x}{2}\right) = \frac{\sin(p+1)\theta}{\sin\theta}. \tag{2.30}$$

Hence, $U_0(x) = 1$, $U_1(x) = 2x$, $U_2(x) = 4x^2 - 1$, $U_3(x) = 8x^3 - 4x$, ... These polynomials may also be generated directly from the differential equation:

$$(1 - x^2)y'' - 3xy' + p(p+2)y = 0. \tag{2.31}$$

(See Exercise 2.11.) They are more suited to approximating functions that take large values in a neighborhood of $x = \pm 1$.

Remark 2.2. *In research on numerical solution of differential equations, (Grosch and Orszag) solve the equation,*

$$u'' + \left(\frac{5}{2} - \frac{1}{4}x^2\right)u = 0, \quad -0 < x < \infty.$$

Because the interval of definition is infinite, an accurate numerical solution is not a simple matter, even for solutions satisfying $u \to 0$, as $x \to \infty$. The approach taken is make a change of independent variable by

$$z = 2\left(\frac{x^2}{x^2 + L^2}\right) - 1,$$

for some constant L. This map transforms $-0 < x < \infty$ to $-1 \le z < 1$. The solution was assumed to be of the form

$$u(x) = \sum_{n=0}^{N} a_n T_n(z),$$

where T_n is the Chebyshev polynomial of degree n. A very accurate solution may be found by taking $L = \sqrt{2}$, and $N = 18$. The comparison with the approximate solution is easy to make in this case because there is an exact solution $u(x) = (x^2 - 1)e^{-x^2/4}$.

Worked Example

We wish to obtain the coefficients for the Chebyshev series for a function $f(t) = 1/(1+t^2)$, that is,

$$\frac{1}{1+t^2} = \sum_{k=0}^{\infty} a_k T_k(t), \quad t \in (-1, 1),$$

where by (2.29), the Chebyshev polynomial is also defined by

$$T_k(t) = \cos(k \cos^{-1}(t)).$$

By the methods we use in Chapter 3, the Chebyshev polynomials may be shown to satisfy the *orthogonality conditions*:

$$\int_{-1}^{1} \frac{T_k T_j dt}{\sqrt{1-t^2}} = \begin{cases} 0, & k \neq j \\ \frac{\pi}{2}, & k = j \neq 0. \\ \pi, & k = j = 0 \end{cases}$$

Hence, the coefficients are given by the integrals,

$$a_k = \frac{2}{\pi} \int_{-1}^{1} \frac{T_k dt}{\sqrt{1-t^2}(1+t^2)}, \quad k > 0 \tag{2.32}$$

$$a_0 = \frac{1}{\pi} \int_{-1}^{1} \frac{dt}{\sqrt{1-t^2}(1+t^2)}.$$

Making the standard change of variable, $t = \cos \psi$, (2.32) can be put in the following form

$$a_k = \frac{1}{\pi} \int_{0}^{2\pi} \frac{\cos(k\psi) d\psi}{1+\cos^2 \psi} = \frac{1}{\pi} \Re\text{eal}\{I\}, \quad k > 0,$$

where the integral I is defined by

$$I(k) = \int_{0}^{2\pi} \frac{e^{ik\psi} d\psi}{1+\cos^2 \psi}. \tag{2.33}$$

We notice that $a_0 = I(0)/2\pi$. Then (2.33) can be understood to be the integral of an analytic function around the unit circle, on which $z = e^{i\psi}$. By this means, it can be transformed into

$$I = \frac{4}{i} \oint_C \frac{z^{k+1} dz}{(z^2+1)^2 + 4z},$$

where C is the unit circle. The poles of the integrand – there are no branch points because k is an integer – are located at the zeroes of the denominator,

$$z_{1,2} = \pm i(\sqrt{2} - 1), \quad z_{3,4} = \pm i(\sqrt{2} + 1).$$

Clearly, only the first set of poles lies inside the unit circle, hence,

$$I = 2\pi i \Big(\text{Res}(z_1) + \text{Res}(z_2) \Big).$$

The formula for the residue in this case is

$$\text{Res} = \frac{1}{i} \frac{z^k}{z^2 + 5}.$$

So,

$$\text{Res}(z_1) = \frac{1}{i} \frac{e^{i\pi k/2}}{2(\sqrt{2}+1)} \left(\sqrt{2}-1\right)^k.$$

The residue at z_2 is almost identical, except that the sign in the exponentiand is changed. Thus, when adding, we get

$$\sum \text{Res} = \frac{1}{i} \frac{\cos(k\pi/2)}{\sqrt{2}+1} \left(\sqrt{2}-1\right)^k.$$

Thus, the integral $I(k)$ has been found to be

$$I(k) = 2\pi \frac{\cos(k\pi/2)}{\sqrt{2}+1} \left(\sqrt{2}-1\right)^k.$$

So, the Chebyshev series has been determined to be

$$\frac{1}{1+t^2} = \frac{1}{\sqrt{2}+1} + \frac{2}{\sqrt{2}+1} \sum_{k=1}^{\infty} \cos(k\pi/2)\left(\sqrt{2}-1\right)^k T_k(t).$$

We notice in passing that this is a strongly convergent series, because the k-term contains the factor

$$\left(\sqrt{2}-1\right)^k = e^{k\log(\sqrt{2}-1)}.$$

2.3.6 Airy Functions

Another differential equation that defines important special functions is:

$$y'' - xy = 0. \tag{2.34}$$

Though $x = 0$ is an ordinary point for this equation, it is also of the form (2.15) with $m = 3$, with $r_3 = p_0 = p_3 = q_0 = 0$, and $q_3 = -1$. If a solution is sought of the form

$$y(x) = \sum_{k=0}^{\infty} c_k x^k,$$

the result is the recurrence formula

$$c_{k+2} = \frac{c_k}{(k+2)(k+1)}, \quad k = 0, 1, \ldots,$$

with $c_2 = 0$. The general solution may be written as

$$y = c_0 \left(1 + \frac{x^3}{2\cdot 3} + \frac{x^6}{(2\cdot 3)(5\cdot 6)} + \cdots\right)$$

$$+ c_1 \left(x + \frac{x^4}{3\cdot 4} + \frac{x^7}{(3\cdot 4)(6\cdot 7)} + \cdots\right). \tag{2.35}$$

Particular values of coefficients c_0 and c_1 lead to the Airy functions of the first and second kinds, $\mathrm{Ai}(x)$ and $\mathrm{Bi}(x)$, respectively. This is clearly a case in which an alternative definition of the special function is desirable, not simply for the values of the functions and their derivatives at the origin but also for large values of their arguments.

2.4 Integral Representations

2.4.1 Physical Problem: Heat Conduction

We have already seen in a couple of instances that alternative representations of special functions are likely to be useful in enhancing some of the ideas involved. The gamma function is an instance, but it does not typically arise as the solution of some differential equation. A more common example arises if one desires to solve the heat equation,

$$\frac{\partial u}{\partial t} = \kappa \frac{\partial^2 u}{\partial x^2}, \quad t > 0, \ 0 < x < \infty,$$

under the conditions

$$u(0, t) = 0, \quad t > 0,$$

$$u(x, t) \to U_0 \ \text{as} \ x \to \infty, \ t > 0.$$

If a solution of the form $u(x, t) = f(\xi)$ is sought, where, for dimensional reasons, $\xi = x/\sqrt{\kappa t}$, an ordinary differential equation for f is found to be

$$f''(\xi) + \frac{1}{2}\xi f'(\xi) = 0.$$

The boundary conditions on f are

$$f(0) = 0, \quad f \to U_0 \ \text{as} \ \xi \to \infty.$$

The solution in terms of an integral is found to be

$$f(\xi) = \frac{U_0}{\sqrt{\pi}} \int_0^{\xi} e^{-s^2/4} ds.$$

A special function which often arises in heat conduction problems is the *error function*, $\mathrm{erf}(z)$, defined as

$$\mathrm{erf}(z) = \frac{2}{\sqrt{\pi}} \int_0^z e^{-s^2} ds. \tag{2.36}$$

So the solution for f may be rewritten as $f(\xi) = U_0 \, \mathrm{erf}(\xi/2)$. The desired solution to the heat equation is

$$u(x, t) = U_0 \, \mathrm{erf}\left(\frac{x}{\sqrt{4\kappa t}}\right).$$

This solution could also have been found by Laplace transform methods as we will show in Chapter 5.

A useful companion integral is the *complementary error function*, erfc(z), defined as

$$\text{erfc}(z) = \frac{2}{\sqrt{\pi}} \int_z^\infty e^{-s^2} ds.$$

Clearly, $\text{erfc}(z) = 1 - \text{erf}(z)$.

2.4.2 Bessel Function Integrals

The integral representation most commonly quoted is

$$J_\nu(x) = \frac{1}{\pi} \int_0^\pi \cos(\nu\theta - x\sin\theta)d\theta - \frac{\sin(\nu\pi)}{\pi} \int_0^\infty e^{-\nu t - x\sinh t} dt, \qquad (2.37)$$

which is valid when $\Re\text{eal}(x) > 0$. This representation may derived from the differential equation by the generalized transform techniques of Chapter 7. From definition (2.21), it follows that

$$Y_\nu(x) = \frac{1}{\pi} \int_0^\pi \sin(\nu\theta - x\sin\theta)d\theta$$

$$-\frac{1}{\pi} \int_0^\infty \left(e^{\nu t} + e^{-\nu t}\cos(\nu\pi)\right) e^{-x\sinh t} dt, \quad \Re\text{eal}(x) > 0. \qquad (2.38)$$

If $\nu = n$, an integer, (2.37) simplifies considerably. The second integral vanishes; the first integral is valid for all finite values of x. Then,

$$J_n(x) = \frac{1}{\pi} \int_0^\pi \cos(n\theta - x\sin\theta)d\theta. \quad n = 0, 1, 2, \ldots. \qquad (2.39)$$

(An alternative derivation of this formula is given in Exercise 2.24.)

Example 2.2. We use the integral formula (2.39) to derive expansions of trigonometric functions in terms of Bessel functions. Thus,

$$J_n(x) = \frac{1}{2\pi} \int_0^\pi \{e^{in\theta - ix\sin\theta} + e^{-in\theta + ix\sin\theta}\}d\theta = \frac{1}{2\pi} \int_{-\pi}^\pi e^{-in\theta + ix\sin\theta} d\theta.$$

This means that $J_n(x)$ determines the coefficients of the complex Fourier series of $e^{ix\sin\theta}$ (See Exercise 3.16.) That is,

$$e^{ix\sin\theta} = \sum_{n=-\infty}^\infty J_n(x)e^{in\theta}.$$

Using Euler's identity $e^{ix\sin\theta} = \cos(x\sin\theta) + i\sin(x\sin\theta)$, even if x is complex, gives

$$\cos(x\sin\theta) = \sum_{n=-\infty}^\infty J_n(x)\cos n\theta,$$

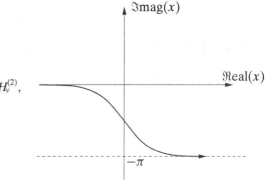

Figure 2.1. Hankel function, $H_\nu^{(2)}$, path.

and

$$\sin(x \sin \theta) = \sum_{n=-\infty}^{\infty} J_n(x) \sin n\theta.$$

Because, as we have already seen in (2.20), $J_n(x) = (-1)^n J_{-n}(x)$, then

$$\cos(x \sin \theta) = J_0(x) + 2 \sum_{n=1}^{\infty} J_{2n}(x) \cos(2n\theta) \qquad (2.40)$$

and

$$\sin(x \sin \theta) = 2 \sum_{n=1}^{\infty} J_{2n-1}(x) \sin\left((2n-1)\theta\right). \qquad (2.41)$$

Hankel Functions

Two other Bessel functions, sometimes called "third kind functions" are the Hankel functions $H_\nu^{(1)}$ and $H_\nu^{(2)}$ defined as

$$H_\nu^{(1)}(x) = J_\nu(x) + i Y_\nu(x)$$

and

$$H_\nu^{(2)}(x) = J_\nu(x) - i Y_\nu(x).$$

From the integral representations developed for $J_\nu(x)$ and $Y_\nu(x)$, because the integrands are entire functions of t,

$$H_\nu^{(1)}(x) = \frac{1}{\pi i} \int_{-\infty}^{\infty + \pi i} e^{-\nu t - x \sinh t} dt, \quad \left|\arg x\right| < \pi/2$$

and

$$H_\nu^{(2)}(x) = -\frac{1}{\pi i} \int_{-\infty}^{\infty - \pi i} e^{-\nu t - x \sinh t} dt, \quad \left|\arg x\right| < \pi/2.$$

See Figure 2.1 for the path for $H_\nu^{(2)}(x)$.

Modified Bessel Function Integrals

Integral representation of the modified Bessel functions are given by

$$I_\nu(x) = \frac{1}{\pi} \int_0^\pi e^{x\cos\theta}\cos(\nu\theta)d\theta - \frac{\sin(\nu\pi)}{\pi}\int_0^\infty e^{-\nu t - x\cosh t}dt,$$

and from definition (2.24),

$$K_\nu(x) = \int_0^\infty e^{-x\cosh t}\cosh(\nu t)dt.$$

2.4.3 Hypergeometric Integrals

The classical statement, derived by Euler for the hypergeometric function, is given by

$$_2F_1(\alpha;\beta;\gamma;x) = \frac{\Gamma(\gamma)}{\Gamma(\beta)\Gamma(\gamma-\beta)}\int_0^1 t^{\beta-1}(1-t)^{\gamma-\beta-1}(1-xt)^{-\alpha}dt. \tag{2.42}$$

This is valid for $\Re\text{eal}(\gamma) > \Re\text{eal}(\beta) > 0$ in the x-plane cut along the real axis from 1 to ∞.

An integral representation for the confluent hypergeometric function (2.27), which is analogous to (2.42) is

$$M(\alpha;\beta;x) = \frac{\Gamma(\beta)}{\Gamma(\alpha)\Gamma(\beta-\alpha)}\int_0^1 t^{\alpha-1}(1-t)^{\beta-\alpha-1}e^{xt}dt, \tag{2.43}$$

valid if $\Re\text{eal}(\beta) > \Re\text{eal}(\alpha) > 0$.

2.4.4 Airy Function Integrals

The Airy functions were defined implicitly by a differential equation (2.34). However, the actual values of the functions at $x = 0$, are not really known by series. The integral representations remove this ambiguity. The original definition is

$$\text{Ai}(x) = \frac{1}{\pi}\int_0^\infty \cos\left(sx + \frac{1}{3}s^3\right)ds.$$

A second, linearly independent solution is

$$\text{Bi}(x) = \frac{1}{\pi}\int_0^\infty \left\{e^{tx-\frac{1}{3}t^3} + \sin\left(tx + \frac{1}{3}t^3\right)\right\}dt.$$

The actual derivation of these integral solutions is best carried out by generalized Laplace transform techniques, as we will show in Chapter 7. However, based on what we have already derived, it is not difficult to show by direct substitution that these integrals both satisfy (2.34) and that in (2.35),

$$c_0 = \text{Ai}(0) = \text{Bi}(0)/\sqrt{3} = 3^{-2/3}/\Gamma(2/3), \tag{2.44}$$

$$c_1 = \text{Ai}'(0) = -\text{Bi}'(0)/\sqrt{3} = 3^{-1/3}/\Gamma(1/3). \tag{2.45}$$

There are interesting integral representations for the functions that arise as polynomial solutions, such as Legendre functions, Chebyshev functions, Hermite functions, and others. We refer the reader to the handbooks and references for further information on these matters.

REFERENCES

M. Abramowitz, and I. A. Stegun, eds., *Handbook of Mathematical Functions*, Dover, New York, 1972.

G. K. Batchelor, Axial Flow in Trailing Vortices, *J. Fluid Mech.* **20**, 65–658 (1964).

P. G. Drazin and W. H. Reid, *Hydrodynamic Stability*, 2nd edition, Cambridge University Press, New York, 2004.

C. E. Grosch and S. A. Orszag, Numerical Solution of Problems in Unbounded Regions: Coordinate Transforms, *J. Comp. Phys.* **25**, 273–295 (1977).

F. B. Hildebrand, *Advanced Calculus for Applications*, 2nd edition, Prentice-Hall, Englewood Cliffs, 1976.

H. Jeffreys and B. S. Jeffreys, *Methods of Mathematical Physics*, 3rd edition, Cambridge University Press, Cambridge, UK, 1953.

D. Lozier, F. Olver, C. Clark, and R. Boisvert, eds., *Digital Library of Mathematical Functions*, National Institute of Standards and Technology, Washington, DC, 2005, available at http://dlmf.nist.gov/.

G. N. Watson, *Theory of Bessel Functions*, 2nd edition, Cambridge University Press, Cambridge, UK, 1966.

E. T. Whittaker and G. N. Watson, *A Course of Modern Analysis*, 4th edition, Cambridge University Press, Cambridge, UK, 1952.

EXERCISES

2.1 Consider the function $P(\eta)$ of Section 2.1.1. It is defined for all $\eta > 0$, where $P'(\eta) < 0$, and we must have $\lim_{\eta \to \infty} P(\eta) = 0$. Show that $\lim_{\eta \to 0^+} P(\eta) = 2\log 2$.

2.2 Verify, by a suitable change of variables that

$$\int_0^\infty x^{p-1} e^{-ax^q} dx = \frac{\Gamma(p/q)}{q a^{p/q}}, \quad a > 0, \ p > 0, q > 0.$$

2.3 Show that

$$\int_0^\infty s^{-n} e^{-k/s^2} ds = \frac{\Gamma\left(\frac{1}{2}n - \frac{1}{2}\right)}{2k^{\frac{1}{2}n - \frac{1}{2}}}, \quad k > 0, \ n > 1.$$

2.4 If m and n are arbitrary but fixed positive integers, show that

$$\int_0^1 x^m (\log x)^n dx = \frac{(-1)^n n!}{(m+1)^{n+1}}.$$

How might this result be generalized if m and n are not integers?

2.5 (a) Use the same method as in finding $\left(\Gamma(\frac{1}{2})\right)^2$ in Section 2.2 to show that

$$(\Gamma(p))^2 = 2^{1-2p} \sqrt{\pi} \frac{\Gamma(2p)\Gamma(p)}{\Gamma\left(p + \frac{1}{2}\right)}.$$

Hint: Note that if $t = \sin^2 \theta$, then

$$\int_0^{\pi/2} \sin^{2p-1} \theta \cos^{2q-1} \theta \, d\theta = \frac{1}{2} \int_0^1 t^{p-1}(1-t)^{q-1} dt = \frac{\Gamma(p)\Gamma(q)}{2\Gamma(p+q)},$$

by (2.11).

(b) Hence show that, if n is a nonnegative integer,

$$\Gamma\left(n + \frac{1}{2}\right) = \frac{\sqrt{\pi}(2n)!}{2^{2n}n!}.$$

2.6 A nonlinear pendulum swings from $\theta = \pi/2$ to $\theta = -\pi/2$. Express its period of oscillation in terms of the beta function

$$B(z, w) = \int_0^1 t^{z-1}(1-t)^{w-1},$$

where $z > 0$, and $w > 0$. Simplify the expression using the identity (2.11).

2.7 The function $\mathrm{li}(x)$, called the "logarithmic integral function" is defined by

$$\mathrm{li}(x) = \int_0^x \frac{dt}{\log t}, \quad x < 1.$$

Show that

$$\mathrm{ei}(x) = \int_x^\infty \frac{e^{-t}}{t} dt = -\mathrm{li}(e^{-x}).$$

2.8 The Associated Legendre Functions, $P_n^\mu(x)$ are given by

$$P_n^\mu(x) = (1 - x^2)^{\mu/2} \frac{d^\mu}{dx^\mu} P_n(x), \quad n > \mu,$$

where $P_n(x)$ is the nth-order Legendre polynomial. Verify that $u = P_n^\mu(x)$ is a solution of the differential equation,

$$((1 - x^2)u')' + \frac{\mu^2 u}{1 - x^2} + n(n+1)u = 0, \quad -1 < x < 1, \quad (2.46)$$

using the fact that $y = P_n(x)$ satisfies (2.16).

2.9 The differential equation,

$$w'' + (a \, \mathrm{sech}^2 z - \kappa)w = 0, \quad -\infty < z < \infty,$$

with the conditions $w \to 0$ as $|z| \to \infty$, occurs in both theoretical and applied physics.

(a) Show that the change of variable $x = \tanh z$, $u(x) = w(z)$, transforms the equation into the associated Legendre equation (2.46).

(b) Hence find three exact solutions for $w(z)$ when $a = 12$ in the cases $\kappa = \mu^2$, $\mu = 1, 2, 3$.

2.10 Given the definition (2.29), the Chebyshev polynomials $T_p(x)$ may be obtained by the expression

$$T_p(x) = \cos(p \cos^{-1} x),$$

where p is zero or a positive integer and $|x| \leq 1$. By considering the expressions for T_{p+1}, T_p, and T_{p-1}, obtain the recurrence relation:

$$T_{p+1}(x) = 2x T_p(x) - T_{p-1}(x).$$

Write out $T_0, T_1, \ldots, T_4$ in explicit polynomial form.

Show that an alternative expression for the functions is given by

$$T_p(x) = \frac{1}{2}\left[(x + i\sqrt{1 - x^2})^p + (x - i\sqrt{1 - x^2})^p\right].$$

2.11 Show that setting $x = 1 - 2t$ and $y(x) = w(t)$ in Equation (2.31) gives a *hypergeometric differential equation*,

$$t(1 - t)w'' + [c - (a + b + 1)t]w' - abw = 0.$$

Express the solution for y of (2.31) in terms of $_2F_1(a, b; c; (1 - x)/2)$ for suitable a, b, c. Argue that this solution is a polynomial if p is a nonnegative integer, which verifies (2.30).

2.12 Given the Bessel equation (2.18), when $\nu = n$ is an integer with series solution (2.19), known to satisfy (2.20), derive the representation

$$e^{\left(x - \frac{1}{x}\right)\frac{t}{2}} = \sum_{n=-\infty}^{\infty} J_n(t)x^n.$$

Method: Show first that

$$e^{\frac{xt}{2}} e^{-\frac{t}{2x}} = \sum_{j=0}^{\infty}\sum_{k=0}^{\infty} \frac{(-1)^k (\frac{t}{2})^{j+k}}{j!k!} x^{j-k}.$$

Then, set $n = j - k$, and apply (2.19) and (2.20).

2.13 Given the power series representations of Kummer's function (2.27) and Bessel's function (2.19), show that

$$J_\nu(x) = \frac{e^{-ix}\left(\frac{x}{2}\right)^\nu}{\Gamma(\nu + 1)} M\left(\nu + \frac{1}{2}, 2\nu + 1, 2ix\right).$$

2.14 Use the change of variable $y = xv(x)$ to show that the general solution of the equation

$$y'' - \frac{1}{x}y' - k^2 y = 0$$

is given by $y = c_1 x I_1(kx) + c_2 x K_1(kx)$.

2.15 (a) Write down the general solution to Bessel's equation in the form
$$x^2 y'' + xy' + (k^2 x^2 - p^2)y = 0,$$
when p is not an integer. *Hint:* Let $z = kx$ and set $u(z) = y(x)$.

(b) Show that the transformation $y = x^p w(x)$ in the equation of part (a) leads to
$$w'' + \frac{A}{x}w' + k^2 w = 0,$$
where A is a constant. Determine the value of A.

(c) Hence, show that the general solution to Bessel's equation in part (a) when $p = \pm 1/2$ is

$$y = \frac{c_1 \sin kx}{\sqrt{x}} + \frac{c_2 \cos kx}{\sqrt{x}}.$$

2.16 If γ is a nonpositive integer, so that the result (2.26) does not apply, transform (2.25) by the substitution $y(x) = x^{1-\gamma} w(x)$ to obtain an equation for $w(x)$. Hence, show that then a basic solution to the hypergeometric defferential equation is

$$y(x) = x^{1-\gamma} {}_2F_1(\alpha - \gamma + 1; \beta - \gamma + 1; 2 - \gamma; x).$$

Note: This is equivalent to taking the second indicial root $s_2 = 1 - \gamma$ in the power series solution.

2.17 Some differential equations may be of hypergeometric type even though they do not obviously appear to be so. Find the general solution of

$$(1 - e^{-x})y'' - \frac{1}{2}y' + e^{-x}y = 0, \quad x > 0,$$

near the singular point $x = 0$ by changing the independent variable to $t = 1 - e^{-x}$. That is, write down two linearly independent solutions in terms of ${}_2F_1$. Show that

$$y_1 = {}_2F_1\left(1, -1, -\frac{1}{2}, 1 - e^{-x}\right)$$

and

$$y_2 = (1 - e^{-x})^{3/2} {}_2F_1\left(\frac{5}{2}, \frac{1}{2}, \frac{5}{2}, 1 - e^{-x}\right).$$

See if you can find a closed form expression for y_1, that is, its series terminates.

2.18 It turns out that the second solution to Legendre's equation is more valuable when $|x| > 1$. Make the change of independent variable $z = x^{-1}$ in (2.16). Show that the resulting differential equation for $w(z) = y(x)$ is

$$(1 - z^2)w'' - 2zw' - n(n + 1)z^{-2}w = 0. \tag{2.47}$$

Hence show that when $n = 0$ there is a solution which is

$$Q_0(z) = \frac{1}{2} \log\left(\frac{1 + z}{1 - z}\right), \quad |z| < 1,$$

the *Legendre function of the second kind* of order zero. The succeeding functions of higher orders may be found from the same recurrence formulas as for Legendre polynomials (Abramowitz & Stegun). However, it may be necessary to express them in closed form. Show that the substitution

$$w(z) = z^{n+1}u(\xi), \quad \xi = z^2,$$

in (2.47) leads to a hypergeometric differential equation, and that a solution to this equation is

$$u(\xi) = {}_2F_1\left(\frac{n + 2}{2}, \frac{n + 1}{2}; n + \frac{3}{2}; \xi\right), \quad |\xi| < 1.$$

The *Q*-functions are thereby defined as

$$Q_n(x) = \frac{\sqrt{\pi} n!}{\Gamma(n + \frac{3}{2})(2x)^{n+1}}\, {}_2F_1\left(\frac{n+2}{2}, \frac{n+1}{2}; n + \frac{3}{2}; \frac{1}{x^2}\right), \quad |x| > 1.$$

2.19 An exact solution of *Rayleigh's stability equation* (Drazin and Reid)

$$(U - c)(\phi'' - \alpha^2 \phi) - U''\phi = 0, \quad 0 < z < \infty.$$

When $U(z) = 1 - e^{-z}$, for $0 < z < \infty$, show that a solution that is bounded as $z \to \infty$ is

$$\Phi(z) = \frac{{}_2F_1(p; q; r; t)t^\alpha}{{}_2F_1(p; q; r; 1)}, \quad 0 \le t \le 1,$$

satisfying $\Phi(z_c) = 1$, where $t = e^{-(z - z_c)}$, given that $U(z_c) = c$. The parameters in the solution are

$$p = \alpha + \sqrt{1 + \alpha^2}, \quad q = \alpha - \sqrt{1 + \alpha^2}, \quad r = 1 + 2\alpha.$$

Method: Change the independent variable in the differential equation to t.

2.20 The *Hermite polynomials $He_n(x)$*, are determined as solutions of the differential equation

$$y''(x) - xy'(x) + ny(x) = 0, \quad -\infty < x < \infty.$$

(a) Use a simple power series $y = \Sigma a_k x^k$ to show that if $n = 0, 1, \ldots$, then $He_0 = 1$, $He_1 = x$, $He_2 = x^2 - 1$, $He_3 = x^3 - 3x$, $He_4 = x^4 - 6x^2 + 3, \ldots$, all satisfying $a_n = 1$.

(b) The polynomials may also be found from the recurrence relation

$$He_{n+1}(x) - x He_n(x) + n He_{n-1}(x) = 0.$$

Verify the expressions in part (a) given $He_0 = 1$ and $He_1 = x$.

2.21 A spherical surface of radius a is kept temperature U_0 and is surrounded by an infinite medium, with initial temperature 0. Show that the temperature at a distance $r > a$ from the center is, at time t,

$$u(r, t) = \frac{aU_0}{r}\left\{\mathrm{erfc}\left(\frac{r - a}{\sqrt{4\kappa t}}\right)\right\}.$$

Hint: The spherically symmetric heat equation is

$$\frac{\partial u}{\partial t} = \frac{\kappa}{r^2}\frac{\partial}{\partial r}\left(r^2 \frac{\partial u}{\partial r}\right),$$

and it reduces to the heat equation by the substitution $u = v/r$.

2.22 Derive (2.38) of Section 2.4.2 from (2.37) and (2.21).

2.23 Use equation (2.19) to show that

$$\frac{1}{x}\frac{d}{dx}[x^\nu J_\nu(x)] = x^{\nu-1}J_{\nu-1}(x)$$

and

$$\frac{1}{x}\frac{d}{dx}\left[x^{-\nu}J_\nu(x)\right] = -x^{-\nu-1}J_{\nu+1}(x).$$

Hence show that

$$J_{\nu+1}(x) + J_{\nu-1}(x) = \frac{2\nu}{x}J_\nu(x).$$

2.24 For each complex t, $J_n(t)$ is defined to be the coefficient of z^n in the Laurent expansion of the *generating function*

$$e^{\left(z-\frac{1}{z}\right)\frac{t}{2}} = \sum_{n=-\infty}^{\infty} J_n(t)z^n.$$

(See Exercise 2.12.)

By integrating around the unit circle centered at the origin in the complex z-plane, show that for $n \geq 0$

$$J_n(t) = \frac{1}{\pi}\int_0^\pi \cos(t\sin\theta - n\theta)d\theta.$$

2.25 Verify that the arclength of the ellipse

$$\frac{x^2}{a^2} + \frac{y^2}{b^2} = 1, \quad 0 < a \leq b,$$

is given by $(2\pi b)_2F_1\left(\frac{1}{2}, -\frac{1}{2}; 1; k^2\right)$, where $k^2 = (b^2 - a^2)/b^2$. This shows that there is a connection here with elliptic functions, which were invented to describe the ellipse!

Find the arclength explicitly, when $k^2 = 1/2$.

2.26 The *incomplete gamma function* is defined as

$$\gamma(b; x) = \int_0^x e^{-t}t^{b-1}dt, \quad \Re\text{eal}(b) > 0.$$

Show that

$$\gamma(b; x) = b^{-1}x^b e^{-x} M(1, 1+b, x)$$
$$= b^{-1}x^b M(b, 1+b, -x).$$

2.27 Given the integral representations of Kummer's function (2.43) and the error function (2.36), show that

$$\text{erf}(x) = \frac{2x}{\sqrt{\pi}}M\left(\frac{1}{2}, \frac{3}{2} - x^2\right)$$
$$= \frac{2x}{\sqrt{\pi}}e^{-x^2}M\left(1, \frac{3}{2} - x^2\right).$$

2.28 Verify the values (2.44) and (2.45) for the functions Ai(x), Bi(x), and their derivatives at $x = 0$.

2.29 Show that the Wronskian of Ai(x) and Bi(x) is

$$W[\text{Ai}(x), \text{Bi}(x)] = \pi^{-1}.$$

Hence, the two solutions are truly linearly independent.

2.30 In (2.34), set $\xi = \frac{2}{3}x^{3/2}$, $y(x) = x^{1/2}w(\xi)$. Show that the resulting differential equation is a modified Bessel's equation of order $\frac{1}{3}$. It follows that

$$\text{Ai}(x) = \pi^{-1}\sqrt{x/3}\,K_{1/3}(\xi)$$

and

$$\text{Bi}(x) = \sqrt{x/3}\,[I_{1/3}(\xi) + I_{-1/3}(\xi)]$$

2.31 Given the special function defined as

$$\text{Hi}(x) = \frac{1}{\pi}\int_0^\infty e^{tx - \frac{1}{3}t^3}\,dt, \quad -\infty < x < \infty,$$

show that $y = \text{Hi}(x)$ satisfies

$$y'' - xy = \frac{1}{\pi}.$$

Evaluate Hi(0) and Hi$'(0)$ in terms of the Γ-function.

2.32 The functions Si(x) and Ci(x), called the "sine integral function" and "cosine integral function," respectively, are defined by

$$\text{Si}(x) = \int_0^x \frac{\sin t}{t}\,dt, \quad \text{Ci}(x) = -\int_x^\infty \frac{\cos t}{t}\,dt.$$

Use variation of parameters to show that $y'' + y = 1/x$ has the particular solution:

$$y = \text{Ci}(x)\sin x - \text{Si}(x)\cos x.$$

3

Eigenvalue Problems and Eigenfunction Expansions

3.1 Preamble

In the twentieth century, there were a number of innovations in applied mathematical techniques. Some had their origins in the years before. Rayleigh seems to have begun many of these innovations that were taken up by others in the twentieth century. Rayleigh's criterion establishes the instability of a rotating fluid. Based on elementary physical arguments, he was able to conclude (loosely) that the flow configuration is stable when the square of the circulation increases outward. In the early years of the twentieth century, work was done to put many of these physical arguments on a stronger mathematical footing. One of the techniques that has proven to be useful is the study of eigenvalue problems for differential equations. This technique relates particularly to the method of normal modes of vibration of a physical system and to discrete-mode instabilities of fluid flows. Many examples of this latter application, which is given some limited discussion in this chapter, may be found in (Chandrasekhar) and (Drazin and Reid), for example.

The general mathematical ideas were first developed by Sturm and Liouville, but Fourier had laid much of the groundwork in his theory of heat conduction. This "continuous" treatment, by means of differential equations, has discrete analogs as well, in the theories of matrices and of particle systems.

The reason for the inclusion of this material in this book is previewed in Section 3.2, and principally emerges throughout this chapter: the solution of partial differential equations by means of expansion of the solution in an eigenfunction series.

3.2 Synge's Setup for Rayleigh's Criterion

In his study of the stability of a rotating heterogeneous liquid, J. L. Synge used the following approach (Synge, 1933). Cylindrical coordinates (r, θ, z) are employed to analyze the rotating motion of an inviscid fluid about the z-axis, with velocity having components (u, v, w); the momentum equations in the radial, circumferential, and axial directions, may be respectively, reduced to

$$\frac{\partial u}{\partial t} + u\frac{\partial u}{\partial r} + w\frac{\partial u}{\partial z} - \frac{v^2}{r} = -\frac{1}{\rho}\frac{\partial p}{\partial r}, \tag{3.1}$$

$$\frac{\partial}{\partial t}(rv) + u\frac{\partial}{\partial r}(rv) + w\frac{\partial}{\partial z}(rv) = 0, \tag{3.2}$$

$$\frac{\partial w}{\partial t} + u\frac{\partial w}{\partial r} + w\frac{\partial w}{\partial z} = -\frac{1}{\rho}\frac{\partial p}{\partial z}. \tag{3.3}$$

Though the density may not be constant, the fluid is assumed to be incompressible so that

$$\frac{\partial \rho}{\partial t} + u\frac{\partial \rho}{\partial r} + w\frac{\partial \rho}{\partial z} = 0. \tag{3.4}$$

The mass-conservation equation then reduces to

$$\frac{\partial}{\partial r}(ru) + \frac{\partial}{\partial z}(rw) = 0, \tag{3.5}$$

so that the velocity field is solenoidal. When the density ρ is assumed to be a function of r only, a solution to the equations of motion may thus be found where

$$u = 0, \quad v = \bar{v}(r), \quad w = 0, \quad \rho = \bar{\rho}(r), \quad p = \bar{p}(r). \tag{3.6}$$

The only stipulation on the equations of motion is a balance between the centrifugal force and the pressure gradient:

$$\frac{\bar{v}^2}{r} = \frac{1}{\bar{\rho}}\frac{d\bar{p}}{dr}. \tag{3.7}$$

Small perturbations are imposed on the motion (3.6) of the form

$$(u, v, w) = (u', \bar{v} + v', w'), \quad \rho = \bar{\rho} + \rho', \quad p = \bar{p} + p'. \tag{3.8}$$

These are substituted into the equations of motion (3.1)–(3.5) and the nonlinear terms are neglected

$$\frac{\partial u'}{\partial t} - \frac{2\bar{v}v'}{r} = -\frac{1}{\bar{\rho}}\frac{\partial p'}{\partial r} + \frac{\bar{\rho}'}{\bar{\rho}^2}\frac{d\bar{p}}{dr}, \tag{3.9}$$

$$\frac{\partial}{\partial t}(rv') + u'\frac{d}{dr}(r\bar{v}) = 0,$$

$$\frac{\partial w'}{\partial t} = -\frac{1}{\bar{\rho}}\frac{\partial p'}{\partial z},$$

$$\frac{\partial \rho'}{\partial t} + u'\frac{d\bar{\rho}}{dr} = 0.$$

It is then possible to eliminate p', ρ', and v', and using the centrifugal balance (3.7), to obtain a single partial differential equation for u' and w'

$$\bar{\rho}\frac{\partial^3 u'}{\partial z\partial t^2} - \frac{\partial}{\partial r}\left(\bar{\rho}\frac{\partial^2 w'}{\partial t^2}\right) + F(r)\frac{\partial u'}{\partial z} = 0,$$

where

$$F(r) = \frac{1}{r^3} \frac{d}{dr} (\overline{\rho} r^2 \overline{v}^2). \tag{3.10}$$

However, because the velocity field is axisymmetric there exists a stream function ψ, such that

$$ru' = -\frac{\partial \psi}{\partial z}, \quad rw' = \frac{\partial \psi}{\partial r},$$

and continuity (3.5) is identically satisfied. Thus, a single equation results, for ψ,

$$\frac{\partial}{\partial r} \left(\frac{\overline{\rho}}{r} \frac{\partial^3 \psi}{\partial r \partial t^2} \right) + \frac{\overline{\rho}}{r} \frac{\partial^4 \psi}{\partial z^2 \partial t^2} + \frac{F}{r} \frac{\partial^2 \psi}{\partial z^2} = 0. \tag{3.11}$$

A *normal mode* solution is sought for traveling waves of the form:

$$\psi = \phi(r) e^{ik(z-ct)}, \quad k > 0 \tag{3.12}$$

where k is a wave number and c is the wave speed. The wave number is taken to be real, thereby localizing the disturbance. The wave speed is allowed to be complex. The flow is said to be *temporally stable* if all disturbances of the form (3.12) decay in time, that is, $c_i < 0$. Likewise, the flow is said to be *temporally unstable*, if there is a disturbance of the form (3.12) that grows in time, that is, $c_i > 0$.

The resulting ordinary differential equation for ϕ is found by substituting (3.12) into (3.11) to yield

$$\frac{d}{dr} \left(\frac{\overline{\rho}}{r} \frac{d\phi}{dr} \right) + \left(c^{-2} \frac{F}{r} - k^2 \frac{\overline{\rho}}{r} \right) \phi = 0. \tag{3.13}$$

Typical boundary conditions for a wall-bounded flow at $r = r_1$ and $r = r_2$ are $\partial \psi / \partial z = 0$, for $r = r_1$ and $r = r_2$, and, therefore,

$$\phi(r_1) = \phi(r_2) = 0. \tag{3.14}$$

This is an example of an eigenvalue problem in which a solution ϕ is obtained for certain values c^{-2} when the wave number and the coefficients are given through the basic flow.

Example 3.1. Suppose that the fluid has the density distribution $\overline{\rho} = \rho_0 r$, ρ_0 constant, and performs solid body rotation, such that $\overline{v} = \omega_0 r$, ω_0 constant. Then, from (3.10), the function

$$F(r) = \frac{1}{r^3} \frac{d}{dr} \left(\rho_0 \omega_0^2 r^5 \right) = 5 \rho_0 \omega_0^2 r. \tag{3.15}$$

The resulting form of the equation for ϕ is, from (3.13)

$$\phi'' + \left(5 \omega_0^2 c^{-2} - k^2 \right) \phi = 0. \tag{3.16}$$

3.3 Sturm–Liouville Problems

The standard form for a Sturm–Liouville problem is a second-order differential equation of the form:

$$\frac{d}{dx}\left(P(x)\frac{dy}{dx}\right) + (\lambda R(x) - Q(x))\,y = 0, \qquad (3.17)$$

defined on a finite interval $a < x < b$, with *separated* boundary conditions:

$$\alpha_0 y(a) + \alpha_1 y'(a) = 0, \qquad (3.18)$$

$$\beta_0 y(b) + \beta_1 y'(b) = 0.$$

The real constants α_0, α_1, β_0, and β_1 are assumed to be fixed, but no two corresponding to one endpoint are both zero. There are well-developed theories of the existence of solutions to the equation. For our purposes, we need only have that $P(x)$, $R(x)$, and $Q(x)$ are continuous real-valued functions and $P(x) > 0$ on $a \le x \le b$. The differential equation and the boundary conditions together define a regular Sturm–Liouville problem. It should be noted that there is no loss of generality in the factored form of the differential equation. An equation in the form:

$$P_2(x)\frac{d^2 y}{dx^2} + P_1(x)\frac{dy}{dx} + P_0(x)y = 0,$$

may be put in factored form by multiplying through the equation by the integrating factor:

$$\mu(x) = \frac{1}{P_2(x)}e^{\int \frac{P_1(x)}{P_2(x)}dx}.$$

There are properties of the solutions of the Sturm–Liouville problem (3.17)–(3.18) that are easy to obtain. Others require a deeper analysis. We will derive the properties as needed for the solution of the stability problem.

Property (i): If throughout the interval (a, b), $R(x)$ is of one sign, then all of the eigenvalues λ are real.

Property (i)′: If throughout the interval (a, b), $Q(x) \ge 0$, and $\alpha_0\alpha_1 \le 0$, and $\beta_0\beta_1 \ge 0$, then all eigenvalues are real. *Note:* $P(x) > 0$, but $R(x)$ may change sign on (a, b).

Property (ii): If throughout the interval (a, b), $R(x)$ is of one sign, $Q(x) > 0$, $\alpha_0\alpha_1 \le 0$, and $\beta_0\beta_1 \ge 0$, then all the eigenvalues λ have the sign of $R(x)$.

Property (ii)′: If all the conditions of Property (ii) hold except that $R(x)$ changes sign in (a, b), then there are both positive and negative eigenvalues λ.

To demonstrate Property (i), suppose that λ *is* complex. Then write

$$\lambda = \lambda_1 + i\lambda_2,$$

in terms of its real and imaginary parts. As would be the case for the eigenvector of a problem involving matrices, we must suppose that the eigenfunction $y(x)$ is also complex. Write

$$y(x) = y_1(x) + iy_2(x)$$

in terms of the real and imaginary parts. Multiply the Sturm–Liouville equation (3.17) by $\overline{y}(x) = y_1(x) - iy_2(x)$ and integrate from $x = a$ to $x = b$. After integration by parts, the result is

$$\int_a^b \left[(\lambda R(x) - Q(x)) |y(x)|^2 - P(x) |y'(x)|^2\right] dx + [P(x)y'(x)\overline{y}(x)]_{x=a}^{x=b} = 0, \quad (3.19)$$

where $|y(x)|^2 = (y_1(x))^2 + (y_2(x))^2$, with a similar meaning for the modulus of the derivative $y'(x)$. Because the boundary conditions are real, $\overline{y}(x)$ satisfies them as does $y(x)$. Thus, evaluating the boundary contributions

$$[P(x)y'(x)\overline{y}(x)]_a^b = -P(b)\frac{\beta_0}{\beta_1} |y(b)|^2 + P(a)\frac{\alpha_0}{\alpha_1} |y(a)|^2,$$

when neither $\alpha_1 = 0$, nor $\beta_1 = 0$. If either of these constants is zero, the boundary contribution at the corresponding endpoint is clearly seen to be zero, since $\overline{y}(x)$ vanishes at that endpoint.

The important observation now is that the boundary contributions are real. Taking the imaginary part of (3.19) gives

$$\lambda_2 \int_a^b R(x) |y(x)|^2 dx = 0. \quad (3.20)$$

Since $R(x)$ does not change sign on (a, b), $\lambda_2 = 0$.

Property (ii) is established by considering the real part of (3.19). It follows that

$$\lambda_1 \int_a^b R(x) |y(x)|^2 dx = \int_a^b P(x) |y'(x)|^2 dx + \int_a^b Q(x) |y(x)|^2 dx$$

$$+ P(b)\frac{\beta_0}{\beta_1} |y(b)|^2 - P(a)\frac{\alpha_0}{\alpha_1} |y(a)|^2. \quad (3.21)$$

By the same reasoning as for the imaginary part (3.20), and the demonstration of Property (i), the boundary contributions are nonnegative under the assumptions $P(x) > 0$, $\alpha_0\alpha_1 \leq 0$ and $\beta_0\beta_1 \geq 0$. Given $P(x) > 0$, $Q(x) > 0$, then $\lambda_1 > 0$ if $R(x) > 0$ on (a, b) or $\lambda_1 < 0$, if $R(x) < 0$ on (a, b), since the integral term multiplying λ_1 in (3.21) has the sign of $R(x)$.

Property (ii) gives sufficient conditions for no eigenvalue to be negative, when $R(x) > 0$, while Property (ii)' gives sufficient conditions for both positive and negative eigenvalues, if $R(x)$ changes sign. However, if $Q(x)$ changes its sign on (a, b) (which violates Property (ii)), while $R(x)$ has one sign, it is possible for there to be only a *finite* number of negative eigenvalues. Property (ii)' however, allows as it turns out, an *infinite* number of eigenvalues of both signs. This process will be simplified with the introduction of Prüfer's method.

Example 3.2. In Section 3.3.1 for the case of solid body rotation, where $\overline{\rho} = \rho_0 r$ (Example 3.1), we can assert, on the basis of Property (i), that all eigenvalues c^{-2} are real, because $P(r) = 1$, $R(r) = 5\omega_0^2 > 0$, and because $Q(r) = k^2 > 0$, the eigenvalues will be positive for conditions like (3.14), where in the notation of (3.18), $\alpha_1 = \beta_1 = 0$.

Unusual phenomena may happen if the boundary conditions are not separated. It is then possible that there may be two eigenfunctions corresponding to an eigenvalue, but clearly not more, since the differential equation is second-order. This, and other possibilities are illustrated in Exercises 3.12, 3.13 and 3.14.

3.3.1 Prüfer's Method

Prüfer's technique examines the eigenvalues and dates from 1926. It is tailored to second-order equations, which, as one would expect, are more amenable to a detailed analysis than those of a higher order. The idea is to write Equation (3.17) in the phase plane, setting

$$z(x) = P(x)y'(x),$$

so that two single, first-order equations result:

$$\frac{dz}{dx} = -(\lambda R(x) - Q(x))y,$$

$$\frac{dy}{dx} = \frac{z}{P(x)}.$$

Now polar coordinates are introduced:

$$y = r(x)\cos\theta(x),$$
$$z = r(x)\sin\theta(x).$$

The variables z and y are then eliminated by differentiation with respect to x:

$$z' = r'\sin\theta + (r\cos\theta)\theta' = -(\lambda R - Q)r\cos\theta,$$

$$y' = r'\cos\theta - (r\sin\theta)\theta' = \frac{r\sin\theta}{P}.$$

Next, solve for r', θ':

$$\frac{dr}{dx} = \left(\frac{1}{P} - (\lambda R - Q)\right)r\sin\theta\cos\theta, \tag{3.22}$$

$$\frac{d\theta}{dx} = -\left(\frac{1}{P}\sin^2\theta + (\lambda R - Q)\cos^2\theta\right). \tag{3.23}$$

The best procedure is to solve Equation (3.23) first for $\theta(x)$ because that equation is independent of $r(x)$. Though, in general, $\theta(x)$, it is difficult to find explicitly, once it is known, say graphically, it is substituted back into equation (3.22) in dr/dx. The equation for $r(x)$ is then easily solved by separation of variables to give

$$r(x) = e^{-\frac{\lambda}{2}\int R\sin 2\theta\, dx} e^{\frac{1}{2}\int(\frac{1}{P}+Q)\sin 2\theta\, dx}. \tag{3.24}$$

Remark 3.1. *There is a periodicity associated with the function $\theta(x)$. That is, if $\theta(x)$ is a solution, so is $\theta(x) + n\pi$, for every integer n. This follows most easily from the definition of $y(x)$ because $y(x) = r\cos\theta(x)$,*

$$r\cos[\theta(x) + n\pi] = r\cos\theta(x)\cos(n\pi) = (-1)^n r\cos\theta(x) = (-1)^n y(x)$$

and

$$r \sin[\theta(x) + n\pi] = r \sin \theta(x) \cos(n\pi) = (-1)^n P y'(x).$$

For even values of n, the same solution is obtained while for odd values of n, the negative of the solution is obtained, which is also a solution. A solution $y(x)$ of (3.17) has a zero at $x = x_1$, if and only if, $\theta(x_1) = \frac{\pi}{2} + n\pi$. The zeros are simple, that is if $y(x_1) = 0$, then $y'(x_1) \neq 0$. This latter condition follows from the fact that $\cos \theta(x_1)$ and $\sin \theta(x_1)$ will not both vanish.

Remark 3.2. *Notice from the equation (3.23) in $d\theta/dx$, for $R(x) > 0$, $\theta(x; \lambda)$ is a decreasing function of λ, and if $\theta(x_n; \lambda) = (n - \frac{1}{2})\pi$, for integer n, then $\theta(x; \lambda) < (n - \frac{1}{2})\pi$ for $x > x_n$. This follows because*

$$\frac{d\theta(x_n; \lambda)}{dx} = \frac{-1}{P(x_n)} < 0.$$

Boundary Conditions. The previous boundary conditions (3.18) are used. Since

$$\tan \theta = \frac{z}{y} = \frac{P y'}{y},$$

take $\theta(a; \lambda) = \gamma$, where γ is the smallest possible number, $-\frac{\pi}{2} \leq \gamma < \frac{\pi}{2}$, such that $\tan \gamma = -\alpha_0 P(a)/\alpha_1$ (if $\alpha_1 = 0$, take $\gamma = -\frac{\pi}{2}$), that is,

$$\theta(a; \lambda) = \tan^{-1}\left(\frac{P(a)y'(a)}{y(a)}\right) = \gamma.$$

Similarly, take $-\frac{\pi}{2} < \delta \leq \frac{\pi}{2}$ so that $\tan \delta = -\beta_0 P(b)/\beta_1$. (If $\beta_1 = 0$, then take $\delta = \frac{\pi}{2}$.) Then the boundary conditions on θ are therefore

$$\theta(a) = \gamma, \tag{3.25a}$$

$$\theta(b) = \delta. \tag{3.25b}$$

The boundary conditions are only important for $\theta(x)$. Once θ is determined, $r(x)$ follows from (3.24).

Remark 3.3. *A solution $y(x)$ of (3.17) is an eigenfunction of the Sturm–Liouville problem (3.17–3.18), if*

$$\theta(a; \lambda) = \gamma, \tag{3.26}$$

$$\theta(b; \lambda) = \delta \pm n\pi, \quad n = 0, 1, 2, \ldots. \tag{3.27}$$

Suppose $\theta(x; \lambda)$ satisfies $\theta(a; \lambda) = \gamma$. Because $\theta(b; \lambda)$ is a decreasing function of λ, it is possible to show, using the conditions of Remark 3.2, that $\theta(b; \lambda)$ is unbounded as $\lambda \to -\infty$. Thus, it must be true that $\theta \to -\infty$. On the other hand, it turns out that as λ increases from $-\infty$, there is a first value λ for which the second of the boundary conditions (3.27) is satisfied. For this eigenvalue, $\theta(b; \lambda_0) = \delta$. As λ

increases, there is an infinite sequence of eigenvalues λ_n for which the second boundary condition is satisfied, namely, those for which $\theta(b; \lambda) = \delta - n\pi$, by Remark 3.1. For each of these,

$$y_n(x) = r_n(x) \cos \theta(x; \lambda_n).$$

Furthermore, the eigenfunction belonging to λ_n has exactly n zeros in the interval $a < x < b$. This follows from counting the zeros of $y_n(x)$ as described in Remark 3.1.

It is now possible to establish Property (ii)', using some of the previous results. The formulation (3.17)–(3.18) remains the same. Remark 3.1 is still valid. However, Remark 3.2 is not valid because $R(x)$ changes sign. It is true though, that on a subinterval $[a_0, b_0]$ on which $R(x) > R_0 > 0$, $P(x) < P_0$, and $Q(x) < Q_0$, so that $\theta(x; \lambda)$ is a decreasing function of λ, and $\theta(x; \lambda) \to -\infty$, as $\lambda \to \infty$. This is readily established by comparing $\theta(x; \lambda)$ with the solution $\omega_0(x; \lambda)$ satisfying an equation like (3.23) with constant coefficients P_0, Q_0, and R_0:

$$\frac{d\omega_0}{dx} = -\frac{1}{P_0} \sin^2 \omega_0 - (\lambda R_0 - Q_0) \cos^2 \omega_0.$$

The successive zeros of $\omega_0(x; \lambda)$ have spacing $\pi [P_0 / (\lambda R_0 - Q_0)]^{\frac{1}{2}}$, which tends to zero as $\lambda \to \infty$. Hence $\omega_0 \to -\infty$ as $\lambda \to \infty$. A standard comparison argument in (Coddington and Levinson) gives that $\theta(x; \lambda) \to -\infty$ as $\lambda \to \infty$. Likewise, on an interval $[a_1, b_1]$ on which $R(x) < R_1 < 0$, $P(x) < P_1$, and $Q(x) < Q_1$, a comparison is made with the solution $\omega_1(x; \lambda)$ satisfying an equation like (3.23) with constant coefficients P_1, Q_1, and R_1:

$$\frac{d\omega_1}{dx} = -\frac{1}{P_1} \sin^2 \omega_1 - (\lambda R_1 - Q_1) \cos^2 \omega_1.$$

In this case, $\theta(x; \lambda) \to -\infty$ as $\lambda \to -\infty$. Thus, in any case, $\theta(b; \lambda) \to -\infty$ as $\lambda \to \pm\infty$. This shows that there are an infinite number of positive and negative eigenvalues.

3.3.2 Orthogonality of Eigenfunctions

The simple Sturm–Liouville problem,

$$y'' + \lambda y = 0, \quad 0 < x < \pi,$$

$$y(0) = 0, \quad y(\pi) = 0,$$

has eigenfunctions $y_n = \sin nx$, where $n = 1, 2, \ldots$. These form the basis of the *Fourier sine series*. The construction of the series is based on the notion that the functions $\sin nx$ are orthogonal on the interval $[0, \pi]$:

$$\int_0^\pi \sin(nx) \sin(mx\,dx) = 0, \quad m \neq n.$$

A similar relation holds for the eigenfunctions of a regular Sturm–Liouville problem, with one important generalization, the introduction of a *weight function*.

Property (iii): When the conditions of Property (i) hold, then the eigenfunctions of the regular Sturm–Liouville system (3.17)–(3.18), which have different eigenvalues are orthogonal with the weight function $R(x)$. That is, given eigenvalues λ_n, λ_m ($\lambda_n \neq \lambda_m$) and corresponding eigenfunctions y_n, y_m,

$$\int_a^b R(x) y_n(x) y_m(x) dx = 0, \quad m \neq n. \tag{3.28}$$

We employ the common notation for the *inner product*,

$$\langle y_n, y_m \rangle = 0, \quad m \neq n.$$

Property (iv): When the conditions of Property (i) hold, where, without loss of generality, it is assumed that $R(x) > 0$, then the eigenfunctions of the regular Sturm–Liouville system (3.17–3.18) may be *normalized* with the weight function $R(x)$, so that

$$\int_a^b R(x) y_n^2(x) dx = 1.$$

This means that

$$\langle y_n, y_n \rangle = 1.$$

We employ the common notation for the *norm*

$$\| y_n \| = \langle y_n, y_n \rangle^{1/2} = 1.$$

Property (iv)′: When the conditions of Property (ii)′ hold, and if the eigenfunctions corresponding to positive and negative eigenvalues are denoted by y_n^+ and y_n^-, respectively, then the eigenfunctions may be normalized as

$$\int_a^b R(x) \left(y_n^+\right)^2 dx = 1$$

and

$$\int_a^b R(x) \left(y_n^-\right)^2 dx = -1.$$

3.3.3 Synge's Proof of Rayleigh's Criterion

We turn our attention again to (3.13), which is now recognizable as a typical Sturm–Liouville problem, in the form (3.17). Synge recognized this and made the following observations.

Since

$$\frac{\bar{\rho}}{r} > 0, \quad \text{and} \quad \frac{k^2 \bar{\rho}}{r} > 0,$$

using Property (i)′, the eigenvalues c^{-2} are real. Furthermore, if

$$F > 0 \quad \text{on} \ (r_1, r_2),$$

then using Property (ii), all values of c^{-2} are positive. Hence, c is real and the oscillation (3.12) is stable. On the other hand, if

$$F < 0 \quad \text{on} \ (r_1, r_2),$$

then, also using Property (ii), c^{-2} is negative. The value of c with positive imaginary part give instability. If F changes sign on (r_1, r_2), then using Property (ii)′, c^{-2} takes both positive and, negative values, and, hence, there is instability. Thus, from (3.10), a necessary and sufficient condition for stability is

$$\frac{d}{dr}(\overline{\rho} r^2 \overline{v}^2) > 0,$$

which is *Rayleigh's criterion*.

Example 3.3. The case treated in Examples 3.1 and 3.2, where $\bar{\rho} = \rho_0 r$ with solid body rotation (3.16), may be solved explicitly under the boundary conditions (3.14). Since the equation has constant coefficients, the eigenfunctions will be expressible as

$$\phi_n = \sin\left[\sqrt{5\omega_0^2 c_n^{-2} - k^2}(r - r_1)\right],$$

and the eigenvalues are found to be

$$c_n^{-2} = \frac{1}{5\omega_0^2}\left[\frac{n^2 \pi^2}{(r_2 - r_1)^2} + k^2\right], \quad n = 1, 2, \ldots.$$

3.4 Expansions in Eigenfunctions

A very important question about a sequence of eigenfunctions y_n of a Sturm–Liouville problem, orthogonal and square integrable with respect to a weight function $R(x)$, is the following: Can every square-integrable function f be expanded into an infinite series $f = \sum c_n y_n$ of the y_n? With a suitable meaning attached to the sense of convergence (in the mean) of the series, the answer to this question is affirmative, and the sequence of eigenfunctions y_n is said to be *complete* (Birkhoff and Rota).

The expansion coefficients for an expansion in the eigenfunctions of the typical problem (3.17)–(3.18) are found by assuming that, for a suitable function $f(x)$,

$$f = \sum c_n y_n. \tag{3.29}$$

Then, by (3.28),

$$\int_a^b f(x) R(x) y_m(x) dx = \sum c_n \int_a^b R(x) y_n(x) y_m(x) dx$$

$$= c_m \int_a^b R(x) y_m^2(x) dx.$$

So mean convergence is defined as $\lim_{N \to \infty} \left\| f - \sum_{n=1}^N c_n y_n \right\|^2 = 0$.

The detailed proof of the mean convergence of (3.29) will not be presented. (See (Birkhoff and Rota).) However, what makes the proof particularly affecting is that it depends on the asymptotic behavior of the eigenfunctions of the Sturm–Liouville problem. Suppose that the equation is written in "Liouville Normal Form"

$$u''(t) + (\lambda - q(t))u(t) = 0, \tag{3.30}$$

which is obtained by setting

$$y(x) = \frac{u(t)}{\sqrt[4]{P(x)R(x)}}, \quad t = \frac{\int_a^x \sqrt{R(x)/P(x)}dx}{\int_a^b \sqrt{R(x)/P(x)}dx} \tag{3.31}$$

in (3.17). Thus, for large n, it may be shown that if in (3.18), $\alpha_1\beta_1 \neq 0$, then

$$u_n(t) \sim \sqrt{2}\cos(n\pi t) + O\left(\frac{1}{n}\right) \quad \text{as } n \to \infty. \tag{3.32}$$

If the condition $\alpha_1\beta_1 = 0$ holds in (3.18), then $\cos(n\pi t)$ is replaced by $\sin(n\pi t)$ in (3.32).

Mathematicians have been engrossed for a long time in determining necessary and sufficient conditions for the *pointwise convergence* of (3.29), that is where convergence holds at each point of the interval $a \leq x \leq b$ (See (Weinberger).) For instance, continuity of $f(x)$ everywhere on the interval is neither necessary nor sufficient. Thus, if the series converges pointwise, and f has a jump discontinuity at $x = x_0$, then it may be proved that the series converges to

$$\frac{f(x_0^+) + f(x_0^-)}{2},$$

at $x = x_0$,

A stronger condition would be *uniform convergence* on the interval. This is true if

$$M_N = \max_{a \leq x \leq b} \left| f - \sum_{n=1}^{N} c_n y_n \right| \to 0,$$

as $N \to \infty$.

Though this is not normally considered the formal definition of uniform convergence it is an equivalent condition, which is easier to apply in practice. However, it is intuitively clear that f would need to satisfy the same boundary conditions as the approximating eigenfunctions, such as (3.18), in order for uniform convergence to be expected; but this is not needed for mean convergence. These ideas are illustrated in Example 3.5.

3.4.1 The Nonhomogeneous Problem; Solvability Condition

Consider the inhomogeneous boundary-value problem:

$$\frac{d}{dx}\left(P(x)\frac{dy}{dx}\right) + (\mu R(x) - Q(x))\, y := Ly + \mu Ry = f(x), \quad a < x < b \tag{3.33}$$

$$B_1 y = \alpha_0 y(a) + \alpha_1 y'(a) = 0, \tag{3.34}$$

$$B_2 y = \beta_0 y(b) + \beta_1 y'(b) = 0.$$

Here μ is a given fixed constant, and f is known. The boundary conditions are the same as (3.18), but μ may or may not be an eigenvalue of (3.17)–(3.18). We examine conditions for (3.33)–(3.34) to have a solution.

Multiply (3.33) by some eigenfunction $\varphi(x)$ of (3.17)–(3.18) and integrate from $x = a$ to $x = b$ to obtain:

$$\int_a^b (Ly + \mu Ry)\varphi dx = \int_a^b f\varphi dx.$$

Employing integration by parts and the boundary conditions (3.34) the left side becomes

$$\int_a^b y(L\varphi + \mu R\varphi)dx$$

and using the fact that φ is an eigenfunction with eigenvalue λ the result is

$$\int_a^b y(-\lambda R\varphi + \mu R\varphi)dx = \int_a^b f\varphi dx.$$

Thus, if μ is the eigenvalue λ, a solution for y depends on whether or not $\int_a^b f\varphi dx = 0$. This is known as the "solvability condition."

Suppose that (3.33)–(3.34) has a solution. Using the eigenfunction expansions based on (3.17)–(3.18), set

$$f = \sum_{n=1}^{\infty} A_n \varphi_n(x), \quad y = \sum_{n=1}^{\infty} B_n \varphi_n(x).$$

Because f is known, so are the coefficients A_n – however, the values B_n have to be determined. Substituting into (3.33) leads to

$$\sum_{n=1}^{\infty} B_n(-\lambda_n + \mu) R\varphi_n(x) = \sum_{n=1}^{\infty} A_n \varphi_n(x).$$

This reinforces the previous result. If $\mu \neq \lambda_n$, for all n, a solution may be found because the eigenfunctions are orthogonal with weight $R(x)$. The coefficients are given by

$$B_n = \frac{A_n \int_a^b (\varphi_n(x))^2 \, dx}{(\mu - \lambda_n) \int_a^b (\varphi_n(x))^2 R(x)dx}, \quad n = 1, 2, \ldots.$$

However, if $\mu = \lambda_{n'}$ for some n', the problem has no solution unless $A_{n'} = 0$, for that n'. In this case, the solution is given by

$$y = \sum_{n=1}^{\infty} B_n \varphi_n(x),$$

where the coefficient $B_{n'}$ is arbitrary. In effect, there are "infinitely many" solutions.

3.5 Worked Examples

3.5.1 Heat Conduction in a Nonuniform Rod

Example 3.4. Consider heat conduction governed by the parabolic differential equation:

$$\rho c \frac{\partial u}{\partial t} = \frac{\partial}{\partial x} \left(\kappa(x) \frac{\partial u}{\partial x} \right), \quad 0 < x < L, \quad t > 0 \tag{3.35}$$

$$u(0, t) = u(L, t) = 0.$$

Here $\rho(x)$ is the density, $c(x)$ the specific heat and $\kappa(x)$ the conductivity (Lin and Segel, Section 4.1). Solving by separation of variables let $u(x, t) = \varphi(x)e^{-\lambda t}$ in (3.35) to find

$$(\kappa(x)\varphi')' + \lambda \rho c \varphi = 0,$$
$$\varphi(0) = \varphi(L) = 0.$$

To be specific, we study the model problem, where $L = 1$ and

$$\kappa(x) = \frac{1}{(1+x)^2} = \rho(x)c(x).$$

The resulting Sturm–Liouville problem looks like

$$\left(\frac{\varphi'}{(1+x)^2} \right)' + \frac{\lambda}{(1+x)^2} \varphi = 0.$$

In order to have the equation look like (3.30), use the substitution (3.31) and set

$$w(x) = \frac{\varphi(x)}{(1+x)},$$

giving

$$w'' + \left[\lambda - \frac{2}{(1+x)^2} \right] w = 0, \tag{3.36}$$

$$w(0) = w(1) = 0. \tag{3.37}$$

Using *Maple* or some other computer algebra systems (CAS), we find that the general solution is

$$w = a \left(\frac{\cos kx}{1+x} + k \sin kx \right) + b \left(\frac{\sin kx}{1+x} - k \cos kx \right), \tag{3.38}$$

where $k = \sqrt{\lambda}$. Applying the left boundary condition,

$$w(0) = a - bk = 0. \tag{3.39}$$

The right boundary condition gives

$$w(1) = a \left(\frac{\cos k}{2} + k \sin k \right) + b \left(\frac{\sin k}{2} - k \cos k \right) = 0. \tag{3.40}$$

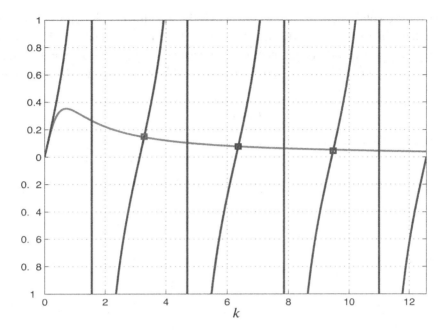

Figure 3.1. Graphical solution of the eigenvalue equation (3.41). Square labels denote solutions where $k = \sqrt{\lambda}$.

The simultaneous solution of (3.39) and (3.40) leads to

$$\begin{vmatrix} 1 & -k \\ \frac{1}{2}\cos k + k \sin k & \frac{1}{2}\sin k - k \cos k \end{vmatrix} = 0.$$

The characteristic equation that determines the eigenvalues is therefore

$$\tan k = \frac{k}{1 + 2k^2}. \tag{3.41}$$

From the graph in Figure 3.1, the lowest eigenvalue is $\lambda_1 \doteq 10.8$. This is because one finds that $\lambda = 0$ is not an eigenvalue since the general solution of (3.36) would be $w = a(1 + x)^{-1} + b(1 + x)^2$, and satisfying (3.37) gives $a = b = 0$. We also observe that from (3.41), $\tan k_n \sim 0$ as $n \to \infty$, and hence $\lambda_n \sim n^2\pi^2$. Furthermore, from (3.38) and (3.39), we have that normalized $w_n(x) \sim \sqrt{2}\sin(n\pi x)$ as $n \to \infty$, all in keeping with the remarks below (3.32) in Section 3.4.

3.5.2 Waves on Shallow Water

Example 3.5. A channel of varying width and bottom topography supports water waves. The governing equation is given by

$$\frac{\partial^2 u}{\partial t^2} = \frac{g}{b}\frac{\partial}{\partial x}\left(bh\frac{\partial u}{\partial x}\right), \tag{3.42}$$

for the displacement $u(x, t)$, where $b(x)$ is the breadth, $h(x)$ is depth of the channel, and g is the acceleration due to gravity (Cochran, Section 2.1). We examine the special case where the channel is measured from $x = 0$ to $x = 1$, $h(x) = (1 + x)^2$, $b(x) = 1 + x$. By separation of variables $u(x, t) = y(x)T(t)$, we obtain the form:

$$u(x, t) = y(x)e^{i\omega t},$$

which gives an equation for the spatial mode shapes,

$$((1 + x)^3 y')' + \lambda(1 + x)y = 0, \tag{3.43}$$

where $\lambda = \omega^2/g$. We take the channel to be closed at both ends so we have the boundary conditions

$$u(0, t) = u(1, t) = 0, \tag{3.44}$$

which lead to

$$y(0) = y(1) = 0. \tag{3.45}$$

Together (3.43) and (3.45) form a Sturm–Liouville problem, so we know that the problem's eigenvalues are real and are nonnegative by Properties (i) and (ii) of Section 3.3. Furthermore, the eigenvalue $\lambda = 0$ is excluded because the general solution of (3.43) would be

$$y = c_1 + c_2(1 + x)^{-2},$$

with which (3.45) would give

$$c_1 + c_2 = 0,$$

$$c_1 + \frac{1}{4}c_2 = 0.$$

For all λ, (3.43) is an Euler–Cauchy-type equation, so it has solutions of the form:

$$y = (1 + x)^s,$$

and by substitution we find

$$s(s + 2) + \lambda = 0.$$

Solving, obtain

$$s = -1 \pm \sqrt{1 - \lambda}.$$

To satisfy $y(0) = 0$, set

$$y = (1 + x)^{-1+\sqrt{1-\lambda}} - (1 + x)^{-1-\sqrt{1-\lambda}}.$$

The condition $y(1) = 0$ gives

$$2^{\sqrt{1-\lambda}} = 1.$$

If $0 < \lambda < 1$ so that $\sqrt{1 - \lambda}$ is real, this equation clearly has no solution. If $\lambda = 1$, the two independent solutions are $(1 + x)^{-1}$ and $(1 + x)^{-1} \log(1 + x)$.

The condition at $x = 0$ is satisfied by the second of these, but not it does not satisfy the condition at $x = 1$. Thus, $\lambda = 1$ is not an eigenvalue.

For $\lambda > 1$, the square root becomes imaginary. We can still obtain two solutions by the method of setting

$$(1 + x)^{-1+i\sqrt{\lambda-1}} = (1 + x)^{-1} e^{i\sqrt{\lambda-1}\log(1+x)}. \tag{3.46}$$

Two independent solutions of (3.43) are given by the real and imaginary parts of (3.46). To make $y(0) = 0$ we need only the imaginary part

$$y(x) = (1 + x)^{-1} \sin\left(\sqrt{\lambda - 1}\log(1 + x)\right).$$

The condition $y(1) = 0$ gives

$$\sin\left(\sqrt{\lambda - 1}\log 2\right) = 0.$$

Thus, $\sqrt{\lambda - 1}\log 2 = n\pi$, or

$$\lambda_n = \frac{n^2\pi^2}{(\log 2)^2} + 1, \quad n = 1, 2, 3, \ldots,$$

which are the eigenvalues. The corresponding eigenfunctions are

$$y_n(x) = (1 + x)^{-1} \sin\left(n\pi\frac{\log(1 + x)}{\log 2}\right).$$

We turn next to the expansion problem. By the results on Sturm–Liouville problems (Birkhoff and Rota), these eigenfunctions are complete. We have the generalized Fourier series

$$f(x) \sim \sum_{n=1}^{\infty} c_n y_n(x),$$

with weight function $R(x) = (1 + x)$ from (3.43). Sufficient Conditions on $f(x)$ that ensure the pointwise convergence of the generalized Fourier series are well-known (Weinberger). We might desire a series solution to the related initial value problem; for instance, to solve (3.42)–(3.44) with $u(x, 0) = f(x)$, $u_t(x, 0) = 0$. The coefficients c_n in the expansion of $f(x)$ will be

$$c_n = \frac{\int_0^1 f(x) \sin\left(n\pi\frac{\log(1+x)}{\log 2}\right) dx}{\int_0^1 (1 + x)^{-1} \sin^2\left(n\pi\frac{\log(1+x)}{\log 2}\right) dx}$$

$$= \frac{2}{\log 2} \int_0^1 f(x) \sin\left(n\pi\frac{\log(1 + x)}{\log 2}\right) dx.$$

Since the equation $T''(t) + \lambda T(t) = 0$, with $T(0) = 1$, $T'(0) = 0$ has the solution $\cos(\sqrt{\lambda}t)$, the inital-boundary problem has the solution

$$u(x, t) = (1 + x)^{-1} \sum_{n=1}^{\infty} c_n \cos\left(\sqrt{\frac{n^2\pi^2}{(\log 2)^2} + 1}\, t\right) \sin\left(n\pi\frac{\log(1 + x)}{\log 2}\right).$$

When the series for $f(x)$ converges uniformly, the series for $u(x, t)$ converges uniformly, and, hence, satisfies the initial and boundary conditions. However, to assure continuous derivatives and satisfaction of the differential equation, we need to have the series for f'' converge uniformly. This can be shown to be true, given that f'' is continuous, that $f(0) = f(1) = 0$, $f''(0) = f''(1) = 0$, and that $\int_0^1 (f''')^2 dx$ is finite. (See (Weinberger, Section 37).)

3.6 Nonstandard Eigenvalue Problems

The theory of Sturm–Liouville problems has been shown to be useful in several contexts. Nevertheless, there are many other eigenvalue problems that arise, not of that particular type. For instance, the order of the equation may be higher than two, and the eigenvalue may occur by multiplying a differential operator. This type of generalized eigenvalue problem was first treated systematically by Kamke. Some authors therefore refer to these as *Kamke problems*. It is to these types of problems we turn next. Rather than elaborate on the complete theory, we will illustrate this theory by several applied problems.

3.6.1 $\mathcal{L}\Phi = \lambda \mathfrak{M}\Phi$

Example 3.6. The study of Rossby waves in a closed ocean illustrates a type of nonstandard eigenvalue problem. A simple example is given by the eigenvalue problem for Rossby normal modes governed by

$$\nabla^2 \psi = -i\lambda \frac{\partial \psi}{\partial x} \quad \text{in } R,$$

$$\psi = 0 \quad \text{on } \partial R,$$

in a closed rectangular basin defined by $R = (0, \pi) \times (0, Y)$, where ψ is the quasigeostrophic stream function, (x, y) are the (eastward, northward) coordinates and ∂R is the boundary of R. This is reduced to a one-dimensional problem using separation of variables by setting

$$\psi(x, y) = \phi(x) \sin ky.$$

Hence, $k = m\pi / Y$, $m = 1, 2m, \ldots$. The resulting ordinary differential equation is

$$\frac{d^2 \phi}{dx^2} - k^2 \phi = -i\lambda \frac{d\phi}{dx},$$

with boundary conditions $\phi(0) = \phi(\pi) = 0$. The resulting eigenvalues and eigenfunctions are

$$\lambda_n^2 = 4(n^2 + k^2),$$

$$\phi_n(x) = e^{-i\lambda_n x/2} \sin nx, \quad n = 1, 2, \ldots.$$

In a notation similar to the heading of this subsection, write $L\phi = \lambda M\phi$, where the operators are defined as

$$L = \frac{d^2}{dx^2} - k^2,$$

$$M = -i\frac{d}{dx}.$$

This notation for the operator M, though less common in Fluid Mechanics, is very prevalent in Quantum Physics. The "completeness and expansion theory" for a class of problems of this type was carried out by (Masuda). (See also Mishoe, who did not have the geophysical application in mind.)

Example 3.7. When the stability of flow between rotating cylinders is studied in the viscous regime, the earlier derivation of the stability equations must be extended. This was carried out by G. I. Taylor in 1923. By an ingenious blend of experiment and theory, he was able to conclude that when the cylinders rotate in the same direction, an instability, observed in the form of vortices, could be described quite accurately mathematically. A companion result, one generalizing Rayleigh's criterion, was later derived by (Synge, 1938), and (Chandrasekhar). The latter author proved stability if

$$\omega_2 r_2^2 > \omega_1 r_1^2,$$

where ω_1 and r_1 are angular velocity and radius, respectively, of the inner cylinder, and ω_2 and r_2 are the corresponding quantities for the outer cylinder. The linearized disturbances he treated were also assumed to be axisymmetric and periodic in the axial direction. The resulting system of ordinary differential equations is

$$M^* Mu + T\alpha^2 \left(\frac{1}{r^2} - \kappa\right) v = -\sigma Mu, \tag{3.47a}$$

$$u + M_0 v = -\sigma v, \tag{3.47b}$$

$$u = u' = v = 0, \quad \text{at } r = \eta, 1,$$

where

$$Mw = \left(-\frac{d^2}{dr^2} - \frac{1}{r}\frac{d}{dr} + \frac{1}{r^2} + \alpha^2\right) w, \quad \eta < r < 1.$$

The parameters occurring in the system are the Taylor number T, wave number α, $\eta = r_1/r_2$, $\mu = \omega_2/\omega_1$, and $0 < \kappa = (1 - \mu/\eta^2)/(1 - \mu) < 1$. The operators M^* and M_0 have the same form as M, but they all have different domains by virtue of the boundary conditions. The important number is σ, which is the temporal growth rate. If $\Re\text{eal}(\sigma) < 0$, the flow is stable.

Synge recognized that by multiplying the first equation (3.47a) by $r\bar{u}$, and integrating from $r = \eta$ to $r = 1$, certain positive definite integrals result. However, one term remains that involves both u and v. For that reason, the second equation (3.47b) must be multiplied by some positive weight function $W(r)$ and v, and the result integrated. When the two expressions are added, the result is

$$\langle M^*Mu, u\rangle + T\alpha^2 \left\langle \left(\frac{1}{r^2} - \kappa\right)v, u\right\rangle + \langle u, Wv\rangle + \langle M_0 v, Wv\rangle$$

$$= -\sigma\left(\langle Mu, u\rangle + \langle Wv, v\rangle\right), \tag{3.48}$$

where

$$\langle f, g\rangle := \int_\eta^1 r f(r)\overline{g}(r)dr.$$

A logical choice for $W(r)$ is $-T\alpha^2(1/r^2 - \kappa)$, since on $\eta \le r \le 1$, $T(1/r^2 - \kappa) < 0$, for the flows under consideration. By taking the real part of (3.48), it is possible to determine $\mathfrak{Real}(\sigma)$. The result is

$$\langle Mu, Mu\rangle + \mathfrak{Real}\left(\langle M_0 v, Wv\rangle\right) = -\mathfrak{Real}(\sigma)(\langle Mu, u\rangle + \langle Wv, v\rangle). \tag{3.49}$$

The second term on the left of (3.49) reduces to

$$\mathfrak{Real}(\langle M_0 v, Wv\rangle) = \mathfrak{Real}\left\{\int_\eta^1 \left[-\frac{d}{dr}\left(r\frac{d}{dr}\right) + \frac{1}{r} + \alpha^2 r\right] v W(r)\overline{v}dr\right\}$$

$$= \int_\eta^1 W(r)\left(r\left|\frac{dv}{dr}\right|^2 + \frac{|v|^2}{r} + \alpha^2 r |v|^2\right)dr$$

$$+ \mathfrak{Real}\left\{\int_\eta^1 r\frac{dv}{dr}\overline{v}W'(r)dr\right\}$$

$$= \int_\eta^1 rW(r)\left(\left|\frac{dv}{dr} - \frac{v}{r}\right|^2 + \alpha^2 |v|^2\right)dr > 0.$$

The last step is determined by the fact that because $W'(r) = 2T\alpha^2/r^3$,

$$\mathfrak{Real}\left\{\int_\eta^1 rW'(r)\frac{dv}{dr}\overline{v}dr\right\} = \frac{1}{2}\int_\eta^1 W'(r)r\frac{d}{dr}|v|^2\,dr = -\int_\eta^1 W(r)\frac{d}{dr}|v|^2\,dr.$$

Consequently, from (3.49), $\mathfrak{Real}(\sigma) < 0$, and the stability result holds.

The structure of the formulation (3.47a), (3.47b) is seen to be of the general form of a matrix operator equation $\mathfrak{L}\Phi = \lambda\mathfrak{M}\Phi$, where

$$\mathfrak{L} = \begin{pmatrix} M^*M & T\alpha^2\left(\frac{1}{r^2} - \kappa\right) \\ 1 & M_0 \end{pmatrix},$$

$$\Phi = \begin{pmatrix} u \\ v \end{pmatrix},$$

$$\mathfrak{M} = \begin{pmatrix} M & 0 \\ 0 & 1 \end{pmatrix},$$

and $\lambda = -\sigma$. The completeness and expansion theory for this problem was first carried out by (DiPrima and Habetler).

Example 3.8. Another important problem of hydrodynamic stability that can be cast in this form occurs in the study of the problem of Rayleigh–Bénard convection. In its most basic form, assuming a Boussinesq fluid, the linearized stability equations for the vertical velocity and temperature perturbations may be reduced to a system with constant coefficients (Joseph)

$$(D^2 - a^2)^2 w - a^2 R_a \theta = \frac{-s}{Pr}(-D^2 + a^2)w, \tag{3.50}$$

$$(-D^2 + a^2)\theta - w = -s\theta, \tag{3.51}$$

where $D = d/dz$. Here R_a is the Rayleigh number and Pr is the Prandtl number. The boundary conditions are taken to be rigid and conducting at the top and bottom:

$$w = Dw = \theta = 0, \text{ at } z = 0, 1.$$

The general formulation makes it clear why the principle of exchange of stabilities holds for this problem. That is, if $R_a < 0$, then $\Re\text{eal}(s) < 0$, so the flow is stable, and if $R_a > 0$, then $\Im\text{mag}(s) = 0$. To see this simply set $w = \Phi_1$ and $\Phi_2 = a^2 R_a \theta$. Then, the system becomes:

$$\begin{pmatrix} M^*M & -1 \\ -1 & \frac{1}{a^2 R_a} M_0 \end{pmatrix} \begin{pmatrix} \Phi_1 \\ \Phi_2 \end{pmatrix} = -s \begin{pmatrix} \frac{1}{Pr} M & 0 \\ 0 & \frac{1}{a^2 R_a} \end{pmatrix} \begin{pmatrix} \Phi_1 \\ \Phi_2 \end{pmatrix}.$$

Since M^*M and M_0 are both positive definite on their domains, if $R_a > 0$, the operator equation,

$$\mathfrak{L}\mathbf{\Phi} = -s\mathfrak{M}\mathbf{\Phi},$$

has the property that $\mathfrak{L} = \mathfrak{L}^*$, $\mathfrak{M} = \mathfrak{M}^*$ and $\langle\mathfrak{M}\mathbf{\Phi}, \mathbf{\Phi}\rangle > 0$ in a suitable inner product, so the eigenvalues s are all real. However if $R_a < 0$, then $\mathfrak{M}$ is no longer a definite operator, though it is symmetric. In that case, complex eigenvalues, $s = \sigma + i\omega$, may occur. However, it is possible to show that $\sigma < 0$. Form the inner product,

$$\langle\mathfrak{L}\mathbf{\Phi}, \mathbf{\Psi}\rangle = -s\langle\mathfrak{M}\mathbf{\Phi}, \mathbf{\Psi}\rangle,$$

where

$$\mathbf{\Psi} = \begin{pmatrix} \Phi_1 \\ -\Phi_2 \end{pmatrix}.$$

Then, since

$$\langle\mathfrak{L}\mathbf{\Phi}, \mathbf{\Psi}\rangle = \int_0^1 \left[|(-D^2 + a^2)\Phi_1|^2 - \Phi_2\bar{\Phi}_1 + \Phi_1\bar{\Phi}_2 - \frac{1}{a^2 R}(|D\Phi_2|^2 + a^2|\Phi_2|^2) \right] dz$$

and $R_a < 0$, $\mathfrak{Real}\,\langle \mathfrak{L}\boldsymbol{\Phi}, \boldsymbol{\Psi}\rangle > 0$. On the other hand,

$$\langle \mathfrak{M}\boldsymbol{\Phi}, \boldsymbol{\Psi}\rangle = \frac{1}{Pr}\int_0^1 \left[|D\Phi_2|^2 + a^2|\Phi_2|^2 - \frac{1}{a^2 R_a}|\Phi_2|^2 \right] dz,$$

which is positive. Consequently $\mathfrak{Real}(s) < 0$.

3.6.2 Sturm–Liouville Problems with Weight Changing Sign (Counterrotating Cylinders)

The consequences of Rayleigh's criterion being violated are usually described simply as an instability; infinitesimal disturbances are predicted to grow in time. This is often evidenced by the emergence of a secondary state. When the cylinders of the problem analyzed by Synge rotate in opposite directions, the growth of the disturbances is known to be different. Even if viscosity is ignored, aspects of this behavior are still evident. Mathematically, this may written as an example of a Sturm–Liouville problem of the type

$$Mg = \lambda w g,$$

with a differential operator M, where the weight function w changes sign. When w is not of one sign, then there are both positive and negative eigenvalues. This is a nonstandard problem because the metric in a space with inner product $\langle w\cdot, \cdot\rangle$ is indefinite. However, this is not completely nonphysical because it corresponds to the case of cylinders rotating in the opposite direction.

In the mathematical counterpart, (Ince) in Sections 10.6.1 and 10.7.1 proves Property (iv)′ above, that for the Sturm–Liouville problem there are infinitely many positive and negative eigenvalues λ_n, such that

$$\lambda_1^+, \lambda_1^+, \ldots, \lambda_m^+ \to +\infty,$$
$$\lambda_1^-, \lambda_1^-, \ldots, \lambda_m^- \to -\infty.$$

The corresponding eigenfunctions are

$$g_1^+, g_2^+, \ldots, g_m^+, \ldots,$$
$$g_1^-, g_2^-, \ldots, g_m^-, \ldots,$$

with normalizations

$$\left\langle wg_i^+, g_j^+\right\rangle = \delta_{ij},$$
$$\left\langle wg_i^-, g_{ij}^-\right\rangle = -\delta_{ij}.$$

Exercises 3.10 and 3.11 illustrate these phenomena.

3.6.3 Singular Sturm–Liouville Problems

An eigenvalue problem is *singular* if the interval (a, b) on which (3.17) is defined is infinite, or if one or more of the coefficients of the equation have singular behavior at $x = a$ or $x = b$. For instance, in

$$\frac{d}{dx}\left(P(x)\frac{dy}{dx}\right) + (\lambda R(x) - Q(x))\, y = 0, \tag{3.52}$$

if $P(x) \to 0$ as $x \to a$ or $x \to b$, the problem is, in general, singular. This happens for the Bessel differential equation:

$$(xy')' + \left(\lambda x - \frac{n^2}{x}\right) y = 0, \ \ 0 < x < b,$$

at $x = 0$. Here the coefficient $P(x) = x$ and $Q(x) = -n^2/x$ are both singular at $x = 0$. Another example of this situation occurs in Exercise 3.4. The issues which arise for singular problems are: What are the correct boundary conditions to apply at a singular point? What happens to the spectrum? That is, when will there still be an infinite sequence of eigenfunctions and eigenvalues? If the nature of the spectrum is different it is necessary to characterize the change. In the case of singular Sturm–Liouville problems, the other possible spectral points are said to be *continuous* as opposed to *discrete*.

Boundary Conditions

In general, the boundary conditions at the singular point $x = a$, say, depend on the behavior of

$$\lim_{x\downarrow a} P(x) \left[u'(x)v(x) - u(x)v'(x)\right].$$

If the differential operator with the boundary conditions is to be well-defined, this limit must vanish ((Holland, p. 198)). It is helpful to know from simply appealing to the coefficients, whether this is the case. The simplest such conditions are those derived by (Kaper et al.). Consider (3.52); they suppose that (i) on (a, b), $P(x) > 0$, but $P(x) \downarrow 0$ as $x \downarrow a$ so that

$$\int_x^b \frac{d\xi}{P(\xi)} = O\left((x - a)^{-\gamma}\right), \ \ \text{as } x \downarrow a, \ \ 0 < \gamma < \frac{1}{2}.$$

(ii) $Q(x)$ is bounded on (a, b).

With these assumptions the authors prove that there are many equivalent boundary conditions at $x = a$, which lead to a well-defined operator. We mention two of these:

$$(\text{B1}) \quad \lim_{x\downarrow a} y(x) \text{ exists and is finite, or}$$

$$(\text{B2}) \quad \lim_{x\downarrow a} (Py')(x) = 0.$$

Example 3.9. (Pipe flow). The following eigenvalue problem occurs in the stability theory of axial flow in a circular pipe.

$$\left(-\frac{d^2}{dr^2} + \frac{1}{r}\frac{d}{dr} + \beta^2\right)\varphi = \lambda\varphi, \quad 0 < r < 1,$$

$$\varphi(1) = 0, \quad \beta \text{ const.}$$

The appropriate boundary condition at $r = 0$ is found by putting the differential equation in factored form:

$$-\frac{d}{dr}\left(\frac{1}{r}\frac{d\varphi}{dr}\right) + \frac{\beta^2}{r}\varphi = \frac{\lambda}{r}\varphi.$$

We expect that $\lim_{r\downarrow 0}\varphi$ is finite, based on (B1). However, from physical considerations, since $\frac{\varphi}{r}$ is proportional to the radial component of the velocity, one would like to apply $\lim_{r\downarrow 0}\frac{\varphi}{r}$ finite. By the equivalence of (B2) it must also be true that $\lim_{r\downarrow 0}\frac{\varphi'}{r} = 0$. For this to be true, one must have $\lim_{r\downarrow 0}\frac{\varphi}{r}$ finite.

Example 3.10. Sometimes a singular problem may be transformed into a regular one, by a suitable transformation. An interesting example is given by the consideration of Rayleigh's equation for the instability of an inviscid plane parallel flow $U(z)\mathbf{i}$ to perturbations with a stream function of the form $\phi(z)e^{i\alpha(x-ct)}$:

$$(U - c)(\phi'' - \alpha^2\phi) - U''\phi = 0, \quad z_1 < z < z_2 \tag{3.53}$$

$$\phi(z) = 0, \quad \text{at} \quad z = z_1, z_2.$$

In this classic problem, like that considered in Section 3.2, c is the wave speed and is the eigenvalue sought to determine the temporal growth rate of a linearized disturbance. In this case, the equation is singular because it is possible that $U(z) = c$. The point $z = z_s$ at which this occurs is usually called the "critical-layer." In a creative insight into the problem, von Mises [p. 294] noticed that if $c = U$, such a wave speed is real, and therefore wrote the equation as

$$\phi'' + K(z)\phi + \lambda\phi = 0, \tag{3.54}$$

where

$$K(z) = \frac{-U''(z)}{U(z) - U(z_s)},$$

given that $U''(z_s) = 0$. This substitution renders $K(z)$ as a regular function, which is assumed positive for the purpose of the investigation. The reconsidered problem is a regular Sturm–Liouville problem, which has an infinite sequence of eigenvalues $\lambda_1, \lambda_2, \ldots, \lambda_n, \ldots, \to \infty$. The conclusion is that there is a smallest value $\lambda_1 = -\alpha_s^2$, for which there is a *neutral mode* with $c = U(z_s)$.

3.7 Fourier–Bessel Series

3.7 Fourier–Bessel Series

As developed in Chapter 2, the Bessel function $J_\nu(x)$ satisfies the differential equation

$$x^2 y'' + xy' + (x^2 - \nu^2)y = 0. \tag{3.55}$$

In solving partial differential equations in curvilinear coordinates, these functions are likely to arise. A typical variant to (3.55) is

$$x^2 y'' + xy' + (\lambda x^2 - \nu^2)y = 0,$$

or in Sturm–Liouville form,

$$(xy')' + \left(\lambda x - \frac{\nu^2}{x}\right) y = 0, \ \ 0 < x < 1, \tag{3.56}$$

whose bounded solution is $y = J_\nu(\sqrt{\lambda}x)$. We have taken the interval to be $[0, 1]$ after any necessary rescaling.

It is desirable to investigate conditions under which one can be sure that

$$\int_0^1 J_\nu(\sqrt{\lambda_k}x)J_\nu(\sqrt{\lambda_j}x)x\,dx = 0. \tag{3.57}$$

For simplicity, we assume that ν is real and that $\nu > -1$ in order that the integral will converge. Relevant boundary conditions are

$$\beta_0 y(1) + \beta_1 y'(1) = 0, \tag{3.58}$$

while at $x = 0$ we allow a condition, such as (B1) or (B2) of Section 3.6.3.

Write (3.56) as

$$Ly + \lambda xy = 0$$

and suppose that $(z(x), \mu)$ is another eigenfunction–eigenvalue pair, satisfying the same boundary conditions as $(y(x), \lambda)$. Evaluate

$$\int_{0+}^1 [z(Ly + \lambda xy) - y(Lz + \mu xz)]\,dx = 0,$$

which, after integration, reduces to

$$x(y'z - z'y)|_{x=1} - \lim_{x\downarrow 0}[x(y'z - z'y)] + (\lambda - \mu)\int_{0+}^1 xyz\,dx = 0. \tag{3.59}$$

By either of the assumed forms:

(B1) $\lim\limits_{x\downarrow 0} y(x), \ \lim\limits_{x\downarrow 0} z(x)$ exist and are finite, or

(B2) $\lim\limits_{x\downarrow 0}(xy')(x) = \lim\limits_{x\downarrow 0}(xz')(x) = 0,$

it follows quite readily that the limits in (3.59) are 0 as $x \downarrow 0$. The conditions at $x = 1$ are symmetric, and consequently, because, in addition to (3.58), we have $\beta_0 z(1) + \beta_1 z'(1) = 0$, and no boundary terms remain.

The desired relation (3.57) follows as long as $\lambda \neq \mu$.

It is important to examine the case where $\lambda = \mu$ because it leads to desirable expressions for normalizing constants in Fourier–Bessel expansions. One way to accomplish this is to relax the boundary condition at $x = 1$, temporarily. We will allow λ and μ to be different and consider the limit as $\lambda \to \mu$, where $\mu > 0$. In (3.59), we substitute $y = J_\nu(\sqrt{\lambda}x)$ and $z = J_\nu(\sqrt{\mu}x)$. The calculations are simpler if we set $\sqrt{\lambda} = s$ and $\sqrt{\mu} = t$. We then obtain

$$\int_{0+}^1 x J_\nu(sx) J_\nu(tx) dx = \frac{s J_\nu'(s) J_\nu(t) - t J_\nu'(t) J_\nu(s)}{t^2 - s^2}.$$

In the limit as $s \to t$, employ L'Hospital's rule and the result is

$$\int_{0+}^1 x \left(J_\nu(tx)\right)^2 dx = \frac{t J_\nu''(t) J_\nu(t) + J_\nu'(t) J_\nu(t) - t J_\nu'(t) J_\nu'(t)}{-2t}.$$

Due to the change of variable, we have

$$t^2 f'' + t f' + (t^2 - \nu^2) f = 0,$$

for $f = J_\nu(t)$. Consequently,

$$t J_\nu''(t) J_\nu(t) + J_\nu'(t) J_\nu(t) = \frac{(\nu^2 - t^2)}{t} \left(J_\nu(t)\right)^2,$$

and the normalizing condition becomes

$$\int_{0+}^1 x \left(J_\nu(tx)\right)^2 dx = \frac{1}{2} \left[\left(J_\nu'(t)\right)^2 + \left(1 - \frac{\nu^2}{t^2}\right) \left(J_\nu(t)\right)^2 \right].$$

In the original notation, it reads

$$\int_{0+}^1 x \left(J_\nu(\sqrt{\mu}x)\right)^2 dx = \frac{1}{2} \left[\left(J_\nu'(\sqrt{\mu})\right)^2 + \left(1 - \frac{\nu^2}{\mu}\right) \left(J_\nu(\sqrt{\mu})\right)^2 \right]. \tag{3.60}$$

The result (3.60) might be further simplified because, according to (3.58) the values $J_\nu'(\sqrt{\mu})$ and $J_\nu(\sqrt{\mu})$ are related. This same technique is also useful in computing normalizing constants for regular Sturm–Liouville problems.

Therefore, one can use such Bessel functions as a basis for a generalized Fourier series, as in the following three examples.

3.7.1 Worked Bessel Function Examples

Example 3.11. Consider the problem of steady heat conduction in a cylindrical rod. If the imposed temperatures are axisymmetric, then the Laplace equation in cylindrical coordinates, is

$$\frac{1}{r} \frac{\partial}{\partial r} \left(r \frac{\partial T}{\partial r} \right) + \frac{\partial^2 T}{\partial z^2} = 0 \quad \text{for} \quad 0 \le r \le a, \quad 0 \le z \le h. \tag{3.61}$$

The boundary conditions are given as

$$T = f(z) \quad \text{on} \quad r = a, \text{ and}$$
$$T = 0 \quad \text{at} \quad z = 0, h.$$

We append the thermal continuity requirement that $f(0) = f(h) = 0$. Based on the homogeneity of the boundary conditions in z, a Fourier sine series seems an obvious choice. For later convenience, we split out the condition on $r = a$, hence,

$$T = (r/a)^2 f(z) + \sum_{n=1}^{\infty} A_n(r) \sin\left(\frac{n\pi z}{h}\right), \tag{3.62}$$

Clearly, then,

$$A_n(a) = 0. \tag{3.63}$$

Substitution of this solution into Equation (3.61) gives

$$\sum_{n=1}^{\infty} \left[r^{-1}(r A_n')' - (n^2\pi^2/h^2) A_n \right] \sin\left(\frac{n\pi z}{h}\right) = -f'' - \frac{4}{a^2} f \equiv F(z). \tag{3.64}$$

The function on the right of (3.64), now denoted as $F(z)$, has a Fourier expansion in z, and hence, if F_n is the Fourier sine series coefficient for $F(z)$, then the orthogonality of the Fourier series leads to

$$r^{-1}(r A_n')' - (n^2\pi^2/h^2) A_n = F_n.$$

Comparison with (3.55) indicates that this is not the Bessel equation, but rather the modified Bessel equation, (2.22). So, the bounded homogeneous solution in this case is I_0, and the particular solution is simple. Hence, we complete the solution with (3.63),

$$A_n(r) = \frac{h^2 F_n}{n^2\pi^2} \left[\frac{I_0(n\pi r/h)}{I_0(n\pi a/h)} - 1 \right].$$

If, for example, $f(z) = T_o z(h - z)/h^2$, a parabolic distribution, then it is easily confirmed that

$$h^2 F_n = 4\left[\frac{2}{n^3\pi^2}\left(\frac{h}{a}\right)^2 - \frac{1}{n\pi} \right]\left[(-1)^n - 1\right].$$

Example 3.12. In underwater and atmospheric applications, scattering of sound waves off objects is a very important problem. Consider the scattering of waves off a circular cylinder in a two-dimensional geometry. The velocity potential obeys the following partial differential equation,

$$\frac{1}{c^2}\frac{\partial^2 \phi}{\partial t^2} = \frac{1}{r}\frac{\partial}{\partial r}\left(r\frac{\partial \phi}{\partial r}\right) + \frac{1}{r^2}\frac{\partial^2 \phi}{\partial \theta^2}, \tag{3.65}$$

where c is the sound speed. The incident wave propagates toward the cylinder at $r = a$ in a form given by

$$\phi_i = e^{i(k(y+ct))}, \quad y = r\sin\theta.$$

The no-penetration boundary condition is on the radial velocity, so

$$\frac{\partial \phi}{\partial r} = 0 \quad \text{on} \quad r = a. \tag{3.66}$$

If we decompose the solution into the incident wave and the rest of the wave field, then

$$\phi = \phi_i + e^{ickt} \Phi(r, \theta),$$

using circular polar coordinates. Substitution into (3.65) and boundary condition (3.66) leads to

$$r^2 \Phi_{rr} + r \Phi_r + \Phi_{\theta\theta} + \lambda^2 r^2 \Phi = 0, \ \lambda \equiv kc$$

$$\Phi_r = -ik \sin\theta e^{ika \sin\theta} \quad \text{at} \quad r = a.$$

This is an obvious problem for the use of Fourier series in angle (Exercise 3.16), and so we write

$$\Phi = \sum_{n=-\infty}^{\infty} e^{in\theta} G_n(r),$$

and its substitution into the equation for Φ gives the differential equation for the Fourier component,

$$r^2 G_n'' + r G_n' + \left(\lambda^2 r^2 - n^2\right) G_n = 0,$$

which is of course a Bessel equation. Instead of the standard two solutions discussed here, we know that an alternative set of linearly independent solutions to this equation is $\left(H_n^{(1)}(\lambda r), \ H_n^{(2)}(\lambda r)\right)$ (Section 2.4.2). By Exercises 8.9–8.10, we may show that the large–r behaviors of these functions are $(\lambda \pi r/2)^{-1/2} \exp(\pm i\lambda r + (2n+1)\pi i/4)$ respectively. Clearly all of the elements of $\Phi \exp(ikct)$ must represent outgoing waves ONLY, so of the two solutions, only $H_n^{(2)}(\lambda r)$ is acceptable. Hence,

$$G_n(r) = c_n H_n^{(2)}(\lambda r).$$

We now turn to the complications of the boundary condition,

$$\Phi_r\big|_{r=a} = -ik \sin\theta e^{ika \sin\theta}. \tag{3.67}$$

In Example 2.2, we found that

$$e^{i\alpha \sin\theta} = \sum_{m=-\infty}^{\infty} e^{im\theta} J_m(\alpha).$$

Using this result, (3.67) may be written as

$$
\Phi_r\big|_{r=a} = -ik\sin\theta \sum_{m=-\infty}^{\infty} e^{im\theta} J_m = -\frac{k}{2}(e^{i\theta} - e^{-i\theta}) \sum_{m=-\infty}^{\infty} e^{im\theta} J_m
$$

$$
= -\frac{k}{2} \sum_{m=-\infty}^{\infty} e^{i(m+1)\theta} J_m + \frac{k}{2} \sum_{m=-\infty}^{\infty} e^{i(m-1)\theta} J_m
$$

$$
= \frac{k}{2} \sum_{n=-\infty}^{\infty} e^{in\theta}[J_{n+1} - J_{n-1}] \tag{3.68}
$$

In this expression, the argument of each of the Bessel functions, which is ka, has been omitted for brevity. The final step, then, is to multiply the above boundary condition (3.68) by $\exp(-in\theta)$ and integrate from zero to 2π, in order to obtain the values for $G'_n(a)$. That process gives the following results:

$$
G'_n(a) = \frac{k}{2}[J_{n+1}(ka) - J_{n-1}(ka)], \quad G'_{-n}(a) = (-1)^n G'_n(a), \quad n \ge 0.
$$

Combining these results gives the completed solution for the reflected and scattered wave field as

$$
\Phi = \frac{1}{2c} \sum_{n=1}^{\infty} \frac{H_n^{(2)}(kr)}{H_n^{(2)'}(ka)}[J_{n+1}(ka) - J_{n-1}(ka)][e^{in\theta} + (-1)^n e^{-in\theta}]
$$

$$
+ \frac{H_0^{(2)}(kr)}{c H_0^{(2)'}(ka)} J_1(ka).
$$

From a Bessel function identity (see Exercise 2.23),

$$
J_{n+1}(z) - J_{n-1}(z) = -2J'_n(z),
$$

the series coefficient may be simplified a bit. Of interest is the short-wave (geometrical acoustics) limit, which may be found using asymptotic results for the various Bessel functions for $k \to \infty$. That is not included here.

Example 3.13. Finally, we consider the sound waves in a circular, two-dimensional region with a forcing at the boundary at a particular frequency, ω. Let $\omega/c \equiv \lambda$, if c is the sound speed. Then, that boundary-value problem is

$$
\frac{1}{r}\frac{\partial}{\partial r}\left(r\frac{\partial p}{\partial r}\right) + \frac{1}{r^2}\frac{\partial^2 p}{\partial \theta^2} + \lambda^2 p = 0, \quad p = f(\theta) \quad \text{on} \quad r = a. \tag{3.69}
$$

The obvious choice here is an exponential Fourier series (see Exercise 3.16), so we write

$$p = (r/a)^2 f(\theta) + \sum_{n=0}^{\infty} A_n(r) e^{in\theta}, \tag{3.70}$$

$$A_n(a) = 0. \tag{3.71}$$

Substitution into (3.69), using the orthogonality of the series and doing a fair bit of algebra leads then to the ordinary differential equation for $A_n(r)$, namely,

$$A_n'' + \frac{1}{r} A_n' - \left(\frac{n}{r}\right)^2 A_n + \lambda^2 A_n = \frac{1}{2\pi} \left[\frac{n^2}{r^2} - \frac{\lambda^2 r^2 + 4}{a^2} \right] \int_0^{2\pi} f(\theta) e^{-in\theta} d\theta$$

$$\equiv H_n(r), \tag{3.72}$$

where the notation $H_n(r)$ has been introduced for convenience. Obviously, this is Bessel's equation for functions of order n, with a rather complicated right-hand side. We could seek a general solution with variation of parameters, but we can avoid the clumsiness of that approach. Worse than that indefinite integrals arise that cannot be written down in closed form. Instead, we note that we do not require the most general solution to (3.72), but one for which (3.71) is satisfied. That being the case, we can expand the solution to (3.72) in a Fourier–Bessel series, namely,

$$A_n(r) = \sum_{k=1}^{\infty} c_{nk} J_n(j_{n,k} r/a), \tag{3.73}$$

where $j_{n,k}$ is the kth zero of the Bessel function of order n. Substitution of (3.73) into (3.72), and use of orthogonality, (3.57), gives the equation

$$\left[\lambda^2 - \frac{j_{n,k}^2}{a^2} \right] c_{nk} \int_0^1 R J_n^2(j_{n,k} R) dR = \int_0^1 R H_n(aR) J_n(j_{n,k} R) dR.$$

Again, use of (3.60) on the left side gives the somewhat simpler form

$$c_{nk} = \frac{2}{(\lambda^2 - j_{n,k}^2/a^2) J_n^2(j_{n,k})} \int_0^1 R H_n(aR) J_n(j_{n,k} R) dR. \tag{3.74}$$

Hence, the final solution is a double Fourier series,

$$p = (r/a)^2 f(\theta) + \sum_{n=0}^{\infty} \sum_{k=1}^{\infty} c_{nk} e^{in\theta} J_n(j_{n,k} r/a).$$

Note, finally, that the R integral that comes into (3.74) through the three terms in $H_n(r)$, cannot be done in closed form.

3.8 Continuous versus Discrete Spectra

It was recognized long ago that if a function $f(x)$ is expanded in a Fourier series on the interval $[0, L]$, and $L \to \infty$, the nature of expansion changes dramatically. For instance, if $f(0) = f(L) = 0$, the appropriate eigenfunction expansion is

$$f(x) = \sum_{n=1}^{\infty} b_n \sin\left(\frac{n\pi x}{L}\right),$$

where

$$b_n = \frac{2}{L} \int_0^L f(\xi) \sin\left(\frac{n\pi\xi}{L}\right) d\xi.$$

It is easily shown heuristically how the series goes over to an integral if $L \to \infty$. Put the integral in the series so that

$$f(x) = \sum_{n=1}^{\infty} \frac{2}{L} \int_0^L f(\xi) \sin\left(\frac{n\pi\xi}{L}\right) \sin\left(\frac{n\pi x}{L}\right) d\xi.$$

Set $n\pi/L = \lambda_n$, so that $\pi/L = \lambda_{n+1} - \lambda_n = \Delta\lambda$. The process $L \to \infty$ gives the formal result:

$$f(x) = \frac{2}{\pi} \int_0^{\infty} d\lambda \int_0^L f(\xi) \sin(\lambda x) \sin(\lambda\xi) d\xi. \tag{3.75}$$

By expanding the notion of integration this can be made rigorous. However, what is not always emphasized is that the expansion functions that are square integrable on the interval $[0, L]$ lose this property when $L = \infty$, and that this is exactly how to identify the occurrence of a continuous spectrum. In particular, the following criterion was enunciated by Rota: "For nth order constant coefficient differential problems, $M\varphi = \lambda\varphi$, defined on the semi–finite interval $[0, \infty)$, the continuous spectrum is identified by the set of values λ such that $\varphi(x; \lambda)$ is not square integrable in x, where $\varphi = e^{i\omega x}$, ω real." Thus, for example, the problem

$$\varphi'' + \lambda\varphi = 0$$
$$\varphi(0) = 0,$$

has the continuous spectrum $\lambda = \omega^2$. When in the differential equation the coefficients are not constant, there are cases where a continuous spectrum will not occur, that is, the spectrum is discrete. Such cases are typified by the following result.

Remark 3.4. *(Titchmarsh, chapter V). Consider the boundary-value problem:*

$$u'' + (\lambda - q(x))u = 0, \quad 0 < x < \infty, \tag{3.76}$$
$$u(0) = 0, \quad |u(\infty)| < K.$$

If the following condition on $q(x)$ is satisfied,

$$\lim_{x \to \infty} q(x) = +\infty,$$

then the spectrum of the problem is all discrete, $\lambda_1 \leq \lambda_2 \leq \ldots$ and $\lambda_n \to \infty$, as $n \to \infty$. The eigenfunction associated with λ_n has n zeros.

If $\lim_{x\to\infty} q(x) < \infty$, there are other results, and depending on the boundary condition at $x = 0$, the spectrum may be both discrete and continuous. Nevertheless, Rota's criterion is always helpful in the identification of the continuous spectrum. So, depending on $\lim_{x\to\infty} q(x) \to q_\infty$, some finite value, the spectrum can be identified from the limiting problem.

Example 3.14. Consider the boundary-value problem:

$$(\cosh^2 xy')' + (\lambda - 2 - \alpha^2 \cosh^2 x)y = 0, \quad 0 < x < \infty,$$
$$y'(0) = 0, \quad y \to 0 \text{ as } x \to \infty$$

where α is a real constant. To put this problem in the form for which the spectral results may be easily found, set $u(x) = \cosh xy(x)$. Then, the problem becomes

$$u'' + (\lambda - 1 - \alpha^2 + 2\,\text{sech}^2 x)u = 0, \quad 0 < x < \infty, \tag{3.77}$$

$$u'(0) = 0, \quad |u(\infty)| < K. \tag{3.78}$$

Here, $q(x) = 1 + \alpha^2 - 2\,\text{sech}^2 x$ and $\lim_{x\to\infty} q(x) = 1 + \alpha^2$. Identify the continuous spectrum in the limiting equation:

$$u'' + (\lambda - 1 - \alpha^2)u = 0,$$

by setting $u = e^{i\omega x}$, which gives $\lambda = 1 + \alpha^2 + \omega^2$, ω real. The continuous spectrum consists of all real numbers $\lambda \geq 1 + \alpha^2$. For the problem (3.77–3.78) one notes that there is also one eigenvalue $\lambda_d = \alpha^2$, corresponding to the discrete spectrum, with eigenfunction $u_d = \text{sech}^2 x$. This eigenfunction is square integrable on $[0, \infty)$.

Also of interest are eigenvalue problems like (3.76) defined on $(-\infty, \infty)$. The conditions for the existence of an all discrete spectrum are similar. For the boundary-value problem,

$$u'' + (\lambda - q(x))u = 0, \quad -\infty < x < \infty, \tag{3.79}$$

if the following conditions on $q(x)$ are satisfied,

$$\lim_{|x|\to\infty} q(x) = \infty,$$

then the spectrum is discrete (Titchmarsh, chapter V).

Example 3.15. An eigenvalue problem that occurs in several areas of mathematical physics is

$$y'' + xy' + \lambda y = 0, \quad -\infty < x < \infty,$$
$$y \to 0, \text{ as } |x| \to \infty.$$

Set in Liouville normal form, the equation is

$$\left(e^{\frac{1}{2}x^2}y'\right)' + \lambda e^{\frac{1}{2}x^2}y = 0.$$

Making the transformation $u(x) = e^{\frac{1}{4}x^2}y(x)$, based on (3.31), leads to

$$u'' + \left(\lambda - \frac{1}{2} - \frac{1}{4}x^2\right)u = 0.$$

Since $q(x) \to \infty$ as $x \to \pm\infty$, by the result just quoted, (Titchmarsh, chapter V) the spectrum is discrete. The eigenfunctions are sometimes called "parabolic cylinder functions." They are given as

$$u_n = e^{-\frac{1}{4}x^2}He_n(x),$$

in which $He_n(x)$ are Hermite polynomials (Lozier) (See Exercise 2.20), defined as

$$He_n(x) = (-1)^n e^{\frac{1}{2}x^2}\frac{d^n}{dx^n}\left(e^{-\frac{1}{2}x^2}\right),$$

that form a complete set with weight function, $w(x) = e^{-x^2/2}$. That is,

$$\int_{-\infty}^{\infty}u_n(x)u_m(x)dx = \int_{-\infty}^{\infty}w(x)He_n(x)He_m(x)dx = 0, \ m \neq n.$$

The eigenfunctions $u(x)$ are square integrable, so it is necessary that

$$\int_{-\infty}^{\infty}e^{\frac{1}{2}x^2}|y(x)|^2dx < \infty.$$

The eigenvalues are $\lambda_n = n = 1, 2, \ldots$.

The continuous spectrum for problems of the form (3.79) is again identified by considering the limiting equation when $|q(x)| \nrightarrow \infty$ as $x \to \pm\infty$.

Example 3.16. In the Rayleigh–Taylor instability of an inviscid stratified fluid (Rayleigh, 1883), density $\bar{\rho} = e^{-\beta z}$, $\beta > 0$, the eigenvalue problem for g, k constant is encountered,

$$\frac{d}{dz}\left(e^{-\beta z}\frac{dw}{dz}\right) + \left(\lambda\beta gk^2e^{-\beta z} - k^2e^{-\beta z}\right)w = 0, \quad -\infty < z < \infty.$$

This equation is also put in the Liouville normal form by setting $u(z) = e^{-\beta z/2}w(z)$. The result is

$$u'' + \left(\lambda\beta gk^2 - \frac{1}{4}\beta^2 - k^2\right)u = 0.$$

Here, if $u = e^{i\omega z}$ is sought, the continuous spectrum is identified as

$$\lambda = \frac{\omega^2 + k^2 + \frac{1}{4}\beta^2}{\beta gk^2}, \quad \omega \text{ is real.}$$

Since the original problem is of Sturm–Liouville form, the spectrum lies on the real line and the continuum eigenfunctions are $w(z) = Ce^{\beta z/2 + i\omega z}$, C is constant.

One of the most useful aspects of the continuum eigenfunctions is in the expansion problem and the development of *transform methods*. For instance, (3.75) may be considered the expansion of $f(x)$ in terms of its sine transform:

$$f_S(\lambda) = \sqrt{\frac{2}{\pi}} \int_0^\infty f(x)\sin(\lambda x)dx.$$

Among the great advantages of integral transform methods, which are discussed in Chapters 5 through 7, is that for problems on $[0, \infty)$ (Laplace transform) and $(-\infty, \infty)$ (Fourier transform), the structure of the spectrum – be it discrete or continuous – emerges naturally when one turns to either the Laplace or Fourier inversion integration. Typically, residues of poles constitute the discrete part of the spectrum, and branch cuts result in a continuous spectrum. Issues of completeness for a discrete spectrum can often be managed best in such a setting. These techniques are developed more thoroughly in Chapter 4.

REFERENCES

G. Birkhoff and G.-C. Rota, On the Completeness of Sturm–Liouville Expansions, *Am. Math. Month.* **67**, 835–841 (1960).

S. Chandrasekhar, *Hydrodynamic and Hydromagnetic Stability*, Oxford University Press, London, 1961.

J. A. Cochran, *Applied Mathematics: Principles, Techniques and Applications,* Wadsworth, Belmont, CA, 1982.

E. A. Coddington and N. Levinson, *Theory of Ordinary Differential Equations*, McGraw-Hill, New York, 1955.

R. C. DiPrima and G. J. Habetler, A Completeness Theory for Non-Selfadjoint Eigenvalue Problems in Hydrodynamic Stability *Arch. Rat. Mech. Anal.* **34**, 218–227 (1969).

P. G. Drazin and W. H. Reid, *Hydrodynamic Stability, 2nd edition*, Cambridge University Press, Cambridge, UK, 2004.

S. S. Holland, Jr., *Applied Analysis by the Hilbert Space Method*, Dekker, New York, 1990.

E. L. Ince, *Ordinary Differential Equations*, Dover, New York, 1956.

D. D. Joseph, *Stability of Fluid Motions, Vols. I, II,* Springer-Verlag, Berlin, 1976.

H. G. Kaper, M. K. Kwong, and A. Zettl, Characterizations of the Friedrich's extensions of singular Sturm–Liouville expressions, *SIAM J. Math. Anal.* **17**, 772–777 (1986).

C.-C. Lin, and L. A. Segel, *Mathematics Applied to Deterministic Problems in the Natural Sciences*, SIAM, Philadelphia, 1988.

D. Lozier, F. Olver, C. Clark, and R. Boisvert, eds., *Digital Library of Mathematical Functions*, National Institute of Standards and Technology, 2005, available at http://dlmf.nist.gov/.

A. Masuda, On the Completeness and the Expansion Theorem for Eigenfunctions of Sturm–Liouville–Rossby Type, *Quart. Appl. Math.* **47**, 435–445 (1989).

L. I. Mishoe, Fourier Series and Eigenfunction Expansions Associated with a Non-Selfadjoint Differential Equation, *Quart. Appl. Math.* **20**, 175–181 (1962).

J. W. S. Rayleigh, Investigation of the Character of the Equilibrium of an Ioncompresible Heavy Fluid of Variable Density, *Proc. Roy. Soc. Lond.* **14**, 170–177 (1883).

G.-C. Rota, On the Spectra of Singular Boundary Value Problems, *J. Math. and Mech.* **10**, 83–90 (1961).

J. L. Synge, The stability of heterogeneous liquids, *Trans. Roy. Soc. Can.* **27**, 1–18 (1933).

J. L. Synge, On the Stability of a Viscous Liquid between Rotating Coaxial Cylinders, *Proc. Roy. Soc. A.* **167**, 250–256 (1938).

E. C. Titchmarsh, Eigenfunction Expansions Associated with Second-Order Differential Equations, Clarendon, Oxford, 1946.

R. von Mises and K. O. Friedrichs, *Fluid Dynamics*, Springer-Verlag, New York, 1971.

H. F. A. Weinberger, First Course in Partial Differential Equations, Dover, New York, 1965.

EXERCISES

3.1 Determine the function $F(r)$ (3.10) for the case of solid body rotation when $\bar{\rho} = \rho_0$, constant. What is the resulting form of the equation for ϕ?

3.2 Prove Property (i)′.

3.3 Show, on the basis of Property (i) that the eigenvalues c^{-2} of Exercise 3.1 are all real.

3.4 Consider the stability equation (Exercise 3.1) for solid body rotation $\bar{v} = \omega_0 r$ (ω_0 const) in a cylinder, where $0 \leq r \leq b$.

(a) Solve explicitly, show that the eigenfunctions are expressible in terms of a Bessel function of integer order, and determine the eigenvalues.

(b) Find an algebraic expression for the smallest eigenvalue $\lambda_1 = 1/c^2$ in terms of ω_0, k and b. Thus, verify that the flow is stable.

3.5 *Rayleigh's criterion for instability of a vertically stratified fluid* (Rayleigh). For an inviscid fluid such as the one in Section 3.2, suppose now that gravity, g, accounts for its acceleration, rather than rotation. The basic state is then

$$\bar{\mathbf{u}} = \mathbf{0}, \quad \rho = \bar{\rho}(z), \quad p = \bar{p}(z),$$

on a layer

$$V = \{(x, y, z) \mid -\infty < x, y < \infty, \ z_1 \leq z \leq z_2\}.$$

Allowing for infinitessimal, three-dimensional perturbations

$$(\mathbf{u}', \rho, p)(x, y, z, t) = (\hat{\mathbf{u}}, \hat{\rho}, \hat{p})(z)e^{i(\alpha x + \beta y) - \sigma t},$$

the linearized disturbance equation for the vertical velocity may be written as

$$\frac{d}{dz}\left(\bar{\rho}\frac{d\hat{w}}{dz}\right) + (\alpha^2 + \beta^2)\left(\frac{g\bar{\rho}}{\sigma^2} - \frac{d\bar{\rho}}{dz}\right)\hat{w} = 0.$$

Suitable boundary conditions are

$$\hat{w}(z_1) = \hat{w}(z_2) = 0.$$

Argue by Sturm–Liouville theory, as in Synge's proof of Rayleigh's criterion for rotating flows (Section 3.3.3), that there is instability if and only if $d\bar{\rho}/dz > 0$ somewhere inside the layer.

3.6 The following Sturm–Liouville problem occurs in the energy stability theory of flow between rotating cylinders (Joseph)

$$(r\phi'(r))' + \frac{\lambda}{r}\phi(r) = 0, \quad 0 < a < r < b,$$

$$\phi = 0, \quad \text{at } r = a, b.$$

Show that the eigenfunctions are

$$\phi_n = \sin[\sqrt{\lambda_n}\ln(r/b)],$$

with eigenvalues

$$\lambda_n = \frac{n^2\pi^2}{[\ln(a/b)]^2}, \quad n = 1, 2, \ldots$$

3.7 Analyze the Sturm–Liouville problem:

$$y'' + (\lambda + 2\,\text{sech}^2 x)y = 0, \quad -a < x < a,$$

$$y(-a) = y(a) = 0.$$

Method: If $\lambda = -k^2$, show that one solution of the differential equation is $e^{kx}(\tanh x - k)$, and find another by replacing k with $-k$; show that the two are usually linearly independent and investigate the exceptional cases. Find the general solution if $\lambda = +\kappa^2$. Then apply the boundary conditions and describe qualitatively the nature of the eigenvalues and eigenfunctions.

3.8 Consider the Sturm–Liouville problem

$$y'' + \left(\lambda - \frac{2}{(1+x)^2}\right)y = 0, \quad 0 < x < 1,$$

$$y(0) = y(1) = 0.$$

(a) Show that the eigenvalues are all positive, based on Property (ii).

(b) If $\lambda = k^2$, the general solution is

$$y = c_1\left(\frac{\cos kx}{1+x} + k\sin kx\right) + c_2\left(\frac{\sin kx}{1+x} - k\cos kx\right).$$

Apply the boundary conditions and derive a transcendental equation to determine the eigenvalues. Find an accurate approximation to the smallest eigenvalue. Describe qualitatively the nature of the eigenfunctions. For a solution, see Section 3.5.

3.9 (a) Show that the differential equation,

$$x^4 y'' + \lambda y = 0, \quad a < x < b,$$

may be solved exactly in terms of Bessel functions, and finally in the elementary form:

$$y = x\left[c_1\cos\left(\frac{k}{x}\right) + c_2\sin\left(\frac{k}{x}\right)\right], \quad k = \sqrt{\lambda}.$$

(b) Apply the boundary conditions, $y(a) = y(b) = 0$, and show that the eigenvalues are

$$\lambda_n = n^2 \pi^2 \left(\frac{ab}{L}\right)^2,$$

where $L = b - a$, with eigenfunctions

$$y_n(x) = x \sin\left[\frac{n\pi b}{L}\left(1 - \frac{a}{x}\right)\right].$$

3.10 Consider the boundary-value problem

$$y'' + \lambda xy = 0, \quad -1 < x < 1,$$
$$y(-1) = y(1) = 0.$$

(a) On the basis of Property (i)′, conclude that all of the eigenvalues are real.

(b) Show implicitly, that if λ is an eigenvalue, so is $-\lambda$.

(c) The equation may be solved explicitly in terms of Airy functions (Chapter 2). Show that the general solution may be written as $y = c_1 Ai(kx) + c_2 Bi(kx)$, for a suitable k. Find the eigenvalue relation, and show that it also gives the same result as part (b). Indicate the locations of the eigenvalues. Use *Maple*, or some computer algebra system, to determine the locations of the first few eigenvalues of smaller magnitudes.

3.11 Consider the boundary-value problem

$$y'' + \lambda \operatorname{sgn}(x)y = 0, \quad -1 < x < \pi, \tag{3.80}$$
$$y(-1) = y(\pi) = 0,$$

where $\operatorname{sgn}(x) = \begin{cases} 1, & \text{if } x > 0 \\ -1, & \text{if } x < 0 \end{cases}$. The coefficient $R(x) = \operatorname{sgn}(x)$ is discontinuous, but continuously differentiable solutions may be found by requiring $y(0^+) = y(0^-)$, and $y'(0^+) = y'(0^-)$. With this proviso, show that (3.80) has all real eigenvalues, but an infinite number of both positive and negative eigenvalues. Find an explicit formula for the corresponding eigenfunctions. Obtain a numerical approximation to the eigenvalue of smallest magnitude.

3.12 Consider the boundary-value problem

$$y'' + \lambda y = 0, \quad 0 < x < 2\pi,$$

with *periodic* boundary conditions

$$y(0) = y(2\pi),$$
$$y'(0) = y'(2\pi).$$

Show that the eigenvalues are $\lambda_n = n^2$, $n = 0, 1, \ldots$, with eigenfunctions $y_0 = 1$, $y_n = \{\cos(nx), \sin(nx)\}$.

3.13 Consider the boundary-value problem:

$$y'' + \lambda y = 0, \quad 0 < x < 1,$$

with

$$y(0) - y(1) = 0,$$
$$y'(0) + y'(1) = 0.$$

Show that every value of λ is an eigenvalue and determine a corresponding eigenfunction.

3.14 Show that the boundary-value problem

$$y'' + \lambda y = 0, \quad 0 < x < 1,$$
$$2y(0) + y(1) = 0, \quad 2y'(0) + y'(1) = 0,$$

has no real eigenvalues. Determine its (complex) eigenvalues and its eigenfunctions.

3.15 Consider the problem of expanding a function $f(x)$ defined on $[0, 2\pi]$ in terms of the eigenfunctions of Exercise 3.12. Assume that

$$f(x) = \sum_{n=0}^{\infty} c_n y_n.$$

Show that the desired expansion becomes

$$f(x) = \frac{a_0}{2} + \sum_{n=0}^{\infty} [a_n \cos(nx) + b_n \sin(nx)],$$

where

$$\frac{a_0}{2} = c_0 = \frac{1}{2\pi} \int_0^{2\pi} f(x)dx,$$

$$a_n = \frac{1}{\pi} \int_0^{2\pi} f(x) \cos(nx)dx,$$

$$b_n = \frac{1}{\pi} \int_0^{2\pi} f(x) \sin(nx)dx.$$

This is called the "full-range Fourier series expansion" of $f(x)$.

3.16 It is often useful to represent Fourier series in terms of complex exponentials. Suppose that a function $f(x)$ is defined on $[0, L]$. Make use of Exercise 3.12 to extend it periodically to give its full-range Fourier series

$$f(x) = \sum_{n=0}^{\infty} \left(a_n \cos \frac{2n\pi x}{L} + b_n \sin \frac{2n\pi x}{L} \right),$$

where

$$a_n = \frac{2}{L} \int_0^L f(x) \cos \frac{2n\pi x}{L} dx$$

$$b_n = \frac{2}{L} \int_0^L f(x) \sin \frac{2n\pi x}{L} dx,$$

$$a_0 = \frac{1}{L} \int_0^L f(x)dx.$$

Show that an *exponential Fourier series* is

$$f(x) = \sum_{n=-\infty}^{\infty} c_n e^{2\pi i n x/L},$$

$$c_n = \frac{1}{L} \int_0^L f(x) e^{-2\pi i n x/L} dx, \tag{3.81}$$

by suitably defining c_n in terms of a_n and b_n. The integral (3.81) is called the "finite Fourier transform" of f.

3.17 Consider the *Sturm–Liouville* problem

$$y'' + \frac{2}{(x+1)} y' + \lambda y = 0, \quad 0 < x < 1, \tag{3.82}$$

$$y'(0) = 0, \quad y(1) = 0.$$

(a) Put the equation in the standard form (3.17).

(b) Show that the substitution $u(x) = (1+x)y(x)$ in (3.82) leads to the differential equation,

$$u'' + \lambda u = 0 \quad (0 < x < 1).$$

What are the corresponding boundary conditions on u?

(c) Determine the eigenfunctions u_n, while finding a transcendental relation giving the eigenvalues λ_n and hence show that

$$y_n = c_n \frac{\sin(\sqrt{\lambda_n}(1-x))}{1+x}, \quad n = 1, 2, \ldots.$$

(d) Write down suitable expressions for the coefficients c_n so that the eigenfunctions are normalized.

3.18 Consider the *Sturm-Liouville* problem

$$y'' + 2y' + (1+\lambda)y = 0, \quad 0 < x < 1, \tag{3.83}$$
$$y'(0) = 0, \quad y'(1) = 0$$

(a) Show that $\lambda_n = n^2\pi^2$ is the nth eigenvalue for (3.83) ($n = 1, 2, \ldots$), and give a corresponding eigenfunction y_n. Is there an eigenvalue λ_0?

(b) Find $R(x)$ such that

$$\langle y_n, y_m \rangle = \int_0^1 R(x) y_n(x) y_m(x) dx = 0,$$

where $m \neq n$.

(c) Determine c_n such that $e^{-x} = \sum_{n=0}^{\infty} c_n y_n(x)$.

3.19 (a) Show that the substitution $u(x) = e^x y(x)$ in (3.83) leads to the differential equation,

$$u'' + \lambda u = 0 \quad (0 < x < 1).$$

What are the corresponding boundary conditions on u?

(b) Normalize the eigenfunctions u_n as $\varphi_n = u_n/\|u_n\|$ and hence show that

$$\varphi_n \sim \sqrt{2}\cos(n\pi x) \quad \text{as } n \to \infty$$

in accordance with (3.32).

3.20 Consider the Sturm–Liouville problem

$$y'' + \lambda y = 0, \quad 0 < x < 1,$$

with the boundary conditions,

$$y(0) = 0, \quad y'(1) = -\beta y(1), \quad \text{where } \beta \text{ is constant.}$$

(a) Show that the eigenfunctions are of the form:

$$y_n = \sin k_n x, \tag{3.84}$$

when $\beta > 0$, where k_n, $n = 1, 2$, are the roots of the equation

$$\tan k = -\frac{k}{\beta}, \tag{3.85}$$

and the eigenvalues $\lambda_n = k_n^2$. Consider separately the cases $\beta > 0$, and $\beta = 0$, and $\beta < 0$. *Note:* If $\beta < 0$, there is the possibility of a nonpositive eigenvalue, since all of the conditions of Property (ii) of Section 3.3 are not met. Determine the corresponding eigenfunction in this special case and the transcendental equation for the eigenvalue.

(b) Show that the eigenfunctions are orthogonal by making use of (3.85), and normalize them as $\varphi_n = y_n/\|y_n\|$. Hence, show that the normalizing constants depend on n if $\beta \neq 0$.

3.21 Consider the Sturm–Liouville problem

$$y'' + \lambda y = 0, \quad 0 < x < 1,$$
$$y'(0) = 0, \quad y(1) + y'(1) = 0.$$

(a) Determine the eigenfunctions and a transcendental equation that gives the eigenvalues. Find a numerical approximation to λ_1, the smallest eigenvalue and show that $\lambda_n \sim n^2\pi^2$ as $n \to \infty$.

(b) Show that the eigenfunctions are orthogonal by making use of the characteristic equation found in part (a). Normalize them as $\varphi_n = y_n/\|y_n\|$, hence verifying that the normalizing constants depend on n. Show that

$$\varphi_n \sim \sqrt{2}\cos(n\pi x) \quad \text{as } n \to \infty,$$

in accordance with (3.32).

3.22 Consider the boundary value problem

$$y'' + 2\operatorname{sgn}(x)y' + \lambda y = 0, \quad y(-\pi) = y(\pi) = 0, \tag{3.86}$$

where $\operatorname{sgn}(x) = -1$, for $x < 0$ and $\operatorname{sgn}(x) = 1$, for $x > 0$.

(a) Obtain the eigenvalues and eigenfunctions for (3.86). *Hint*: Make use of the proviso of Problem 3.11.

(b) Show that (3.86) can be written in the Sturm–Liouville form:

$$\frac{d}{dx}\left[e^{2|x|}\frac{dy}{dx}\right] + \lambda e^{2|x|}y = 0, \quad y(-\pi) = y(\pi) = 0.$$

What is the orthogonality condition for a pair of eigenfunctions, $y_n(x)$ and $y_m(x)$?

3.23 Consider the Sturm–Liouville problem

$$y'' - 2y' + (1+\lambda)y = 0, \quad 0 < x < 1$$
$$y(0) = 0, \quad y(1) + y'(1) = 0.$$

(a) If λ_n is the n_{th} eigenvalue for the problem $(n = 1, 2, \ldots)$, obtain a transcendental equation determining λ_n and give a corresponding eigenfunction y_n.

(b) Find $R(x)$, such that

$$\langle y_n, y_m \rangle = \int_0^1 R(x)y_n(x)y_m(x)dx = 0,$$

where $m \neq n$.

(c) Determine c_n, such that $e^x = \sum_{n=0}^{\infty}c_n y_n(x)$.

3.24 Consider the eigenvalue problem given in Exercise 3.6 when $a < 1$ and $b = 1$.

(a) If $f(r)$ is defined on $[a, 1]$, show that

$$f(r) = \sum_{n=1}^{\infty}c_n \sin\left(\frac{n\pi \ln r}{\ln a}\right),$$

where

$$c_n = \frac{\int_a^1 f(r)\sin\left(\frac{n\pi \ln r}{\ln a}\right)\frac{dr}{r}}{\int_a^1 \sin^2\left(\frac{n\pi \ln r}{\ln a}\right)\frac{dr}{r}}.$$

Introduce an appropriate change of variables and show that this reduces to

$$c_n = 2\int_0^1 f(a^t)\sin(n\pi t)dt.$$

(b) Evaluate, c_n when $f(r) = 1$ and when $f(r) = r$.

3.25 Suppose the nonhomogeneous problem of interest is

$$\frac{d}{dx}\left(P(x)\frac{dy}{dx}\right) + (\mu R(x) - Q(x))\,y := Ly + \mu Ry = 0, \quad a < x < b \qquad (3.87)$$

$$B_1 y := \alpha_0 y(a) + \alpha_1 y'(a) = \gamma_1,$$

$$B_2 y := \beta_0 y(b) + \beta_1 y'(b) = \gamma_2. \qquad (3.88)$$

Here, γ_1 and γ_2 are nonzero constants.

Derive analogous conditions to those in Section 3.4.1, for the existence of a solution. Consider the alternatives if μ is or is not an eigenvalue of (3.17)–(3.18).

(a) Argue why if μ is an eigenvalue then if $\gamma_1\gamma_2 \neq 0$, there is a solution only under special conditions. Show this by multiplying (3.87) by an eigenfunction φ of (3.17)–(3.18) and integrating from $x = a$ to b. Note however, that the solution is not unique since any eigenfunction may added to give another solution. Hence, there are "infinitely many" solutions.

(b) If μ is not an eigenvalue argue why if $\gamma_1\gamma_2 \neq 0$, there is a unique solution. *Hint*: Examine the general solution $y = c_1\varphi_1 + c_2\varphi_1$ to (3.87) where the basic solutions φ_1 and φ_2 are chosen so that $B_1\varphi_1 = 0$ and $B_2\varphi_2 = 0$. This permits the analysis of the more general boundary-value problem of (3.33) and (3.88) by writing the solution by superposition as the sum of each of the two types.

3.26 As an illustration of the results of Exercise 3.25, consider the problem of finding the steady-state temperature distribution $T(r)$ in a long cylindrical annulus $a \leq r \leq b$ as the solution to the boundary-value problem

$$\frac{d}{dr}\left(r\frac{dT}{dr}\right) = 0,$$

$$\alpha_0 T(a) + \alpha_1 \frac{dT}{dr}(a) = \gamma_1, \quad \beta_0 T(b) + \beta_1 \frac{dT}{dr}(b) = \gamma_2.$$

The boundary conditions indicate that the heat exchange is prescribed on the walls, and that they may not be perfectly insulated.

(a) Suppose $\alpha_1 = \beta_1 = 0$. Show that the homogeneous problem has only the trivial solution. Hence, find the unique solution to the nonhomogeneous problem.

(b) Suppose $\alpha_0 = \beta_0 = 0$. Show that the homogeneous problem has the solution $T = T_0$, constant. This corresponds to an eigenvalue $\lambda = 0$, as a Sturm–Liouville problem. Integrate the differential equation to find conditions on the fluxes in order to find out whether a solution exists.

3.27 Consider the buckling of a column with constant stiffness EI. Suppose that the column is of length L and is cantilevered at its base, that is $y'(0) = 0$, where $y(x)$ is the deflected shape. For a uniform column, Euler beam theory says that

$$EIy'' = P[y(L) - y(x)], \quad 0 < x < L,$$

$$y(0) = 0, \quad y'(0) = 0.$$

Differentiating to get rid of $y(L)$, obtain the eigenvalue problem

$$y''' + \lambda y' = 0, \quad 0 < x < L,$$

$$y(0) = y'(0) = y''(L) = 0,$$

where $\lambda = P/EI$.

Show that all of the eigenvalues are real and positive without solving completely for all the eigenfunctions.

Find the eigenvalues and corresponding eigenfunctions or "buckling modes." In particular, show that the critical (minimum) buckling load is

$$P_{cr} = \frac{\pi^2 EI}{4L^2}.$$

Sketch the critical buckling mode.

3.28 Find the eigenvalues and eigenfunctions of the boundary value problem:

$$y'''' + \lambda y'' = 0, \quad 0 < x < 1,$$
$$y(0) = y'(0) = y(1) = y''(1) = 0.$$

3.29 Solve the system (3.50)–(3.51) with stress-free conducting boundary conditions

$$w = D^2 w = \theta = 0, \quad \text{at } z = 0, 1.$$

Show that the system may be reduced to

$$(D^2 - a^2)(D^2 - a^2 - s/\mathrm{Pr})(D^2 - a^2 - s)w = -a^2 R_a w.$$

Hence, show that the eigenfunctions are

$$w_n(z) = \sin(n\pi z), n = 1, 2, \ldots,$$

with eigenvalues

$$s_n = -\frac{1}{2}(1 + \mathrm{Pr})(n^2\pi^2 + a^2)$$

$$\pm \left[\frac{1}{4}(\mathrm{Pr} - 1)^2(n^2\pi^2 + a^2)^2 + a^2 R_a \mathrm{Pr}/(n^2\pi^2 + a^2) \right]^{1/2}.$$

Notice that if $R_a < 0$, complex eigenvalues may indeed occur.

3.30 Find the eigenfunctions and eigenvalues of the pipe flow problem, Example 3.9 of Section 3.6.3. *Hint*: See Exercise 2.14 in Chapter 2. Use the change of variable $\varphi = rv(r)$. Show that the eigenvalues satisfy the condition

$$(\lambda_n - \beta^2)^{1/2} \sim \left(n + \frac{1}{4} \right) \pi \text{ as } n \to \infty.$$

3.31 Consider the singular boundary-value problem

$$y'' - \frac{1}{x}y' + \lambda y = 0, \quad 0 < x < 1,$$

$$\lim_{x \downarrow 0} \frac{y}{x} \text{ finite}, \quad 2y(1) - y'(1) = 0.$$

Show that $\lambda_0 = 0$ is an eigenvalue and find a corresponding eigenfunction.

Convert this to a Sturm–Liouville problem and, hence, conclude that all eigenvalues are real and nonnegative. *Note*: You do not need to find all of the eigenvalues to answer this question.

3.32 Express in terms of Bessel functions, the eigenvalues and eigenfunctions of the singular Sturm–Liouville problem

$$\phi''(r) - \frac{1}{r}\phi'(r) + \lambda\phi(r) = 0, \quad 0 < r < 1,$$

$$\lim_{r \to 0^+} \frac{\phi}{r} \text{ finite, } \phi(1) - \phi'(1) = 0.$$

What is the orthogonality relation between the eigenfunctions? Estimate numerically the smallest nonzero eigenvalue.

3.33 A homogeneous cord of mass, m, and of length, L, is fixed by its upper end ($x = L$) to a vertical axis and rotates about the axis with a constant angular velocity ω_0. The equation of lateral vibrations of the cord about its vertical position of equilibrium with the displacement given by $u(x, t)$ is

$$\frac{\partial^2 u}{\partial t^2} = g\frac{\partial}{\partial x}\left(x\frac{\partial u}{\partial x}\right) + \omega_0^2 u,$$

where g is the acceleration due to gravity, with the boundary conditions

$$u(0, t) \text{ finite, } u(L, t) = 0.$$

(a) Look for free vibrations of the form

$$u(x, t) = \phi(x)e^{i\omega t}.$$

(b) Show that the resulting ordinary differential equation for $\phi(x)$ may be transformed into a Bessel equation of order zero by changing the independent variable from x to $z = 2\sqrt{\lambda x/g}$, where $\lambda = \omega^2 + \omega_0^2$.

(c) Hence, represent the free vibration solutions satisfying the boundary conditions.

Note: The problem of the vibrating attached cord with no rotation, that is, $\omega_0 = 0$, was the first instance in which Bessel functions are known to have arisen in applied mathematics.

3.34 Example 3.11 of Section 3.7 contrasts to a problem with variable temperature at the ends which cannot be done by a z-sine series. Suppose that we wish to solve

$$\frac{1}{r}\frac{\partial}{\partial r}\left(r\frac{\partial T}{\partial r}\right) + \frac{\partial^2 T}{\partial z^2} = 0, \quad \text{for} \quad 0 \le r \le a, \quad 0 \le z \le h, \tag{3.89}$$

but, now with boundary conditions,

$$T = 0 \quad \text{at} \quad z = 0, \qquad T = 0 \quad \text{on} \quad r = a, \quad T = g(r) \quad \text{on} \quad z = h. \tag{3.90}$$

As before, thermal continuity demands that $g(a) = 0$. In contrast to what we did in (3.62), this time write

$$T = \sum_{n=1}^{\infty} A_n(z)J_0(j_{0,n}r/a), \tag{3.91}$$

where $\{j_{0,n}\}$ are the zeros of $J_0(x)$. Hence, the homogeneous condition at $r = a$ is guaranteed. Show that substitution of (3.91) into (3.89) this time gives

$$A_n'' - \left(j_{0,n}^2/a^2\right) A_n = 0.$$

The homogeneous condition at $z = 0$ requires the use of the sinh function solution, so show that

$$A_n = c_n \sinh\left(\frac{j_{0,n}z}{a}\right)$$

and imposing the final condition in (3.90) gives

$$\sum_{n=1}^{\infty} c_n \sinh(j_{0,n}h/a) J_0(j_{0,n}r/a) = g(r).$$

Make use of the orthogonality condition (3.57), multiply by r and J_0, and integrate. This leads to a determination of the constants,

$$c_n = \frac{\int_0^1 Rg(aR)J_0(j_{0,n}R)dR}{\sinh(j_{0,n}h/a)\int_0^1 RJ_0^2(j_{0,n}R)dR}.$$

Use (3.60) and the fact that $J_0' = -J_1$, to conclude that

$$c_n = \frac{2}{J_1^2(j_{0,n})\sinh(j_{0,n}h/a)}\int_0^1 Rg(aR)J_0(j_{0,n}R)dR.$$

3.35 Show that if in Rayleigh's equation (3.53),

$$U(z) = \tanh z, \quad -1 < z < 1,$$

then $z_s = 0$ and (3.54) reduce to

$$\phi'' + (\lambda + 2\operatorname{sech}^2 z)\phi = 0, \quad -1 < z < 1,$$

$$\phi(-1) = \phi(1) = 0.$$

Use the results of Exercise 3.7 to calculate $\lambda_1 = -\alpha_s^2$.

3.36 Under the Boussinesq approximation, the stability equation in Example 3.16 reduces to

$$\frac{d^2w}{dz^2} + \left(\lambda\beta gk^2 - k^2\right)w = 0, \quad -\infty < z < \infty.$$

Show that its continuous spectrum is given by

$$\lambda = \frac{\omega^2 + k^2}{\beta gk^2}, \quad \omega \text{ is real.}$$

What are the corresponding continuum eigenfunctions?

4

Green's Functions for Boundary-Value Problems

4.1 Preamble

One of the more remarkable stories in the history of modern applied mathematics is the origin of Green's functions. They are named for a mostly self-taught mathematician, George Green, who introduced them in his studies of potential theory as it occurs in electricity and magnetism. As a result of working independently, his first work on the subject in 1828 was published at his own expense! This turned out to be justified as he also derived his famous theorem in the same work, and Green's theorem has certainly stood the test of time. Later, he did work on elasticity in which he introduced Green's tensors. A history is available to the interested reader in (Challis and Sheard).

4.1.1 Sources and Fundamental Solutions

As we noted in Chapter 1, the problem of potential flow in two-dimensional fluid mechanics involves the solutions of Laplace's equation $\nabla^2 \phi = 0$. An axisymmetric solution, that depends on r only, the distance from the origin satisfies

$$\frac{\partial}{\partial r}\left(r\frac{\partial \phi}{\partial r}\right) = 0.$$

After integrating twice we arrive at the solution $\phi(r) = c_1 \ln r + c_2$, the constant solution being expected. In a similar way, in three dimensions, a radially symmetric solution satisfies

$$\frac{\partial}{\partial r}\left(r^2\frac{\partial \phi}{\partial r}\right) = 0$$

and is given by $\phi(r) = c_1/r + c_2$. These solutions are useful in practice, and we note that for $c_1 \neq 0$, they are singular at $r = 0$. This property makes them *source functions*. A common way of representing this in differential equations is typified by

$$\frac{\partial}{\partial r}\left(r\frac{\partial \phi_s}{\partial r}\right) = \frac{\delta_+(r)}{2\pi},$$

in the two-dimensional case, where $\phi_s(r) = \ln r/2\pi$. The factor 2π appears because polar coordinates are tacitly assumed and integration of the all-polar angles will occur, while in the three-dimensional case

$$\frac{\partial}{\partial r}\left(r^2 \frac{\partial \phi_s}{\partial r}\right) = \frac{\delta_+(r)}{4\pi},$$

with $\phi_s(r) = 1/4\pi r$. Here the factor 4π appears, because if spherical coordinates are appropriate, then integration over all angles is allowed.

The expression $\delta_+(r)$ obeys the property,

$$\int_0^\infty \delta_+(r)\varphi(r)dr = \varphi(0),$$

for a suitable test function $\varphi(r)$.

In the problem of potential flow past a sphere of radius a, moving with speed U, it is useful to introduce the Stokes stream function $\psi(r, \theta)$ to represent the desired three-dimensional axisymmetric result (Batchelor). The velocity components are

$$u_r = \frac{1}{r^2 \sin\theta}\frac{\partial \psi}{\partial \theta}, \quad u_\theta = -\frac{1}{r \sin\theta}\frac{\partial \psi}{\partial r},$$

where $r = \sqrt{x^2 + y^2 + z^2}$ is the radial distance from the origin and $\theta = 0$ in the x–y plane, which shows the direction of the motion. The solution is found to be given by

$$\psi = -\frac{1}{2}Ur^2 \sin^2\theta\left(1 - \frac{a^3}{r^3}\right),$$

which satisfies the conditions $\psi = 0$ on $r = a$ and $\psi \to -\frac{1}{2}Ur^2 \sin^2\theta$ as $r \to \infty$. Part of this result,

$$\gamma = \frac{1}{2}Ua^3 \frac{\sin^2\theta}{r},$$

is referred to as a "source doublet" at the origin. It is a *fundamental solution* of the equation,

$$\frac{\partial^2 \psi}{\partial r^2} + \frac{(1 - \mu^2)}{r^2}\frac{\partial^2 \psi}{\partial \mu^2} = 0,$$

where $\mu = \cos\theta$.

We see then that a fundamental solution plays a significant role; it satisfies the differential equation except at the one point where it has a singularity of the same order as the spherically symmetric source.

In our discussion in this chapter, we will extend these ideas to handle cases where boundaries are present, and therefore the expressions for the source functions are more involved. For the most part, we will restrict the treatment to one variable. In subsequent chapters, we will see how, after integral transformations, the dimensions may be reduced and only one variable with boundary conditions will remain. However, as the following example illustrates, often because of symmetry, only one direction has boundary conditions to be satisfied and once the boundary conditions are satisfied, the solution is completed.

4.1.2 Conduction of Heat in a Spherical Shell with Sources

The early time behavior of the heat diffusion process is of great interest in nuclear science and engineering. Here we consider a simple relevant model of heat conduction in a spherical shell or hollow sphere, with a spherically symmetric source. (See (Özişik) and (Wilkins).) It might represent the heating of the pressure shell in a water moderated and cooled reactor.

Due to symmetry the heat conduction problem, is described by the partial differential equation for the temperature $T(r, t)$ as

$$\frac{\partial T}{\partial t} = \frac{\partial^2 T}{\partial r^2} + \frac{2}{r} \frac{\partial T}{\partial r} + \frac{f(r)}{r}, \quad t > 0, \quad r_1 < r < r_2, \tag{4.1}$$

with the initial condition

$$T(r, 0) = 0. \tag{4.2}$$

To start, we suppose that the inner and outer surfaces are kept at some zero reference temperature implying the following boundary conditions

$$T(r_1, t) = T(r_2, t) = 0 \quad t > 0.$$

Here the thermal diffusivity is scaled to unity. If we make the transformation $u(r, t) = r T(r, t)$, the one-dimensional heat-conduction problem results

$$\frac{\partial u}{\partial t} = \frac{\partial^2 u}{\partial r^2} + f(r); \tag{4.3}$$

the initial condition remains

$$u(r, 0) = 0,$$

and the boundary conditions are still

$$u(r_1, t) = 0, \tag{4.4}$$
$$u(r_2, t) = 0. \tag{4.5}$$

We consider the possibility that a steady-state is reached so that $\partial u / \partial t = 0$. Then, the governing equation (4.3) reduces to

$$\frac{d^2 u_s}{dr^2} + f(r) = 0. \tag{4.6}$$

The boundary conditions for u_s remain the same, (4.4–4.5). To further simplify the analysis, we make changes of variables in (4.6), setting

$$x = r - r_1, \ b = r_2 - r_1, \ y(x) = u_s(r), \ h(x) = f(r). \tag{4.7}$$

Then, we solve

$$y''(x) + h(x) = 0, \ 0 < x < b, \tag{4.8}$$

where

$$y(0) = 0, \tag{4.9a}$$
$$y(b) = 0. \tag{4.9b}$$

Because of the simplicity of the equation, (4.8) may be integrated (twice). However, the form of the source may render the integral difficult to evaluate. Nevertheless, in principle, it may be written down. Integrating once, we obtain

$$y' = - \int_0^x h(t)dt + c_1,$$

and then after the second integration, we obtain

$$y = - \int_0^x d\xi \int_0^\xi h(t)dt + c_1 x + c_2.$$

Apply the boundary condition (4.9a) at $x = 0$:

$$[y]_{x=0} = c_2 = 0.$$

Using integration by parts this may be written as

$$y = -x \int_0^x h(t)dt + \int_0^x \xi h(\xi)d\xi + c_1 x. \tag{4.10}$$

(Note that t is a dummy variable.) Apply the boundary condition at $x = b$ to find that

$$[y]_{x=b} = -b \int_0^b h(\xi)d\xi + \int_0^b \xi h(\xi)d\xi + c_1 b = 0,$$

which gives

$$c_1 = \frac{1}{b} \int_0^b (b - \xi)h(\xi)d\xi. \tag{4.11}$$

Inserting (4.11) into (4.10) and recognizing that $0 \le x \le b$, the solution may be written as

$$y = \frac{1}{b} \int_0^x \xi(b - x)h(\xi)d\xi + \frac{1}{b} \int_x^b x(b - \xi)h(\xi)d\xi.$$

This last expression is more compactly written as

$$y(x) = \int_0^b G(x, \xi)h(\xi)d\xi, \tag{4.12}$$

where

$$G(x, \xi) = \begin{cases} x(b - \xi)/b, & x < \xi \\ \xi(b - x)/b, & x > \xi \end{cases}$$

and is called the "Green's function for the boundary-value problem" (4.8), (4.9a–4.9b).

To complete the solution, one should return to the original variables (4.7) to obtain steady temperature $T_s(r) = u_s/r$ as

$$T_s(r) = \int_{r_1}^{r_2} K(r, \rho) f(\rho) d\rho,$$

where $K(r, \rho) = G(r - r_1, \rho - r_1)/r$. The form (4.12) has the advantage that, for different sources, it provides the solution for the steady distribution in an input–output format. The method of determining G depends on being able to integrate twice and would have to be generalized to work in other cases. The purpose of this chapter is to look at some of these cases. We begin next with second-order equations and move on to higher equations and systems later in this chapter.

4.2 Green's Function

4.2.1 Second-Order Problems

Though variation of parameters is a method by which a particular solution to a nonhomogeneous linear ordinary differential equation (see Appendix 4.7.2, Equation (4.51)) may be found, when boundary conditions are appended or the inhomogeneous term varies, it is expedient to find another method to express the general form of the solution. We illustrate this first using the variation of parameters, and then we discuss the means by which this method may be applied to higher-order problems. Consider then the problem,

$$(p(x)y')' + q(x)y = f(x), \quad a < x < b, \tag{4.13}$$

with the simple boundary conditions,

$$y(a) = 0, \quad y(b) = 0. \tag{4.14}$$

We suppose that $\lambda = 0$ is not an eigenvalue of the boundary-value problem,

$$(p(x)y')' + (\lambda r(x) + q(x)) y = 0, \quad a < x < b,$$

with (4.14). Then there is fundamental pair of solutions u_1, u_2 such that by (4.53), a particular solution is given by $y = \chi_1(x)u_1(x) + \chi_2(x)u_2(x)$, so that

$$\chi_1' u_1 + \chi_2' u_2 = 0,$$
$$\chi_1' u_1' + \chi_2' u_2' = f/p.$$

Consequently, the general solution of (4.13) is

$$y = -u_1(x) \int_a^x \frac{f(\xi)u_2(\xi)}{pW} d\xi + u_2(x) \int_a^x \frac{f(\xi)u_1(\xi)}{pW} d\xi + c_1 u_1 + c_2 u_2, \tag{4.15}$$

where $W(u_1, u_2) = u_1 u_2' - u_1' u_2$. Suppose that $u_1(a) = 0$ and $u_2(b) = 0$. This is certainly possible because u_1, u_2 are linearly independent, and $\lambda = 0$ is not an eigenvalue. Satisfying (4.14), we find that in (4.15), $c_2 = 0$ and

$$-u_1(b) \int_a^b \frac{f(\xi)u_2(\xi)}{pW} d\xi + c_1 u_1(b) = 0.$$

Again, since $u_1(b) \neq 0$, this gives

$$c_1 = \int_a^b \frac{f(\xi)u_2(\xi)}{pW} d\xi.$$

The solution to the boundary-value problem (4.13)–(4.14) is written as

$$y = u_1(x) \left[-\int_a^x \frac{f(\xi)u_2(\xi)}{pW} d\xi + \int_a^b \frac{f(\xi)u_2(\xi)}{pW} d\xi \right] + u_2(x) \int_a^x \frac{f(\xi)u_1(\xi)}{pW} d\xi.$$

The compact and symmetric version of this result is

$$y = \int_a^b G(x, \xi) f(\xi) d\xi,$$

where

$$G(x, \xi) = \begin{cases} u_1(x)u_2(\xi)/(p(\xi)W(\xi)), & x < \xi \\ u_2(x)u_1(\xi)/(p(\xi)W(\xi)), & x > \xi \end{cases}. \qquad (4.16)$$

Example 4.1. As a specific instance of this result, let $p(x) = -1$, $q(x) = 0$, $a = 0$; then the problem to be solved is

$$-y'' = f, \quad 0 < x < b,$$
$$y(0) = y(b) = 0.$$

It suffices to take $u_1 = x$, $u_2 = b - x$, so that the boundary conditions are satisfied. Then, $W = -b$, and

$$y = \int_0^b G(x, \xi) f(\xi) d\xi,$$

where

$$G(x, \xi) = \begin{cases} x(b - \xi)/b, & x < \xi \\ \xi(b - x)/b, & x > \xi \end{cases}. \qquad (4.17)$$

We can make certain observations about the Green's function (4.16), which carry over to the higher-order case. As a function of x,

  (i) $G(x, \xi)$ is continuous on $a \leq x \leq b$.
 (ii) Its derivative $\partial G / \partial x$ has an upward jump at $x = \xi$, that is,

$$\left[\frac{\partial G}{\partial x} \right]_{x=\xi^-}^{x=\xi^+} = \frac{1}{p(\xi)}.$$

(iii) G satisfies $L(G) = 0$, except at $x = \xi$.
(iv) G satisfies the boundary conditions.

Example 4.2. An important special case of (4.13)–(4.14) may be used to illustrate these results. That is, suppose $q(x) = 0$. Then the general homogeneous solution is determined by direct integration to be

$$u_h(x) = c_1 \int \frac{dx}{p(x)} + c_2.$$

To construct the Green's function satisfying (4.14), take

$$u_1(x) = \int_a^x \frac{dt}{p(t)}, \quad u_2(x) = -\int_x^b \frac{dt}{p(t)}.$$

These two solutions differ by a constant. The required Green's function by (4.16) is

$$G(x, \xi) = \frac{u_1(x)u_2(\xi)}{\int_a^b \frac{dt}{p(t)}}, \quad x < \xi,$$

$$G(\xi, x) = G(x, \xi).$$

The method of constructing the Green's function by variation of parameters is suggestive that whenever the boundary conditions are separated, the form (4.16) is to be expected (Hildebrand). This will be demonstrated later in the context of higher-order equations, but now we turn to another example.

Example 4.3. Consider the problem

$$-y'' + k^2 y = f, \quad 0 < x < 1,$$

$$2y(0) - y'(0) = 0,$$

$$y(1) = 0.$$

Here, we may take

$$u_1(x) = 2\sinh(kx) + k\cosh(kx) \text{ and } u_2(x) = \sinh(k(1 - x)),$$

which satisfy the left and right boundary conditions, respectively. Then the form (4.16) gives the Green's function

$$G(x, \xi) = \frac{1}{pW} \begin{cases} (2\sinh(kx) + k\cosh(kx))\sinh(k(1 - \xi)), & x < \xi \\ (2\sinh(k\xi) + k\cosh(k\xi))\sinh(k(1 - \xi)), & x > \xi \end{cases},$$

where $pW = k^2 \cosh(k) + 2k \sinh(k)$.

There is another observation that follows, as we will see, from (4.55) in Appendix 4.7, which is that $pW = $ constant. Refer to a classic reference (Courant and Hilbert p. 371) for more examples of this type of Green's function.

Before we turn to higher-order cases, we consider one other example in which the attendant boundary conditions are not separated, and, hence, the structure of Green's function is slightly more complicated.

Example 4.4. Construct the Green's function for the following problem. Is the Green's function symmetric? Express the solution in terms of the Green's function

$$-u'' = f(x) \ \ 0 < x < 1,$$
$$u(0) = 2u(1), \quad 2u'(0) = u'(1).$$

Since the boundary conditions are not separated, we write the Green's function as

$$G(x, \xi) = \begin{cases} c_1 u_1 + c_2 u_2, & x < \xi \\ d_1 u_1 + d_2 u_2, & x > \xi \end{cases}.$$

It suffices to take $u_1 = 1$ and $u_2 = x$. Conditions (i)–(iv) are enough to determine G. They give

(i) G is continuous at $x = \xi$. Thus,

$$c_1 + c_2 \xi = d_1 + d_2 \xi.$$

(ii)
$$\left[\frac{\partial G}{\partial x} \right]_{x=\xi^-}^{x=\xi^+} = \frac{1}{p(\xi)} \Longrightarrow$$
$$d_2 - c_2 = -1.$$

(iii) $L(G) = 0$, except at $x = \xi$.
(iv) Boundary conditions

$$u(0) = 2u(1) \ \Rightarrow \ c_1 = 2(d_1 + d_2)$$
$$2u'(0) = u'(1) \ \Rightarrow \ 2c_2 = d_2.$$

Solving for those four values c_1, c_2, d_1, and d_2 we find that

$$G(x, \xi) = \begin{cases} -x - 2\xi + 4, & x < \xi \\ -2x - \xi + 4, & x > \xi \end{cases}.$$

Since $G(x, \xi) = G(\xi, x)$, the Green's function is symmetric. The solution is given by

$$y(x) = \int_0^1 G(x, \xi) f(\xi) d\xi.$$

The fact that $G(x, \xi) = G(\xi, x)$, in the last example, is dependent on the boundary conditions. Exercise 4.12 illustrates that such symmetry need not hold. In the next subsection conditions for symmetry will be derived, and this will be formalized in Theorem 4.1 of Section 4.3.

4.2.2 Higher-Order Problems

Now our intent is to solve

$$Ly = p_0(x) \frac{d^n y}{dx^n} + p_1(x) \frac{d^{n-1} y}{dx^{n-1}} + \cdots + p_{n-1}(x) \frac{dy}{dx} + p_n(x) y = r(x),$$

with n homogeneous boundary conditions $B_i(y) = 0$, $i = 1, \ldots, n$, involving the values $y, y', \ldots, y^{(n-1)}$ at $x = a$ and $x = b$. We suppose that these boundary conditions are such that the problem,

$$Lu = 0, \tag{4.18}$$

$$B_i(u) = 0, \quad i = 1, \ldots, n,$$

has only the trivial solution. In this case, a unique Green's function may be determined. Otherwise a *generalized* Green's function may be constructed, as will be shown later. The Green's function $G(x, \xi)$ for (4.18) as a function of x (Ince, chapter XI)

(i) Is continuous on $a \le x \le b$, and has continuous derivatives up to order $(n-2)$.
(ii) Its $(n-1)^{st}$ derivative has an upward jump at $x = \xi$, that is,

$$\left[\frac{\partial^{(n-1)}}{\partial x^{(n-1)}} G \right]_{x=\xi^-}^{x=\xi^+} = \frac{1}{p_0(\xi)}.$$

(iii) G satisfies $L(G) = 0$, except at $x = \xi$.
(iv) G satisfies the boundary conditions.

Example 4.5. As a simple example we consider the third-order problem for which to find the Green's function,

$$u''' = 0, \quad 0 < x < 1,$$

$$u(0) = u'(0) = u''(1) = 0.$$

Once again, we assume different forms for $x \lessgtr \xi$:

$$G(x, \xi) = \begin{cases} c_1 + c_2 x + c_3 x^2, & x < \xi, \\ d_1 + d_2 x + d_3 x^2, & x > \xi. \end{cases}$$

Applying the boundary conditions first, to reduce the number of constants, we find

$$G(0, \xi) = G_x(0, \xi) = 0 \quad \Rightarrow \quad c_1 = c_2 = 0,$$

and

$$G_{xx}(1, \xi) = 0 \quad \Rightarrow \quad d_3 = 0.$$

From the continuity of G and of G_x at $x = \xi$, we have

$$c_3 \xi^2 = d_1 + d_2 \xi$$

and

$$2 c_3 \xi = d_2.$$

Lastly the jump condition leads to

$$0 - 2 c_3 = 1.$$

Therefore, the required Green's function is

$$G(x, \xi) = \begin{cases} -\frac{1}{2}x^2, & x < \xi, \\ \frac{1}{2}\xi^2 - x\xi, & x > \xi \end{cases}.$$

Note the lack of symmetry.

4.2.3 Adjoint and Self-Adjoint Problems

The boundary-value problem $\{Lu = 0, B_i(u) = 0, i = 1, \ldots, n\}$ defined in (4.18) will be denoted by $L_x u = 0$. The adjoint problem will be called $L_x^* v = 0$ described by $\{L^* v = 0, B_i^*(v) = 0, i = 1, \ldots, n\}$. The functional form of $L_x^* v$ is given by

$$L^* v = (-1)^n \frac{d^n}{dx^n}(p_0 v) + (-1)^{n-1}\frac{d^{n-1}}{dx^{n-1}}(p_1 v) + \cdots - \frac{d}{dx}(p_{n-1}v) + p_n v. \quad (4.19)$$

The boundary conditions $B_i^*(v)$ are determined by the Lagrange identity[†]

$$vLu - uL^*v = \frac{d}{dx}P(u, v). \quad (4.20)$$

That is, if v satisfies the adjoint equation and u satisfies the equation $L_x u = 0$, then requiring

$$P(u, v) = 0 \quad (4.21)$$

determines the adjoint boundary conditions.

When $L^* = L$, the operator is *formally self-adjoint* and if the boundary conditions agree, the operator is said to be *self-adjoint*. Examples of these calculations follow in the next two examples. It should be understood that certain differentiability requirements are necessary on the functions involved so that the integrations by parts may be performed, all within the realm of square integrable functions (Holland).

Example 4.6. Consider the operator L_x defined by

$$Lu = u''', \quad 0 < x < 1,$$
$$u(0) = u'(0) = u''(1) = 0.$$

We simply integrate by parts to illustrate (4.20). From

$$\int_0^1 u'''\bar{v}dx = [u''\bar{v} - u'\bar{v}' + u\bar{v}'']_0^1 - \int_0^1 u\bar{v}'''dx,$$

we have admitted the possibility that u and v are complex-valued. Then $L^* v = -v'''$ and the complementary boundary conditions needed so that the boundary terms vanish simultaneously are $v(0) = v'(1) = v''(1) = 0$. In summary, the adjoint problem L_x^* is defined by

$$\{L^* v = -v''', \ 0 < x < 1, \ v(0) = v'(1) = v''(1) = 0\}.$$

[†] The expressions (4.19) and (4.20) are derived in the Appendix Section 4.7.3.

If the boundary conditions on L are not separated, the adjoint conditions need not be separated either. However, this does not preclude a self-adjoint operator.

Example 4.7. Consider the second-order problem

$$Lu = -u'', \ 0 < x < 1,$$

$$u(0) = 2u(1), \quad 2u'(0) = u'(1).$$

Here, it is clear that

$$\int_0^1 -u'' \bar{v} dx = [-u'\bar{v} + u\bar{v}']_0^1 - \int_0^1 u\bar{v}'' dx,$$

so L is formally self-adjoint. Applying the boundary conditions on u leads to the requirement

$$u'(0)\left(-2\bar{v}(1) + \bar{v}(0)\right) + u(1)\left(\bar{v}'(1) - 2\bar{v}'(0)\right) = 0.$$

Thus the conditions on v are identical to those on u and the operator L_x is self-adjoint; $L_x = L_x^*$.

Separated Boundary Conditions

Though an expression exists in principle for the form of the Green's function of the general boundary-value problem with mixed boundary conditions, it lacks the symmetry that occurs when the boundary conditions are separated. For the rest of the development of this section, we make that assumption (see Appendix, Section 4.7.3 and (Herron)).

Again, refer to the homogeneous differential equation (4.18), but temporarily relax the number of boundary conditions, while assuming that L is formally self-adjoint. Let there be k boundary conditions involving values at $x = b$. Then there will be $n - k$ linearly independent solutions satisfying the conditions there. Let these solutions be $v_1(x), v_2(x), \ldots, v_{n-k}(x)$. On the other hand, suppose that there are $n - k$ boundary conditions at $x = a$. Then there will be k linearly independent solutions satisfying these, which we call $u_1(x), \ldots, u_k(x)$. The construction of G proceeds as before. First, write

$$G(x, \xi) = \begin{cases} \sum_{i=1}^{k} c_i(\xi)u_i(x), & x < \xi \\ \sum_{i=1}^{n-k} d_i(\xi)v_i(x), & x > \xi \end{cases}.$$

The continuity and jump conditions are applied at $x = \xi$ to obtain

$$\mathbf{W}(\mathbf{\Phi}(\xi))\mathbf{\Gamma}(\xi) = \frac{1}{p_0(\xi)}\mathbf{e}_n, \tag{4.22}$$

where $\mathbf{W}$ is the Wronskian matrix for the solutions defined above, $\mathbf{e}_n$ is the nth unit vector and

$$\boldsymbol{\Gamma}(\xi) = \begin{pmatrix} d_1(\xi) \\ \vdots \\ d_{n-k}(\xi) \\ -c_1(\xi) \\ \vdots \\ -c_k(\xi) \end{pmatrix}, \quad \boldsymbol{\Phi}(\xi) = \begin{pmatrix} v_1(\xi) \\ \vdots \\ v_{n-k}(\xi) \\ u_1(\xi) \\ \vdots \\ uc_k(\xi) \end{pmatrix}.$$

Then the solution of (4.22) is

$$\Gamma(\xi) = \frac{1}{p_0(\xi)} \mathbf{W}^{-1}(\boldsymbol{\Phi}(\xi)) \mathbf{e}_n. \tag{4.23}$$

In the Appendix, we have examined in more detail the case where L is formally self-adjoint. In that case it is known that when the coefficients are real,

$$\mathbf{W}^T(\boldsymbol{\Phi}(\xi)) \mathbf{R} \mathbf{W}(\boldsymbol{\Phi}(\xi)) \equiv \mathcal{P}, \tag{4.24}$$

where $\mathbf{R}$ is defined by (4.57) and $\mathcal{P}$ is a constant matrix. Furthermore, if the number of boundary conditions required at each endpoint is the same; then the order of the differential operator is even and $\mathcal{P}$ has a special structure (see Appendix, Section 4.7.3) given by

$$\mathcal{P} = \begin{pmatrix} \mathbf{O} & \mathbf{P} \\ -\mathbf{P}^T & \mathbf{O} \end{pmatrix}. \tag{4.25}$$

Here $\mathbf{O}$ (the zero matrix) and $\mathbf{P}$ are square matrices of size $\left(\frac{n}{2} \times \frac{n}{2}\right)$. If these conditions hold the problem is self-adjoint.

We employ (4.24) to rewrite (4.23) as

$$\Gamma(\xi) = \frac{1}{p_0(\xi)} \mathcal{P}^{-1} \mathbf{W}^T(\boldsymbol{\Phi}(\xi)) \mathbf{R} \mathbf{e}_n.$$

Note that by (4.57), $\mathbf{R}\mathbf{e}_n = p_0 \mathbf{e}_1$. Hence,

$$\Gamma(\xi) = \mathcal{P}^{-1} \mathbf{W}^T(\boldsymbol{\Phi}(\xi)) \mathbf{e}_1.$$

An inspection of $\mathbf{W}^T(\boldsymbol{\Phi}(\xi))$ shows that

$$\mathbf{W}^T(\boldsymbol{\Phi}(\xi)) \mathbf{e}_1 = \boldsymbol{\Phi}(\xi).$$

The unknown elements of the Green's function are therefore

$$d_i(\xi) = \left[\mathcal{P}^{-1}\boldsymbol{\Phi}(\xi)\right]_i, \quad i = 1, \dots, \frac{n}{2}$$

$$c_i(\xi) = -\left[\mathcal{P}^{-1}\boldsymbol{\Phi}(\xi)\right]_i, \quad i = 1, \dots, \frac{n}{2}.$$

The form of the Green's function for the self-adjoint problem of even order, with separated boundary conditions, reads

$$G(x, \xi) = \begin{cases} -\mathbf{U}^T(x)\mathbf{P}^{-1}\mathbf{V}(\xi), & x < \xi, \\ -\mathbf{V}^T(x)\left(\mathbf{P}^T\right)^{-1}\mathbf{U}(\xi), & x > \xi, \end{cases} \tag{4.26}$$

where

$$\mathbf{U}(x) = \begin{pmatrix} u_1(x) \\ \vdots \\ u_{\frac{n}{2}}(x) \end{pmatrix}, \quad \mathbf{V}(x) = \begin{pmatrix} v_1(x) \\ \vdots \\ v_{\frac{n}{2}}(x) \end{pmatrix}.$$

With the choice of the solutions $\mathbf{U}$ and $\mathbf{V}$, we may determine the entries of $\mathbf{P}$ as

$$[\mathbf{P}]_{ij} = P(u_j, v_i), \ i, j = 1, \ldots, \frac{n}{2}$$

by means of Lagrange's identity. Observe that if $n = 2$, then $\mathbf{P}$ is a scalar, $p_0 W$.

A Fourth-Order Example

Example 4.8. A mathematical model of a uniform vibrating beam uses the problem

$$L\phi = \phi'''' - \beta^4 \phi, \ 0 < x < 1. \tag{4.27}$$

The boundary conditions for a cantilevered beam are

$$\phi(0) = \phi'(0) = \phi''(1) = \phi'''(1) = 0.$$

The concomitant is given by

$$P(u, v) = u'''v - uv''' - (u''v' - u'v'').$$

The functions satisfying the boundary conditions are chosen as

$$u_1(x) = \cosh(\beta x) - \cos(\beta x),$$
$$u_2(x) = \sinh(\beta x) - \sin(\beta x),$$

at the left endpoint, and

$$v_1(x) = \cosh(\beta - \beta x) + \cos(\beta - \beta x),$$
$$v_2(x) = \sinh(\beta - \beta x) + \sin(\beta - \beta x),$$

at the right endpoint.

The elements of P needed are obtained by evaluating the concomitants at $x = 0$:

$$P_{11} = P(u_1, v_1)(0) = 2\beta^3(\sinh \beta - \sin \beta)$$
$$P_{12} = P(u_2, v_1)(0) = 2\beta^3(\cosh \beta + \cos \beta)$$
$$P_{21} = P(u_1, v_2)(0) = 2\beta^3(\cosh \beta + \cos \beta)$$
$$P_{22} = P(u_2, v_2)(0) = 2\beta^3(\sinh \beta + \sin \beta).$$

From (4.26) Green's function for any self-adjoint fourth-order system is given by

$$G(x, \xi) = \begin{cases} \frac{1}{\det(P)} \{u_1(x)[-P_{22}v_1(\xi) + P_{12}v_2(\xi)] \\ +u_2(x)[P_{21}v_1(\xi) - P_{11}v_2(\xi)]\}, & x < \xi \end{cases},$$

$$G(\xi, x) = G(x, \xi).$$

Notice that $P_{12} = P_{21}$ is symmetric because of the symmetric choice of the functions which satisfy the boundary conditions. Thus, when the computations are completed, the result for the cantilevered boundary condition is

$$G(x, \xi; \beta) = \frac{1}{4\beta^3(1 + \cosh \beta \cos \beta)} \begin{cases} g(x, \xi; \beta), & x < \xi \\ g(\xi, x; \beta), & x > \xi \end{cases},$$

where

$$g(x, \xi; \beta) = u_1(x)[(\sinh \beta + \sin \beta)v_1(\xi) + (\cosh \beta + \cos \beta)v_2(\xi)]$$
$$+ u_2(x)[(\sinh \beta - \sin \beta)v_2(\xi) + (\cosh \beta + \cos \beta)v_1(\xi)].$$

4.3 Connections with Distributions

We have shown that the solution of the nonhomogeneous boundary-value problem may be written as an integral

$$y = \int_a^b G(x, \xi) f(\xi) d\xi. \tag{4.28}$$

The boundary conditions $B_i(y) = 0$, $i = 1, \ldots, n$, involve only $y, y', \ldots, y^{(n-1)}$, at $x = a$ and $x = b$. Thus we have

$$B_i(y) = \int_a^b B_i(G(x, \xi)) f(\xi) d\xi = 0,$$

because G satisfies the same boundary conditions.

Operating on (4.28) with the operator L, we are able to differentiate the expression under the integral sign with L denoted by L_x to find

$$Ly = \int_a^b L_x G(x, \xi) f(\xi) d\xi \overset{?}{=} f(x).$$

Examine in a little more detail the case where $n = 2$ and where the operator L is given by (4.13). We have

$$y(x) = \int_a^x G(x, \xi) f(\xi) d\xi + \int_x^b G(x, \xi) f(\xi) d\xi.$$

Differentiating, noting that since G is continuous at $x = \xi$,

$$y'(x) = \int_a^x \frac{\partial G}{\partial x}(x, \xi) f(\xi) d\xi + \int_x^b \frac{\partial G}{\partial x}(x, \xi) f(\xi) d\xi.$$

However, $\partial G/\partial x$ has a jump at $\xi = x$, so for the second derivative we find

$$y''(x) = \int_a^x \frac{\partial^2 G}{\partial x^2}(x,\xi) f(\xi) d\xi + \lim_{\varepsilon \downarrow 0} \left[\frac{\partial G}{\partial x}(x,\xi) f(\xi) \right]_{\xi = x - \varepsilon}$$

$$+ \int_x^b \frac{\partial^2 G}{\partial x^2}(x,\xi) f(\xi) d\xi - \lim_{\varepsilon \downarrow 0} \left[\frac{\partial G}{\partial x}(x,\xi) f(\xi) \right]_{\xi = x + \varepsilon}$$

$$= \int_a^x \frac{\partial^2 G}{\partial x^2}(x,\xi) f(\xi) d\xi + \int_x^b \frac{\partial^2 G}{\partial x^2}(x,\xi) f(\xi) d\xi + \frac{f(x)}{p(x)}.$$

The result is

$$Ly = \int_a^x L_x G(x,\xi) f(\xi) d\xi + \int_x^b L_x G(x,\xi) f(\xi) d\xi + f(x)$$

$$= f(x),$$

because $L_x G(x,\xi) = 0$, $x \neq \xi$. Thus, we see that $L_x G(x,\xi)$ must play the role of the Dirac function or *distribution* $\delta(x - \xi)$, so that

$$\int_a^b \delta(x - \xi) f(\xi) d\xi = f(x),$$

in order that f is to be recovered (see (Friedman) and (Keener)).

A more general argument may be successfully applied to the nth order problem given that (4.28) holds where

$$\left[\frac{\partial^{n-1} G}{\partial x^{n-1}} \right]_{x = \xi^-}^{x = \xi^+} = \frac{1}{p_0(\xi)}.$$

Alternatively, as in Section 4.2.3, we think of the collection

$$\{ Lu, \ B_i(u) = 0, \ i = 1, \dots, n \}$$

as *defining* the differential operator L_x. Then, when G exists, we have the inverse operator L_x^{-1} so that

$$y = L_x^{-1} f. \tag{4.29}$$

It is important to recognize that the transposed Green's function and the adjoint boundary-value problem are related. We state this as a theorem (Friedman, p. 173).

Theorem 4.1. *Let $G(x,\xi)$ be the Green's function for the operator L_x as defined in Section 4.2.3, and let $H(x,\xi)$ be the Green's function for the adjoint operator L_x^* also defined there; then*

$$G(x,\xi) = H(\xi, x).$$

In particular, if L_x is self-adjoint, then $G(x,\xi) = G(\xi, x)$.

Proof. Start with the definitions

$$L_x G(x, \xi) = \delta(x - \xi)$$

and

$$L_x^* H(x, z) = \delta(x - z).$$

Using the properties of the adjoint operator, it follows that

$$\int_a^b H(x, z) L_x G(x, \xi) dx = \int_a^b L_x^* H(x, z) G(x, \xi) dx.$$

From the definitions, we have

$$\int_a^b H(x, z)\delta(x - \xi) dx = \int_a^b \delta(x - z) G(x, \xi) dx,$$

and

$$H(\xi, z) = G(z, \xi).$$

Replacing z with x gives the desired result. If L_x is self-adjoint, then $L_x = L_x^*$ and $G(x, \xi) = G(\xi, x)$. □

We notice how many of the Green's functions we have constructed have the symmetry property. For the second-order differential operator (4.13) with separated boundary conditions, this is to be expected; and for higher-order problems with separated boundary conditions, this has been demonstrated in (4.26). However, for problems with more general boundary conditions, such as Example 4.4 of Section 4.2.1, it is comforting to know that our intuition was correct.

4.4 First-Order System: Green's Matrices

For systems of differential equations, the desired representation of the solution may be expressed in terms of a *Green's matrix*. The system

$$\frac{d\mathbf{u}}{dt} = \mathbf{P}(t)\mathbf{u} + \mathbf{f}(t), \quad a < t < b, \tag{4.30}$$

where

$$\mathbf{u}(t) = \begin{pmatrix} u_1(t) \\ \vdots \\ u_k(t) \end{pmatrix}, \quad \mathbf{f}(t) = \begin{pmatrix} f_1(t) \\ \vdots \\ f_k(t) \end{pmatrix},$$

and $\mathbf{P}(t)$ is a $k \times k$ nonsingular matrix, plays an important role because many higher-order systems may be reduced to this form. Suggestively, the independent variable is t because the system in (4.30) is sometimes referred to as evolution equations.

We suppose that boundary conditions are given, such as

$$\mathbf{A}\mathbf{u}(a) + \mathbf{B}\mathbf{u}(b) = \mathbf{0}, \tag{4.31}$$

with $\mathbf{A}$ and $\mathbf{B}$ are $(k \times k)$ constant matrices. The homogeneous counterpart of (4.30) will have a $k \times k$ *fundamental (nonsingular) matrix* $\mathbf{U}(t)$, such that

$$\frac{d\mathbf{U}}{dt} = \mathbf{P}(t)\mathbf{U},$$

with

$$\mathbf{A}\mathbf{U}(a) + \mathbf{B}\mathbf{U}(b) \equiv \mathbf{D} \neq \mathbf{O},$$

otherwise $\lambda = 0$ is an eigenvalue of (4.30)–(4.31) with $\mathbf{f}(t) = \mathbf{0}$. Thus, taking

$$\mathbf{G}(t, \tau) = \begin{cases} -\mathbf{U}(t)\mathbf{D}^{-1}\mathbf{B}\mathbf{U}(b)\mathbf{U}^{-1}(\tau), & t < \tau \\ \mathbf{U}(t)\mathbf{D}^{-1}\mathbf{A}\mathbf{U}(a)\mathbf{U}^{-1}(\tau), & t > \tau \end{cases}, \tag{4.32}$$

we have the desired Green's function.

Notice that $\mathbf{G}$ has a jump at $t = \tau$, in fact,

$$[\mathbf{G}]_{t=\tau^-}^{t=\tau_+} = \mathbf{I}. \tag{4.33}$$

And when $t \neq \tau$, $\partial\mathbf{G}/\partial t - \mathbf{P}(t)\mathbf{G} = \mathbf{O}$, so that

$$\frac{\partial\mathbf{G}}{\partial t} - \mathbf{P}(t)\mathbf{G} = \delta(t - \tau)\mathbf{I}.$$

The conditions (4.31) are satisfied for $\mathbf{G}$ as a function of t. The solution to (4.30) with (4.31) is

$$\mathbf{u}(t) = \int_a^b \mathbf{G}(t, \tau)\mathbf{f}(\tau)d\tau. \tag{4.34}$$

If either $\mathbf{B} = \mathbf{O}$ (or $\mathbf{A} = \mathbf{O}$), the Green's function that results is for the initial-value problem. However, the representation (4.34) is particularly useful when a periodic solution is sought, so that $\mathbf{A} = -\mathbf{B}$.

The derivation of the representation (4.32) is outlined in Exercise 4.21. The general theory for nth order systems has also been developed. (See (Reid).)

4.5 Generalized Green's Functions

The ability to solve the second-order, boundary-value problem

$$Lu = (pu')' + qu = f, \ a < x < b,$$

$$B_1u(a) = 0,$$

$$B_2u(b) = 0,$$

when $\lambda = 0$ is an eigenvalue of $Lu + \lambda ru = 0$ is the motivation for this section. We have shown in Section 3.4.1 that a solution to this boundary-value problem exists only if

$$\int_a^b f(x)\varphi(x)dx = 0,$$

where φ is the corresponding eigenfunction. This solvability condition must be employed in order to find a *generalized Green's function* $G^\dagger(x, \xi)$ (sometimes called a "modified" Green's function (Stakgold)) to solve the nonhomogeneous problem, whose solution *will not be unique*. So, if

$$u(x) = \int_a^b G^\dagger(x, \xi)f(\xi)d\xi$$

is the representation of the solution, $G^\dagger(x, \xi)$ is defined by

$$L_x G^\dagger(x, \xi) = \delta(x - \xi) + cr(x)\varphi(x), \qquad (4.35)$$

where c is independent of x, and φ is the eigenfunction.

The conditions to be satisfied are as follows. As a function of x:

(i) $G^\dagger$ satisfies the boundary conditions.
(ii) $G^\dagger$ is continuous at $x = \xi$.
(iii)

$$\left.\frac{\partial G^\dagger}{\partial x}\right|_{x=\xi^-}^{x=\xi^+} = \frac{1}{p(\xi)}\int_a^b r\varphi^2 dx.$$

(iv)

$$LG^\dagger = cr\varphi, \quad x \neq \xi.$$

The value of c is determined from (4.35). Multiply both sides by $\varphi(x)$ and integrate to obtain

$$\int_a^b \varphi(x)L_x G^\dagger(x, \xi)dx = \int_a^b \delta(x - \xi)\varphi(x)dx + c\int_a^b r\varphi^2 dx = 0,$$

after integration by parts. Hence, it suffices to take

$$c = -\varphi(\xi) \Big/ \int_a^b r\varphi^2 dx .$$

Condition (iv) is met with this value for c.

To restrict the nonuniqueness of the solution, one can eliminate the "component" of the solution u in the direction of φ. This requirement is that

$$\int_a^b u(x)r(x)\varphi(x)dx = 0.$$

Consequently, there is the accompanying condition

(i)′

$$\int_a^b \varphi(x)r(x)G^\dagger(x, \xi)dx = 0.$$

This replaces condition (i), because φ is an eigenfunction. For the boundary-value problem with separated boundary conditions, this is all that is needed to find a unique generalized Green's function. The solution thus obtained may be written as

$$u = L_x^\dagger f,$$

in keeping with (4.29), where $L_x^\dagger$ denotes the *generalized inverse* of the operator L_x defined by $\{Lu, \, B_i(u) = 0, \, i = 1, 2\}$, which is in complete analogy to the well-known generalized inverse defined for singular symmetric matrices (Noble and Daniel).

For those problems with mixed (i.e., nonseparated) boundary conditions, two independent eigenfunctions may occur. Then it is necessary to assume

$$L_x G^\dagger = \delta(x - \xi) + r(x)(c_1 \varphi_1 + c_2 \varphi_2).$$

Choosing the eigenfunctions as orthogonal, solving for c_1 and c_2 in a like manner as before, we have

$$c_i = -\varphi_i(\xi) \left/ \int_a^b r \varphi_i^2 dx, \quad i = 1, 2. \right.$$

For higher-order, *self-adjoint* boundary-value problems, the procedure is easily extended, since the eigenfunctions corresponding to $\lambda = 0$ may be used in the same way to define and construct a symmetric $G^\dagger(x, \xi) = G^\dagger(\xi, x)$. If the problem

$$\{Lu = 0, \, B_i(u) = 0, \, i = 1, \ldots k\},$$

as in (4.18), is not self-adjoint, then the adjoint problem

$$\{L^* v = 0, \, B_i^*(v) = 0, \, i = 1, \ldots, 2n - k\},$$

as defined in (4.19), must be introduced. A careful analysis of this most general situation has been carried out in the literature (Loud). The interested reader will find that Exercises 4.26 and 4.27 typify this situation.

Example 4.9. We consider the important operator L_x as defined by

$$Lu = -u'', \quad 0 < x < 1,$$

$$u'(0) = u'(1) = 0.$$

This problem occurs in one-dimensional heat conduction with insulated end conditions.

The eigenfunction is $\varphi(x) = 1$, and it suffices to take $r(x) = 1$ so we are called on to solve

$$-G_{xx} = \delta(x - \xi) - 1.$$

We may take

$$G = \begin{cases} c_1 + c_2 x + \frac{1}{2} x^2, & x < \xi, \\ d_1 + d_2 x + \frac{1}{2} x^2, & x > \xi. \end{cases}$$

(i) Satisfying the boundary conditions $c_2 = 0$ and $d_2 = -1$.
(ii) Continuity at $x = \xi$ requires $d_1 - c_1 = \xi$.
(iii) The jump condition at $x = \xi$ is identically satisfied.
(i)′ The condition

$$\int_a^b G^\dagger(x, \xi) dx = 0$$

leads to

$$\int_0^\xi \left(c_1 + \frac{1}{2}x^2 \right) dx + \int_\xi^1 \left(d_1 - x + \frac{1}{2}x^2 \right) dx = 0.$$

This provides another condition which along with condition (ii) gives $c_1 = \frac{1}{3} - \xi + \frac{1}{2}\xi^2$ and $d_1 = \frac{1}{3} + \frac{1}{2}\xi^2$. The desired generalized Green's function is

$$G^\dagger(x, \xi) = \begin{cases} \frac{1}{3} - \xi + \frac{1}{2}\xi^2 + \frac{1}{2}x^2, & x < \xi \\ \frac{1}{3} - x + \frac{1}{2}\xi^2 + \frac{1}{2}x^2, & x > \xi \end{cases}$$

$$= \frac{1}{3} + \frac{1}{2}(x^2 + \xi^2) - \min(x, \xi).$$

4.6 Expansions in Eigenfunctions

The purpose of this section is to derive alternative expressions for Green's functions using the expansions from Chapter 3. We begin with the second-order case and analyze the operator,

$$Ly = (py')' - qy, \quad a < x < b,$$

with Sturm–Liouville boundary conditions at $x = a, b$.

Suppose $\lambda \neq \lambda_n$, where λ_n is one of the eigenvalues of the operator (with weight $r(x)$). Then, since $Ly + \lambda_n r y_n = 0$,

$$Ly + \lambda r y_n = -\lambda_n r y_n + \lambda r y_n = (\lambda - \lambda_n) r y_n. \tag{4.36}$$

Let $G(x, \xi; \lambda)$ be the Green's function of (4.36). Then, we obtain an *integral equation* for y_n as

$$y_n(x) = (\lambda - \lambda_n) \int_a^b G(x, \xi; \lambda) r(\xi) y_n(\xi) d\xi. \tag{4.37}$$

A key idea is to use the expansion we know to be true from Chapter 3,

$$G(x, \xi; \lambda) = \sum_k \alpha_k y_k(\xi), \tag{4.38}$$

where, assuming that the eigenfunctions are normalized,

$$\alpha_k = \int_a^b G(x, \xi; \lambda) r(\xi) y_k(\xi) d\xi.$$

Substituting the series expansion (4.38) into (4.37) yields

$$y_n(x) = (\lambda - \lambda_n) \int_a^b \sum_k \alpha_k y_k(\xi) r(\xi) y_n(\xi) d\xi,$$

which gives, since only the term with $k = n$ in the series survives,

$$\alpha_n(x) = \frac{y_n(x)}{\lambda - \lambda_n}.$$

The resulting expansion for G is

$$G(x, \xi; \lambda) = \sum_n \frac{y_n(x) y_n(\xi)}{\lambda - \lambda_n}. \tag{4.39}$$

Example 4.10. As an illustration, we consider the expansion problem using

$$Ly = -y'', \quad -1 < x < 1,$$

with boundary conditions $y = 0$ at $x = \pm 1$. The eigenvalues and eigenfunctions are easily shown to be

$$\lambda_n = (n\pi/2)^2, \quad y_n(x) = \sin(\sqrt{\lambda_n}(x+1)), \quad n = 1, 2, \dots.$$

So, if we are interested in the expansion of $G(x, \xi; \lambda)$ defined by

$$LG + \lambda G = -G_{xx} + \lambda G,$$

$$G(-1, \xi) = G(1, \xi) = 0,$$

we make use of (4.39) to write

$$G(x, \xi; \lambda) = \sum_{n=1}^{\infty} \frac{1}{(\lambda - \frac{n^2\pi^2}{4})} \sin\left[\frac{n\pi}{2}(x+1)\right] \sin\left[\frac{n\pi}{2}(\xi+1)\right].$$

This expansion will converge uniformly on $-1 \le x, \xi \le 1$, for $\lambda \ne n^2\pi^2/4$. However, since $\partial G/\partial x$ has a jump at $x = \xi$, the same is not true of this derivative and indeed a *Gibb's phenomenon* will occur at this point.

4.6.1 Delta Function Representation

We use the operational definition that we developed in Section 4.3:

$$L_x G + \lambda r G = \delta(x - \xi).$$

Hence, by applying this to (4.39), we obtain, formally,

$$\delta(x - \xi) = (L_x + \lambda r) \sum_n \frac{y_n(x) y_n(\xi)}{\lambda - \lambda_n}$$

$$= \sum_n \frac{(L_x + \lambda r) y_n(x) y_n(\xi)}{\lambda - \lambda_n}.$$

Making use of (4.36), we have

$$\delta(x - \xi) = \sum_n r(x) y_n(x) y_n(\xi)$$

$$= \sum_n r(\xi) y_n(x) y_n(\xi), \qquad (4.40)$$

since $\delta(x - \xi) = \delta(\xi - x)$. This is an operational form whose justification we will not thoroughly pursue. However, by seeking alternative views of it, we get a better picture of not only its verity but also its utility. So, on the other hand, we can view $G(x, \xi; \lambda)$ as an analytic function of λ, except for poles at the eigenvalues (assumed simple), where $\lambda = \lambda_n$. If C is a closed contour in the complex λ-plane, containing all the poles of G, then, from (4.39),

$$\frac{1}{2\pi i} \oint_C G(x, \xi; \lambda) d\lambda = \frac{1}{2\pi i} \oint_C \sum_n \frac{y_n(x) y_n(\xi)}{\lambda - \lambda_n} d\lambda$$

$$= \sum_n y_n(x) y_n(\xi), \quad \text{since} \oint_C \frac{d\lambda}{\lambda - \lambda_n} = 2\pi i.$$

Thus, we have the expansion of δ

$$\delta(x - \xi) = \frac{1}{2\pi i} \oint_C r(\xi) G(x, \xi; \lambda) d\lambda. \qquad (4.41)$$

Now a few of these details will be applied to specific problems involving discrete and continuous spectra.

4.6.2 Worked Examples

Example 4.11. We study the problem,

$$y'' + \lambda y = 0, \quad 0 < x < 1,$$
$$y(0) = 0, \quad y'(1) - y(1) = 0.$$

From our work in Chapter 3, we know that the problem has the eigenfunctions,

$$y_0 = 1, \quad \lambda_0 = 0,$$
$$y_n = \sin\left(\sqrt{\lambda_n} x\right), \quad n = 1, 2, \ldots,$$

where

$$\sqrt{\lambda_n} = \tan \sqrt{\lambda_n}. \qquad (4.42)$$

We seek an alternative representation of the form (4.40). The Green's function (see Exercise 4.3) is given by

$$G(x, \xi; k) = \frac{\sin(kx) \left[\sin(k(1 - \xi)) - k \cos(k(1 - \xi))\right]}{k(k \cos k - \sin k)}, \quad x < \xi,$$

$$G(\xi, x; k) = G(x, \xi; k), \quad \text{where } k = \sqrt{\lambda}.$$

We examine the poles of G in k. First, we notice the eigenvalue at $\lambda = 0$. Since $\lambda = k^2$, we can evaluate

$$\lim_{k \to 0} k^2 G = \lim_{k \to 0} \frac{k(kx - \cdots)(k(1 - \xi) - \cdots - k + \cdots)}{k(1 - \frac{1}{2}k^2 + \cdots) - (k - \frac{1}{6}k^3 + \cdots)}$$

$$= 3x\xi.$$

At the other eigenvalues,

$$\lim_{k \to k_n} \frac{(k^2 - k_n^2)}{k \cos k - \sin k} = \frac{2k_n}{-k_n \sin k_n},$$

so summing, the complete expansion is for $x < \xi$:

$$\delta(x - \xi) = 3x\xi - 2\sum_{n=1}^{\infty} \frac{\sin(\sqrt{\lambda_n}x)\left[\sin\left(\sqrt{\lambda_n}(1 - \xi)\right) - \sqrt{\lambda_n}\cos\left(\sqrt{\lambda_n}(1 - \xi)\right)\right]}{\sqrt{\lambda_n}\sin\sqrt{\lambda_n}}.$$

This expression may be simplified by the eigenvalue relation (4.42) to

$$\delta(x - \xi) = 3x\xi + 2\sum_{n=1}^{\infty} \frac{\sin\left(\sqrt{\lambda_n}x\right)\sin\left(\sqrt{\lambda_n}\xi\right)}{\sin^2\sqrt{\lambda_n}}.$$

Likewise, when $x > \xi$, the same expansions result.

This technique works with some modifications for non–self-adjoint, boundary-value problems. (See (Cole), (Friedman), and (Ince).) It was such an approach that led to one of the earliest expansion methods for the linearized hydrodynamic stability problem of channel flows (Schensted). We will not pursue this further here. Instead, we turn to self-adjoint examples in which a continuous spectrum arises.

Example 4.12. Consider then the familiar problem (Friedman, p. 217)

$$u'' + \lambda u = 0, \quad 0 < x < \infty,$$

$$u(0) = 0, \quad \int_0^{\infty} u^2 dx < \infty.$$

Its Green's function satisfies (4.41)

$$G_{xx} + \lambda G = -\delta(x - \xi)$$

and is given by

$$G(x, \xi; \lambda) = \begin{cases} \frac{\sin(\sqrt{\lambda}x)}{\sqrt{\lambda}}e^{i\sqrt{\lambda}\xi}, & x < \xi \\ \\ \frac{\sin(\sqrt{\lambda}\xi)}{\sqrt{\lambda}}e^{i\sqrt{\lambda}x}, & x > \xi \end{cases},$$

where $\lambda \notin [0, \infty)$. Note that $\sin(\sqrt{\lambda}x)/\sqrt{\lambda}$ is an analytic function of λ without a branch point. However, $e^{i\sqrt{\lambda}x}$ has a branch point at $\lambda = 0$. Integrate over $\{\lambda \mid \Re\text{eal}(\lambda) \geq 0, \Im\text{mag}(\lambda) = 0\}$ given as the branch cut. Put $\lambda = k^2$, so that $d\lambda = 2kdk$. On the integrals from ∞, $k = -\sqrt{\lambda}$, while on the integrals to ∞, $k = -\sqrt{\lambda}$. Using (4.41), we find

$$\delta(x - \xi) = \frac{1}{2\pi i} \left[2H(\xi - x) \left\{ \int_0^\infty \sin(kx)e^{ik\xi} dk + \int_\infty^0 \sin(kx)e^{-ik\xi} dk \right\} \right.$$

$$\left. + 2H(x - \xi) \left\{ \int_0^\infty \sin(k\xi)e^{ikx} dk + \int_\infty^0 \sin(k\xi)e^{-ikx} dk \right\} \right]$$

$$= \frac{2}{\pi} \int_0^\infty \sin(kx) \sin(k\xi) dk.$$

This particular representation gives the *Fourier sine transform*. We see this by multiplying both sides of the equation by $f(\xi)$ and integrating with respect to ξ, to obtain

$$f(x) = \frac{2}{\pi} \int_0^\infty f(\xi)d\xi \int_0^\infty \sin(kx) \sin(k\xi) dk.$$

Define

$$F_s(k) = \sqrt{\frac{2}{\pi}} \int_0^\infty f(\xi) \sin k\xi \, d\xi,$$

and

$$f(x) = \sqrt{\frac{2}{\pi}} \int_0^\infty F_s(k) \sin(kx) dk,$$

to obtain the famous transform pair.

Example 4.13. A more involved example is,

$$u'' + (\lambda - 1 - \alpha^2 + 2 \operatorname{sech}^2 x)u = 0 \quad -\infty < x < \infty,$$

with $u \to \infty$ as $x \to \pm\infty$. We assume that α is real.

The equation has exact solutions given by

$$u(x) = c_1 e^{kx}(\tanh x - k) + c_2 e^{-kx}(\tanh x + k),$$

where $k = (1 + \alpha^2 - \lambda)^{\frac{1}{2}}$ is the principal square root. As long as λ is not on the nonnegative real line, we have the Green's function

$$G(x, \xi; \lambda) = \frac{1}{2k(1 - k^2)} \begin{cases} e^{k(x-\xi)}(\tanh x - k)(\tanh \xi + k), & x < \xi \\ e^{-k(x-\xi)}(\tanh x + k)(\tanh \xi - k), & x > \xi \end{cases}.$$

The spectrum of the corresponding operator L_x is both discrete and continuous. There is one eigenvalue at $\lambda = \alpha^2$. The continuous spectrum is defined parametrically by

$$\lambda = \omega^2 + \alpha^2 + 1, \quad 0 \le \omega < \infty.$$

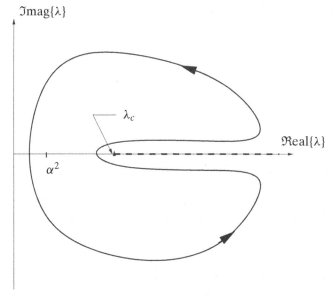

Figure 4.1. $\lambda_c = 1 + \alpha^2$.

The counterpart of (4.41) is performed over the contour C shown in Figure 4.1. It begins at $+\infty$, enclosing all poles and branch cuts, and terminates at $+\infty$. This substitution is made in the integral. Note that if $\lambda_0 \in C$, then, for some ω_0, we have

$$\lim_{\lambda \to \lambda_0^+} k(\lambda) = k(\lambda_0^+) = i\omega_0$$

on the "top" of the continuous spectrum, and

$$\lim_{\lambda \to \lambda_0^-} k(\lambda) = k(\lambda_0^-) = -i\omega_0$$

on the "bottom" of the continuous spectrum. Furthermore, $d\lambda = 2\omega d\omega$, so the integrals become

$$\delta(x - \xi) = \text{Res}\left[G(x, \xi; \lambda)\right]_{k=1}$$

$$+ \frac{1}{2\pi i} H(\xi - x) \left\{ \int_0^\infty \frac{e^{i\omega(x-\xi)}(\tanh x - i\omega)(\tanh \xi + i\omega)}{i(\omega^2 + 1)} d\omega \right.$$

$$+ \int_\infty^0 \frac{e^{-i\omega(x-\xi)}(\tanh x + i\omega)(\tanh \xi - i\omega)}{-i(\omega^2 + 1)} d\omega \right\}$$

$$+ \frac{1}{2\pi i} H(x - \xi) \left\{ \int_0^\infty \frac{e^{-i\omega(x-\xi)}(\tanh x + i\omega)(\tanh \xi - i\omega)}{i(\omega^2 + 1)} d\omega \right.$$

$$+ \int_\infty^0 \frac{e^{i\omega(x-\xi)}(\tanh x - i\omega)(\tanh \xi + i\omega)}{-i(\omega^2 + 1)} d\omega \right\}, \quad x < \xi.$$

There is a similar expression if $x > \xi$. The required formula is

$$\delta(x - \xi) = \frac{1}{2}\operatorname{sech} x \operatorname{sech} \xi - \frac{1}{2\pi}\int_{-\infty}^{\infty}\frac{e^{i\omega(x-\xi)}(\tanh x - i\omega)(\tanh \xi + i\omega)}{\omega^2 + 1}d\omega.$$

This expression may be further simplified after significant algebra (compare with Exercise 4.29), giving

$$\delta(x - \xi) = \frac{1}{2\pi}\int_{-\infty}^{\infty}e^{i\omega(x-\xi)}d\omega.$$

Example 4.14. We present another example that leads to a classical transform. We consider the Bessel equation of order zero

$$(xu')' + \lambda xu = 0, \quad 0 < x < \infty,$$

where

$$\lim_{x\downarrow 0} u \text{ is finite,} \quad \int_0^\infty x\,|u|^2\,dx < \infty,$$

because the weight function is $r(x) = x$. This Green's function satisfies

$$(xG_x)_x + \lambda xG = -\delta(x - \xi),$$

the boundary and integrability conditions as a function of x and is given by

$$G(x, \xi; \lambda) = \begin{cases} -\frac{\pi}{2}J_0\left(\sqrt{\lambda}x\right)Y_0\left(\sqrt{\lambda}\xi\right), & x < \xi \\ -\frac{\pi}{2}J_0\left(\sqrt{\lambda}\xi\right)Y_0\left(\sqrt{\lambda}x\right), & x > \xi \end{cases},$$

when $\lambda \notin [0, \infty)$. Once again, we perform a contour integral over the branch cut along the nonnegative real axis, on the bottom of which $k = -\sqrt{\lambda}$, while on the top $k = \sqrt{\lambda}$. As the Bessel functions are continued in the complex plane, $J_0(z) = J_0(-z)$, while $Y_0(-z) = Y_0(z) + 2iJ_0(z)$. Making use of (4.41), we find that

$$\delta(x - \xi) = \frac{1}{2\pi i}\oint_C \xi G(x, \xi; \lambda)d\lambda$$

$$= -\frac{1}{2\pi i}\frac{\pi}{2}\xi\left[H(\xi - x)\left\{\int_\infty^0 J_0\left(kx\right)\left(Y_0\left(k\xi\right) + 2iJ_0(k\xi)\right)2k\,dk\right.\right.$$

$$\left. + \int_0^\infty J_0\left(kx\right)Y_0\left(k\xi\right)2k\,dk\right\}$$

$$+ H(x - \xi)\left\{\int_\infty^0 J_0\left(k\xi\right)\left(Y_0\left(kx\right) + 2iJ_0(kx)\right)2k\,dk\right.$$

$$\left.\left. + \int_0^\infty J_0\left(k\xi\right)Y_0\left(kx\right)2k\,dk\right\}\right].$$

Hence,

$$\delta(x - \xi) = \int_0^\infty \xi J_0(kx) J_0(k\xi) k\,dk.$$

This leads to the *Hankel transform* of the order zero. By multiplying by $f(\xi)$ and integrating, we obtain

$$f(x) = \int_0^\infty f(\xi)d\xi \int_0^\infty \xi J_0(kx) J_0(k\xi) k\,dk.$$

Then, we define

$$F_h(k) = \int_0^\infty f(\xi) J_0(k\xi) \xi\,d\xi,$$

so that

$$f(x) = \int_0^\infty F_h(k) J_0(kx) k\,dk.$$

REFERENCES

G. Batchelor, *An Introduction to Fluid Dynamics*, Cambridge University Press, Cambridge, UK, 1967.

L. Challis and F. Sheard, The Green of Green's Functions, *Physics Today*, Dec., 41–46 (2003).

E. A. Coddington and N. Levinson, *Theory of Ordinary Differential Equations*, McGraw-Hill, New York, 1955.

R. H. Cole, *Theory of Ordinary Differential Equations*, Appleton-Century-Crofts, New York, 1968.

R. Courant and D. Hilbert, *Methods of Mathematical Physics*, vol. I, Wiley-Interscience, New York, 1953.

B. Friedman, *Principles and Techniques of Applied Mathematics*, Dover, New York, 1990.

I. H. Herron, A Method of Constructing Green's Functions for Ordinary Differential Systems, *Bull. Inst. Math. Appl.* **25**, 233–237 (1989).

F. B. Hildebrand, *Methods of Applied Mathematics*, Dover, New York, 1992.

S. S. Holland, Jr., *Applied Analysis by the Hilbert Space Method*, Dekker, New York, 1990.

E. L. Ince, *Ordinary Differential Equations*, Dover, New York, 1956.

J. P. Keener, *Principles of Applied Mathematics: Transformation and Approximation*, 2nd Edition, Perseus Books, Cambridge, MA, 2000.

W. S. Loud, Some Examples of Generalized Green's Functions and Generalized Green's Matrices, *SIAM Rev.* **12**, 194–210 (1970).

B. Noble and J. W. Daniel, *Applied Linear Algebra*, 3rd edition, Prentice-Hall, Englewood Cliffs, NJ, 1988.

M. N. Özişik, *Boundary Value Problems in Heat Conduction*, Dover, New York, 1968.

W. T. Reid, *Ordinary Differential Equations*, Wiley, New York, 1971.

I. V. Schensted, *Contributions to the Theory of Hydrodynamic Stability*, Ph.D. dissertation, University of Michigan, 1960.

I. Stakgold, Green's Functions and Boundary-Value Problems, 2nd edition, Wiley-Interscience, New York, 1998.

J. E. Wilkins, Jr., Conduction of Heat in an Insulated Spherical Shell with Arbitrary Spherically Symmetric Sources, *SIAM Rev.* **1**, 149–153 (1959).

EXERCISES

4.1 Derive the Green's function for the boundary-value problem

$$-y'' = f(x), \quad 0 < x < 1,$$

$$y(0) = 0, \quad 2y(1) - y'(1) = 0.$$

Show that it is given by

$$G(x, \xi) = \begin{cases} x(1 - 2\xi), & x < \xi \\ \xi(1 - 2x), & x > \xi \end{cases}.$$

4.2 Consider the problem (4.1) with initial conditions (4.2), but with insulated boundary conditions

$$\frac{\partial T}{\partial r}(r_1, t) = \frac{\partial T}{\partial r}(r_2, t) = 0. \tag{4.43}$$

(a) Recognize that the equation may be written as

$$\frac{\partial T}{\partial t} = \frac{1}{r}\frac{\partial^2}{\partial r^2}(rT) + \frac{f(r)}{r}.$$

(b) Looking for a steady solution T_s, $\partial T_s/\partial t = 0$, show that this is possible with (4.43) only if

$$\int_{r_1}^{r_2} r f(r) dr = 0. \tag{4.44}$$

(c) However, even when (4.44) holds, show that no unique steady solution of the form

$$T_s(r) = \int_{r_1}^{r_2} K(r, \rho) f(\rho) d\rho$$

is possible. That is, no Green's function exists for (4.6) with boundary conditions (4.43).

4.3 Analyze the problem

$$-y'' - k^2 y = f(x), \quad 0 < x < 1,$$

$$y(0) = 0, \quad y'(1) - y(1) = 0.$$

Show that the Green's function is given by

$$G(x, \xi) = \frac{\sin(kx)\left[\sin(k(1 - \xi)) - k\cos(k(1 - \xi))\right]}{k(k\cos k - \sin k)}, \quad x < \xi$$

$$G(\xi, x) = G(x, \xi).$$

Are there values of k for which the Green's function does not exist?

4.4 Consider the boundary-value problem

$$-\frac{d^2 y}{dx^2} + k^2 y = f(x), \quad a < x < b$$

with the boundary conditions

$$\frac{dy}{dx}(a) - y(a) = 0, \tag{4.45}$$

and

$$\frac{dy}{dx}(b) - y(b) = 0. \tag{4.46}$$

Show that this boundary-value problem may be solved explicitly to give

$$y(x) = \int_a^b G(x, \xi; k) f(\xi) d\xi, \tag{4.47}$$

where

$$G(x, \xi; k) = \begin{cases} u(x; k)v(\xi; k)/W(k), & x < \xi \\ v(x; k)u(\xi; k)/W(k), & x > \xi \end{cases},$$

is the Green's function, with

$$u(x; k) = ka \cosh(k(x - a)) + \sinh(k(x - a)),$$
$$v(x; k) = kb \cosh(k(b - x)) - \sinh(k(b - x)),$$

and

$$W(k) = k(k^2 ab - 1) \sinh(k(b - a)) + k^2(b - a) \cosh(k(b - a)).$$

4.5 Show that the solution of the boundary-value problem,

$$x^2 y'' - 2xy' + 2y = -f(x),$$
$$y(1) = 0, \quad y(2) = 0,$$

is given by

$$y(x) = -\int_1^2 f(\xi)\xi^{-4} G(x, \xi) d\xi.$$

where

$$G(x, \xi) = \begin{cases} (x^2 - x)(\xi^2 - 2\xi), & x < \xi \\ (x^2 - 2x)(\xi^2 - \xi), & x > \xi \end{cases}.$$

Use the expression just obtained to find $y(x)$, when

$$f(x) = \begin{cases} -1, & 1 \le x < \frac{3}{2} \\ 1, & \frac{3}{2} < x \le 2 \end{cases}.$$

4.6 Consider the singular boundary-value problem

$$-\frac{d^2y}{dx^2} + \frac{1}{x}\frac{dy}{dx} = f(x), \quad 0 < x < 1,$$

$$\lim_{x\downarrow 0}\frac{y(x)}{x} \text{ finite}, \quad y(1) = 0.$$

Use a Green's function to show that the solution may be written as

$$y(x) = \int_{0^+}^1 \frac{f(\xi)}{\xi} G(x, \xi)\,d\xi,$$

where

$$G(x, \xi) = \begin{cases} \frac{1}{2}x^2(1 - \xi^2), & x < \xi \\ \frac{1}{2}\xi^2(1 - x^2), & x > \xi \end{cases}.$$

4.7 Find the Green's function for the singular boundary-value problem

$$-2xy'' - y' = f(x), \quad 0 < x < 1,$$

$$\lim_{x\to 0^+}(\sqrt{x}\,y') = 0, \quad y(1) = 0.$$

4.8 Construct the Green's function for the boundary-value problem

$$-u'' + k^2 u = f(x), \quad -a < x < a, \quad k > 0,$$

$$u(-a) = u(a) = 0.$$

Show that as $a \to \infty$, the Green's function becomes

$$G(x, \xi) = \frac{1}{2k}e^{-k|x-\xi|}.$$

4.9 Consider the boundary-value problem

$$-u'' + k^2(x)u = f(x), \quad -\infty < x < \infty,$$

$$u(x) \to \infty \text{ as } x \to \pm\infty,$$

where $k_1 > 0, k_2 > 0$, and

$$k(x) = \begin{cases} k_1, & x < 0 \\ k_2, & x > 0 \end{cases}.$$

Show that its Green's function is given by

$$G(x, \xi) = \begin{cases} u_1(x)u_2(\xi)/P(\xi), & x < \xi \\ u_2(x)u_1(\xi)/P(\xi), & x > \xi \end{cases},$$

where

$$u_1(x) = e^{k_1 x}H(-x) + e^{k_2 x}H(x),$$

$$u_2(x) = e^{-k_1 x}H(-x) + e^{-k_2 x}H(x),$$

and

$$P(\xi) = 2k_1 H(-\xi) + 2k_2 H(\xi).$$

4.10 Consider the infinite boundary-value problem

$$Ly = x^2 y'' + 2xy' + x^2 y = f(x), \quad 0 < x < \infty,$$

$$y(0) = 0, \quad y(\infty) = 0.$$

(a) We know that the solutions to $Ly = 0$ are

$$\frac{\sin x}{x} \quad \text{and} \quad \frac{\cos x}{x}.$$

Using this fact, obtain the Green's function and show that the general solution in this case is

$$y(x) = \frac{1}{x} \int_0^x \frac{\sin(x - \xi)}{\xi} f(\xi) d\xi. \tag{4.48}$$

(b) Show that (4.48) is valid *only* for $f(0) = 0$. Explain.

4.11 *The initial-value problem by variation of parameters.*
Consider Equation (4.13) with only the initial conditions

$$y(a) = y'(a) = 0.$$

By variation of parameters, its general solution is given by (4.15), where u_1 and u_2, are any two linearly independent solutions to the homogeneous equation. Now apply the initial conditions, to show that

$$y = -u_1(x) \int_a^x \frac{f(\xi)u_2(\xi)}{pW} d\xi + u_2(x) \int_a^x \frac{f(\xi)u_1(\xi)}{pW} d\xi,$$

that is, $c_1 = c_2 = 0$. Hence, one can conclude that the Green's function is

$$G(x, \xi) = \begin{cases} 0, & x < \xi \\ [u_2(x)u_1(\xi) - u_1(x)u_2(\xi)] / (p(\xi)W(\xi)), & x > \xi \end{cases}. \tag{4.49}$$

4.12 Find the Green's function for the boundary-value problem

$$-y'' = f(x), \quad 0 < x < 1,$$

$$2y(0) + y(1) = 0, \quad 2y'(0) + y'(1) = 0.$$

Show that it may be written as

$$G(x, \xi) = \begin{cases} \frac{2}{9} - \frac{1}{3}(x - \xi), & x < \xi \\ \frac{2}{9} + \frac{2}{3}(\xi - x), & x > \xi \end{cases}.$$

4.13 Given the differential operator defined by

$$Lu = -u'' - u \quad 0 < x < \pi,$$

$$u'(0) = u'(\pi),$$

$$u(0) - u(\pi) + u'(\pi) = 0,$$

show that its Green's function is

$$G(x, \xi) = -\frac{1}{4} \cos x \cos \xi - \frac{1}{2} \sin |x - \xi|.$$

4.14 If a beam of length b, of uniform cross-section is built into a wall, is unsupported at the right end, and carries a distributed load $w(x)$ along its length, the displacement $u(x)$ of its centerline satisfies the boundary-value problem:

$$u'''' = \frac{1}{EI} w(x), \quad 0 < x < b,$$
$$u(0) = u'(0) = u''(b) = u'''(b) = 0,$$

where E is Young's modulus, and I is the second moment of cross-section.

(a) Show that the Green's function is given by

$$G(x, \xi) = \begin{cases} \frac{1}{2} x^2 \xi - \frac{1}{6} x^3, & x < \xi \\ \frac{1}{2} \xi^2 x - \frac{1}{6} \xi^3, & x > \xi \end{cases},$$

and write the solution $u(x)$ as an integral.

(b) Evaluate the integral explicitly when $w(x)$ is given by

$$w(x) = \begin{cases} 0, & 0 < x < \frac{b}{2}, \\ w_0, & \frac{b}{2} < x < b, \end{cases}$$

w_0 constant.

4.15 Construct the Green's function for the third-order operator

$$Lu = u''', \quad 0 < x < 1,$$
$$u(0) = u'(0) = u''(1) + \sigma u(1) = 0, \quad \sigma \neq -2.$$

4.16 Consider the vibrating beam model (4.27) with simply supported boundary conditions:

$$\phi(0) = \phi''(0) = 0,$$
$$\phi(1) = \phi''(1) = 0.$$

Show that its Green's function is given by

$$G(x, \xi; \beta) = \frac{\sinh \beta u_2(x) v_2(\xi) - \sin \beta u_1(x) v_1(\xi)}{2\beta^3 \sin \beta \sinh \beta}, \quad x < \xi$$
$$G(\xi, x; \beta) = G(x, \xi; \beta), \quad x > \xi,$$

where

$$u_1(x) = \sinh(\beta x),$$
$$u_2(x) = \sin(\beta x)$$

at the left endpoint and

$$v_1(x) = \sinh(\beta - \beta x),$$
$$v_2(x) = \sin(\beta - \beta x)$$

at the right endpoint.

4.17 Find the Green's function for the vibrating beam model (4.27) with clamped boundary conditions:

$$\phi(0) = \phi'(0) = 0,$$
$$\phi(1) = \phi'(1) = 0.$$

This Green's function also has applications in viscous flow problems.

4.18 *The initial-value problem by distributions.*
Consider Equation (4.13) with only initial conditions

$$y(a) = y'(a) = 0.$$

Suppose that u_1 and u_2 are any two linearly independent solutions to the homogeneous equation. To determine the Green's function assume that

$$G(x, \xi) = \begin{cases} 0, & x < \xi \\ c_1 u_1(x) + c_2 u_2(x), & x > \xi \end{cases},$$

so that the initial conditions are satisfied. Require that (a) $G(x, \xi)$ is continuous at $x = \xi$. (b) Its derivative $\partial G / \partial x$ has an upward jump at $x = \xi$, that is,

$$\left[\frac{\partial G}{\partial x} \right]_{x=\xi^-}^{x=\xi^+} = \frac{1}{p(\xi)}.$$

Hence, one can conclude that the Green's function is the same as given in Exercise 4.11

4.19 Consider the boundary-value problem

$$Lu = u''', \quad 0 < x < 1,$$
$$u(0) = u'(0) = u''(1) = 0,$$

which is Example 4.5 in Section 4.2.2. Its Green's function $G(x, \xi)$ is determined there. In Example 4.6 of Section 4.2.3, the adjoint boundary problem is determined to be

$$L^* v = -v''', \ 0 < x < 1, \ v(0) = v'(1) = v''(1) = 0.$$

Compute the Green's function $H(x, \xi)$ for this problem and, hence, verify that $H(x, \xi) = G(\xi, x)$ in accordance with Theorem 4.1 of Section 4.3.

4.20 Given the matrix differential equation,

$$\frac{d\mathbf{u}}{dt} = \begin{pmatrix} -\cos^2 t & \sin^2 t \\ \cos^2 t & -\sin^2 t \end{pmatrix} \mathbf{u} + \mathbf{f}(t),$$

with initial conditions $\mathbf{u}(0) = \mathbf{0}$, determine the Green's matrix function explicitly and write the solution in terms of it. *Method*: Show that

$$\mathbf{U}(t) = \begin{pmatrix} 5 - \cos 2t - 2\sin 2t + 6e^{-t} & 5 - \cos 2t - 2\sin 2t - 4e^{-t} \\ 5 + \cos 2t + 2\sin 2t - 6e^{-t} & 5 + \cos 2t + 2\sin 2t + 4e^{-t} \end{pmatrix}$$

is a suitable *fundamental matrix*. That is, show that $\mathbf{U}'(t) = \mathbf{P}(t)\mathbf{U}(t)$ and $\det \mathbf{U} \neq 0$. The use of *Maple* or some other CAS is acceptable in this problem.

4.21 Derive the representation (4.32). *Method*: Though it may seem a little backward, begin with the representation,

$$\mathbf{G}(t, \tau) = \begin{cases} \mathbf{U}(t)\mathbf{C}_1, & t < \tau \\ \mathbf{U}(t)\mathbf{C}_2, & t > \tau \end{cases},$$

where $\mathbf{U}$ is a fundamental matrix and $\mathbf{C}_1$ and $\mathbf{C}_2$ are matrices independent of t. Require that $\mathbf{G}$ satisfy (4.31) and (4.33).

4.22 Consider the operator L_x defined by

$$Lu = -u'', \quad -1 < x < 1,$$
$$u(-1) = u(1)$$
$$u'(-1) = u'(1).$$

Show that $\varphi(x) = 1$ is an eigenfunction corresponding to $\lambda = 0$ and derive the generalized Green's function

$$G^\dagger(x, \xi) = -\frac{1}{2}|x - \xi| + \frac{1}{4}(x - \xi)^2 + \frac{1}{6}.$$

4.23 Show that $\lambda = 0$ is an eigenvalue for the homogeneous form of the boundary-value problem

$$-y'' = f(x), \quad 0 < x < 1,$$
$$y(0) = 0, \quad y(1) - y'(1) = 0.$$

Find the generalized Green's function for the problem.

4.24 Compute the generalized Green's function for the operator L_x defined by

$$Lu = -u'', \quad 0 < x < 1,$$
$$u(0) - u(1) + u'(1) = 0,$$
$$u'(0) - u'(1) = 0.$$

Notice that there are two eigenfunctions that correspond to the eigenvalue $\lambda = 0$.

4.25 Determine the generalized Green's function for the singular problem

$$-\left(\frac{1}{x}u'\right)' = f(x), \quad 0 < x < 1,$$

$$\lim_{x \to 0^+} \frac{u}{x} \text{ finite}, \quad 2u(1) - u'(1) = 0.$$

Hint: Due to the nature of the Sturm–Liouville eigenfunctions, take $r(x) = 1/x$.

4.26 Consider the operator L_x defined by

$$Lu = u''', \quad 0 < x < 1,$$
$$u(0) = u'(0) = u''(1) - 2u(1) = 0. \tag{4.50}$$

(a) Show that $\lambda = 0$ is an eigenvalue of the problem $Lu = \lambda u$ with (4.50), while finding the eigenfunction $\varphi(x)$.

(b) Derive the adjoint problem for L_x^* defined by

$$L^*v = -v''', \quad 0 < x < 1,$$
$$v(0) = v'(1) = v''(1) + 2v(1) = 0,$$

and show that the problem $L^*v = \mu v$ has an eigenfunction $\psi(x)$ corresponding to $\mu = 0$.

(c) Define the generalized Green's function $G^\dagger$ by using

$$L_x G^\dagger = \delta(x - \xi) + c\psi.$$

Choose

$$c = -\psi(\xi) \bigg/ \int_0^1 \psi^2 d\xi.$$

Solve for $G^\dagger$ and satisfy the three boundary conditions of L_x at $x = 0, 1$. Employ the conditions on the continuity of $G^\dagger$ and $G_x^\dagger$, and the jump on $G_{xx}^\dagger$ at $x = \xi$.

(d) Make the additional requirement that

$$\int_0^1 \varphi(x) G^\dagger(x, \xi) dx = 0,$$

to find that the desired generalized Green's function is

$$G^\dagger(x, \xi) = \frac{1}{2}(x - \xi)^2 H(x - \xi) + x^2 \left(\frac{1}{12}\xi^5 - \frac{5}{6}\xi^2 + \frac{5}{4}\xi - \frac{1}{2} \right)$$
$$- \frac{15}{8}(\xi^2 - 2\xi) \left(\frac{1}{60}x^5 - \frac{1}{12}x^4 + \frac{11}{224}x^2 \right),$$

where

$$H(x - \xi) = \begin{cases} 0, & x < \xi \\ 1, & x > \xi \end{cases}.$$

4.27 A generalized Green's function is needed for the operator L_x defined by

$$Lu = -u'' + k^2 u, \quad k > 0, \quad 0 < x < \infty,$$

where

$$u(0) = u'(0) = 0, \quad \int_0^\infty u^2 dx < \infty.$$

(a) Show that the adjoint operator L_x^* is given by

$$L^*v = -v'' + k^2 v, \quad 0 < x < \infty, \quad \int_0^\infty v^2 dx < \infty.$$

That is, no boundary conditions are needed on v.

(b) Argue that $\lambda = 0$ is a simple eigenvalue for $L^* v = \lambda v$. Call its eigenfunction $\psi(x)$.

(c) Define the generalized Green's function $G^\dagger$ by using

$$L_x G^\dagger = \delta(x - \xi) + c\psi.$$

Choose

$$c = -\psi(\xi) \Big/ \int_0^\infty \psi^2 d\xi.$$

to solve for $G^\dagger$ and satisfy the two boundary conditions of L_x at $x = 0$. Require that $\int_0^\infty |G^\dagger|^2 dx < \infty$. Employ the conditions on the continuity of $G^\dagger$ and the jump on $G_x^\dagger$ at $x = \xi$.

(d) Then find that the desired generalized Green's function is

$$G^\dagger(x, \xi) = \frac{1}{2k} \left(e^{-k|x-\xi|} - e^{-k(x+\xi)} \right) - xe^{-k(x+\xi)}.$$

Notice that no additional requirement such as condition (i)′ of Section 4.5 is needed because $\lambda = 0$ is not an eigenvalue of L_x.

4.28 Use the eigenvalue problem

$$u'' + \lambda u = 0, \quad 0 < x < \infty$$

$$u'(0) = 0, \quad \int_0^\infty u^2 dx < \infty,$$

to derive the expansion

$$\delta(x - \xi) = \frac{2}{\pi} \int_0^\infty \cos(kx) \cos(k\xi) dk,$$

which gives the *Fourier cosine transform*. Obtain a suitable transform pair.

4.29 Use the eigenvalue problem

$$u'' + \lambda u = 0, \quad -\infty < x < \infty$$

$$\int_{-\infty}^\infty u^2 dx < \infty,$$

to derive the expansion

$$\delta(x - \xi) = \frac{1}{2\pi} \int_{-\infty}^\infty e^{ik(x-\xi)} dk,$$

which leads to the *Fourier integral theorem*. Obtain a suitable transform pair. *Method*: Show first that the appropriate Green's function is

$$G(x, \xi) = \frac{i}{2\sqrt{\lambda}} e^{i\sqrt{\lambda}|x-\xi|},$$

where $\lambda \notin [0, \infty)$.

4.30 Use the eigenvalue problem

$$u'' + \lambda u = 0, \quad 0 < x < \infty$$

$$u'(0) - \alpha u(0) = 0, \quad \int_0^\infty u^2 dx < \infty,$$

where α is a real parameter, to derive an expansion for $\delta(x - \xi)$.

Show that if $\alpha < 0$, the spectrum is both continuous and discrete with a single eigenvalue and a corresponding eigenfunction.

4.7 Appendix: Linear Ordinary Differential Equations

4.7.1 Fundamental Solutions and the Wronskian Matrix

Consider the equation

$$p_0(x)\frac{d^n y}{dx^n} + p_1(x)\frac{d^{n-1}y}{dx^{n-1}} + \cdots + p_{n-1}(x)\frac{dy}{dx} + p_n(x)y = r(x), \qquad (4.51)$$

which we will abbreviate as

$$Ly \equiv (p_0 D^n + p_1 D^{n-1} + \cdots + p_{n-1}D + p_n)y = r.$$

Suppose that $p_0, p_1, \ldots, p_n, r$ are continuous on $a \leq x \leq b$, and $p_0(x) \neq 0$, then $u_2(x_0)$ has at least one solution $y \in C^n[a, b]$. The equation

$$Lu = 0 \qquad (4.52)$$

is the homogeneous equation corresponding to (4.51). We will assume that most of the superposition properties of (4.52) are known. There are several definitions that we introduce here (Cole).

The *Wronskian vector* of a C^{n-1} function u is defined as

$$\mathbf{w}(u) = \begin{pmatrix} u \\ u' \\ \vdots \\ u^{(n-1)} \end{pmatrix}.$$

Thus if $u_1, u_2, \ldots, u_n$ are linearly independent solutions of (4.52), they comprise a *fundamental set* and we may form a *fundamental vector* $\mathbf{\Phi}$ whose components are these solutions. The *Wronskian matrix* for this fundamental set or for the fundamental vector is

$$\mathbf{W}(\mathbf{\Phi}) = \mathbf{W}(u_1, u_2, \ldots, u_n) \equiv \begin{pmatrix} u_1 & u_2 & \cdots & u_n \\ u_1' & u_2' & \cdots & u_n' \\ \vdots & \vdots & \vdots & \vdots \\ u_1^{(n-1)} & u_2^{(n-1)} & \cdots & u_n^{(n-1)} \end{pmatrix};$$

whose determinant is $W = \det |\mathbf{W}|$. Observe that

$$\frac{dW}{dx} = \begin{vmatrix} u_1 & u_2 & \cdots & u_n \\ u_1' & u_2' & \cdots & u_n' \\ \vdots & \vdots & \vdots & \vdots \\ u_1^{(n-2)} & u_2^{(n-2)} & \cdots & u_n^{(n-2)} \\ u_1^{(n)} & u_2^{(n)} & \cdots & u_n^{(n)} \end{vmatrix},$$

for all the other determinants that arise in the differentiation have two alike rows and will therefore vanish. Then because

$$p_0 u_i^{(n)} = -p_1 u_i^{(n-1)} - \cdots - p_{n-1} u_i' - p u_i, \quad i = 1, \ldots, n;$$
$$\frac{dW}{dx} = -\frac{p_1}{p_0} W.$$

Hence,

$$W = W_0 \exp\left(-\int_{x_0}^{x} \frac{p_1(t)}{p_0(t)} dt\right).$$

This is the *Abel identity*. It follows that if $p_0(x)$ does not vanish in the interval $[a, b]$, then if W vanishes at x_0, it will be identically zero.

The Initial-Value Problem

There is in general an infinite number of fundamental sets, but there is a particularly simple system. Take one set, say $\{v_k(x), \ k = 1, \ldots, n\}$ and manipulate this set so that

$$v_k^{(j-1)}(x_0) = \delta_{jk} = \begin{cases} 1, & k = j \\ 0, & k \neq j \end{cases}.$$

Define $u_i(x) = v_i(x)$, $i = 1, \ldots, n$. Then $u_1, u_2, \ldots, u_n$ forms a fundamental set with the property that $W(u_1, u_2, \ldots, u_n)(x_0) = 1$. This shows that for a fundamental set, W will never vanish. Furthermore, the unique solution of (4.52) that satisfies the initial conditions

$$u(x_0) = y_0, u'(x_0) = y_0', \ldots, u^{(n-1)}(x_0) = y_0^{(n-1)}$$

is

$$u(x) = y_0 u_1(x) + y_0' u_2(x) + \cdots + y_0^{(n-1)} u_n(x).$$

4.7.2 Variation of Parameters

Now we proceed to our main objective for this chapter – to solve (4.51). We will do so when a fundamental set of solutions of (4.52) is known. We generalize the standard method for second-order equations by assuming a solution of the form

$$y = \chi_1 u_1 + \chi_2 u_2 + \cdots + \chi_n u_n,$$

where the "constants" $\chi_k(x)$, $k = 1, \ldots, n$ are allowed to vary. In particular, take

$$\left. \begin{array}{l} \chi_1' u_1 + \chi_2' u_2 + \cdots + \chi_n' u_n = 0, \\[1mm] \chi_1' u_1' + \chi_2' u_2' + \cdots + \chi_n' u_n' = 0, \\[1mm] \qquad\qquad\vdots \\[1mm] \chi_1' u_1^{(n-2)} + \chi_2' u_2^{(n-2)} + \cdots + \chi_n' u_n^{(n-2)} = 0, \\[1mm] \chi_1' u_1^{(n-1)} + \chi_2' u_2^{(n-1)} + \cdots + \chi_n' u_n^{(n-1)} = r(x)/p_0(x). \end{array} \right\} \tag{4.53}$$

Written in matrix notation, this is

$$\mathbf{W} \begin{pmatrix} \chi_1' \\ \vdots \\ \chi_n' \end{pmatrix} \equiv \mathbf{W}\chi' = \begin{pmatrix} 0 \\ \vdots \\ \frac{r(x)}{p_0(x)} \end{pmatrix},$$

and solving for the vector of parameters

$$\chi' = \mathbf{W}^{-1} \begin{pmatrix} 0 \\ \vdots \\ r(x)/p_0(x) \end{pmatrix}.$$

Once the antiderivative is found, a particular solution to (4.51) can be expressed as $y_p = \chi^T \mathbf{w}(u)$.

4.7.3 The Adjoint Equation; Lagrange's Identity

The results outlined here are standard and are discussed in more detail in (Coddington and Levinson), (Cole), (Herron), and (Ince). This begins with the search for an integrating factor for the differential expression,

$$Lu = p_0 u^{(n)} + \cdots + p_n u,$$

defined on $[a, b]$. Suppose that $v(x)$ is a function with n continuous derivatives. Then,

$$\int_a^x v\,Lu\,dx = \int_a^x \sum_{j=0}^{n-1} p_j u^{(n-j)} v\,dx + \int_a^x p_n u v\,dx$$

$$= \sum_{j=0}^{n-1} \int_a^x p_j v\,d\left(u^{(n-j-1)} \right) + \int_a^x p_n u v\,dx,$$

in order to prepare for integration by parts.

This is carried out successively to obtain

$$\int_a^x v\, Lu\, dx = \sum_{j=0}^{n-1} \left[p_j v u^{(n-j-1)} \right]_x^a - \sum_{j=0}^{n-1} \int_a^x \frac{d}{dx}(p_j v)\, dx + \int_a^x p_n uv\, dx$$

$$= \cdots$$

$$= \left[\sum_{l=0}^{n-1} \sum_{j=0}^{n-1-l} (-1)^l (p_j v)^l u^{(n-j-l-1)} \right]_a^x + \sum_{l=0}^n \int_a^x (-1)^l u \frac{d^l}{dx^l}(p_{n-l} v)\, dx.$$

The differential operator on v under the integral sign,

$$L^* v = (-1)^n \frac{d^n}{dx^n}(p_0 v) + (-1)^{n-1} \frac{d^{n-1}}{dx^{n-1}}(p_1 v) + \cdots - \frac{d}{dx}(p_{n-1} v) + p_n v \quad (4.54)$$

is called the "formal adjoint of L." The expression in the square brackets is called the "bilinear concomitant of u and v" and is written, $P(u, v)$. Sometimes the notation $[uv]$ is used to denote $P(u, v)$. An important result is that by differentiating, with respect to x,

$$v\, Lu - u\, L^* v = \frac{d}{dx} P(u, v). \quad (4.55)$$

The relation (4.55) is known as *Lagrange's identity*. If v satisfies the adjoint equation and u satisfies the equation, then

$$P(u.v) = C. \quad (4.56)$$

Alternatively, the Lagrange identity may be written as

$$v\, Lu - u\, L^* v = \frac{d}{dx} \left(\mathbf{w}^T(v) \mathbf{R} \mathbf{w}(u) \right),$$

where the elements of the matrix $\mathbf{R}$ are

$$R_{ij} = \begin{cases} \sum_{h=i}^{n-j+1} (-1)^{h-1} \binom{h-1}{i-1} p_{n-h-j+1}^{(h-i)}, & i \leq n-j+1 \\ 0, & i > n-j+1 \end{cases}.$$

So, in this notation, the concomitant is

$$P(u, v) = \mathbf{w}^T(v) \mathbf{R} \mathbf{w}(u). \quad (4.57)$$

If $L = L^*$, the differential operator is said to be *formally self-adjoint*. Observe that

$$Lu = \frac{d^n}{dx^n} \left(p_0 \frac{d^n}{dx^n} u \right) + \frac{d^{n-1}}{dx^{n-1}} \left(p_1 \frac{d^{n-1}}{dx^{n-1}} u \right) + \cdots + \frac{d}{dx} \left(p_{n-1} \frac{d}{dx} u \right) + p_n u$$

is the most general $(2n)$th order formally self-adjoint expression.

The structure of (4.57) is illuminating.

Theorem 4.2. *Cole's Representation (Cole). If $\mathbf{\Phi}$ and $\mathbf{\Psi}$ are fundamental vectors for $Lu = L^*v = 0$, respectively, then*

$$\mathbf{W}^*(\mathbf{\Psi})\mathbf{R}\mathbf{W}(\mathbf{\Phi}) = \mathcal{P},$$

where $\mathcal{P}$ is a nonsingular constant matrix.

The following theorem and its corollary are proved in (Herron).

Theorem 4.3. *If $L = L^*$, then $\mathbf{R} = -\mathbf{R}^T$.*

Corollary 4.1. *If $\mathbf{\Phi} = \mathbf{\Psi}$, then $\mathcal{P} = -\mathcal{P}^T$.*

Separated Boundary Conditions

We look more closely at the case where separated boundary conditions are appended to L and L^*. There is an important conclusion that follows, which we state in the form of a theorem. First, we state a preliminary result.

Lemma 4.4. *((Ince), (Coddington and Levinson)). If the Lagrange identity (4.55) is integrated, Green's identity is obtained. It may be written as*

$$P(u, v)(b) - P(u, v)(a) = A_1(u)A_n^*(v) + A_2(u)A_{n-1}^*(v) + \cdots$$
$$+ A_n(u)A_1^*(v) + B_1(u)B_n^*(v) + \cdots + B_n(u)B_1^*(v). \tag{4.58}$$

Here, the set $\{A_s\}$ consists of n, linearly independent, homogeneous expressions in $u(a), u'(a), \ldots, u^{(n-1)}(a)$, and the members of the set $\{B_s\}$ are such expressions in $u(b), u'(b), \ldots, u^{(n-1)}(b)$. For the fixed sets $\{A_s\}$ and $\{B_s\}$, there exists a unique set of linear combinations of $v(a), v'(a), \ldots, v^{(n-1)}(a)$ called $\{A_s^*\}$ and another set of functions $v(b), v'(b), \ldots, v^{(n-1)}(b)$ called $\{B_s^*\}$. The sets $\{A_s^*\}$ and $\{B_s^*\}$ are called the "adjoint boundary expressions."

We now ask, under what conditions does $P(u, v)$ vanish? First, suppose that

$$A_s(u(a)) = 0, \quad s = 1, \ldots, n - k, \tag{4.59}$$

with the result of (4.56), we see that (4.58) becomes

$$P(u, v)(a) = A_{n-k+1}A_k^* + \cdots + A_n A_1^*$$

and

$$P(u, v)(b) = B_1 B_n^* + \cdots + B_n B_1^*.$$

Thus, if also

$$A_s^*(v(a)) = 0, \quad s = 1, \ldots, k, \tag{4.60}$$

then $P(u, v) = 0$. The same result is obtained if, instead of (4.59), it is assumed that

$$B_s(u(b)) = 0, \quad s = 1, \dots, k. \tag{4.61}$$

Suppose on the other hand that $P(u, v)$ vanishes. We know that the expressions $\{A_s\}$ are independent as are the the expressions $\{B_s\}$. Therefore, if u satisfies the boundary conditions (4.59), then (4.60) must hold. Likewise, if the conditions (4.61) are met then

$$B_s^*(v(b)) = 0, \quad s = 1, \dots, n - k.$$

We have therefore proved the following theorem.

Theorem 4.4. *Let u and v be solutions of the equations $Lu = 0$ and $L^*v = 0$, respectively. Suppose u satisfies $n - k$ and v satisfies k separated boundary conditions. Then $P(u, v)$ vanishes if and only if the respective conditions are all satisfied at the same point.*

Corollary 4.2. *When the boundary value problem defined by (4.52) with boundary conditions*

$$A_i(u(a)) = B_j(u(b)) = 0, \ i = 1, \dots, n - k; \ j = 1, \dots k,$$

is self-adjoint as in Section 4.2.3, $P(u, u) = 0$.

Proof. If the problem is self-adjoint, n is even and the number of boundary conditions at each point is equal. Specifically, they are of the form:

$$A_s(u(a)) = A_s^*(u(b)) = 0,$$
$$B_s((u(b)) = B_s^*(U(b)) = 0, \quad s = 1, \dots, n/2.$$

The use of the preceding theorem for $u = v$ gives the result. $\qquad\square$

We may now observe that *Cole's representation*,

$$\mathbf{W}^T(\mathbf{\Phi}(\xi))\mathbf{R}\mathbf{W}(\mathbf{\Phi}(\xi)) \equiv \mathcal{P},$$

given by Theorem 4.2 and stated in (4.24), of which (4.56) defines each entry by virtue of (4.57), leads to the special structure of (4.25).

5

Laplace Transform Methods

5.1 Preamble

In this chapter, we use the Laplace transform to solve a series of partial differential equations. Because of the character of the transform, it is well-suited to the analysis of initial-value problems, in particular. The approach here is to build on the foundation of complex analysis to carefully examine the inversion integral (the Bromwich integral) in the (complex) s-plane. In many areas of engineering, linear systems are designed and analyzed by working directly in the s-plane, without explicitly doing the inversion, and this is done, in general, with transforms that are rational functions. However, in problems that arise as a result of the solution of partial differential equations, the transforms are almost always quite complicated, and, in general, include both classical transcendental functions and special functions. The behavior of those functions and their oft-occurring branch point singularities are key to obtaining the solution in the original, physical variables.

5.2 The Laplace Transform and Its Inverse

If $f(t)$ is a function defined in $t \geq 0$, then its Laplace transform is given by

$$F(s) = \mathcal{L}(f(t)) \equiv \int_0^\infty f(t)e^{-st}dt. \tag{5.1}$$

There are conditions usually stated for the existence of (5.1); the main one is that $f(t)$ be of *exponential order*, that is, there are constants M and c so that $|f(t)| < M\exp(ct)$, when t, sufficiently large. The function f may be complex-valued. However, not every common function has a Laplace transform. For example, $f(t) = e^{t^2}$ has no Laplace transform. In our applications, most of the Laplace transforms will arise when one is searching for solutions to differential equations. Usually, it will be clear if a solution has unusual growth properties, which might be handled by a change of variable, for instance, before the Laplace transform is applied.

We will allow the transform variable s to be complex. Then it becomes useful to know that $F(s)$ is analytic on the region in which (5.1) converges. More precisely, we have the following theorem.

Theorem 5.1. *(Widder). If (5.1) converges for $\Re\text{eal}(s) > c$, then $F(s)$ is analytic in the half-plane $\Re\text{eal}(s) > c$.*

The inversion, from the Laplace domain to the t-domain is given by the complex integral,

$$f(t) = \mathcal{L}^{-1}(F(s)) = \frac{1}{2\pi i} \int_{\Gamma} F(s) e^{st} ds, \tag{5.2}$$

where Γ is a path lying wholly in $\Re\text{eal}\{s\} > c > \gamma$, in the complex s-plane, with end-points at $\gamma - i\infty$ and $\gamma + i\infty$; Γ is chosen so that (5.1) converges absolutely on the line $\Re\text{eal}(s) = \gamma$, and $F(s)$ is analytic to the right of Γ (Widder). This discussion is completed in Section 5.2.2. We will first outline some of the important properties of the Laplace transform. Following that discussion is a series of examples of solutions of partial differential equations by means of Laplace transforms.

Property (i): Transform of a derivative.

An important property for the use of Laplace transforms is to show how to do the transform of a t-derivative.

If $F(s)$ is the Laplace transform of a function $f(t)$, then

$$\mathcal{L}\{f'\} = sF(s) - f|_{t=0+}. \tag{5.3}$$

Begin with

$$\mathcal{L}\{f'\} = \int_0^\infty f'(t) e^{-st} dt.$$

Using integration by parts, this may be rewritten as

$$\mathcal{L}\{f'\} = f(t) e^{-st} \big|_{t=0+}^\infty + s \int_0^\infty f(t) e^{-st} dt. \tag{5.4}$$

Since $f(t)$ is of exponential order,

$$\lim \left| f(t) e^{-st} \right| < \lim \left| Me^{-(s-c)t} \right| \to 0 \quad \text{as } t \to \infty.$$

The formula (5.3) then follows. We seek to use this method to find the Laplace transform of higher derivatives. As long as the derivatives are of exponential order, the formula may be extended. Again, for the types of problems we will consider, generally this will be the case. The resulting formulas are

$$\mathcal{L}\{f''\} = s^2 F(s) - s f(0+) - f'(0+)$$

$$\vdots$$

$$\mathcal{L}\{f^{(n)}\} = s^n F(s) - s^{n-1} f(0+) - \cdots - f^{(n-1)}(0+), \quad n = 1, 2, \ldots.$$

Limiting Behavior: Initial-Value and Final-Value Theorems

We investigate the limiting behavior of Laplace transforms. Notice that if $f(t)$ is piecewise continuous on $0 \leq t \leq T$ and is of exponential order

$$|f(t)| < Me^{ct} \quad \text{for } t > T, \tag{5.5}$$

for fixed c, M, then $F(s) \to 0$ as $s \to \infty$, as long as $\Re\text{eal}(s) > c$. Furthermore, $sF(s)$ is bounded as $s \to \infty$. The demonstration of this follows.

By (5.5)

$$\left|e^{-st} f(t)\right| < Me^{-(s_1-c)t}, \quad \text{for } t > T,$$

with $s = s_1 + is_2$,

$$|F(s)| = \left|\int_0^\infty e^{-st} f(t)dt\right| \leq \int_0^\infty e^{-s_1 t} |f(t)| \, dt$$

$$\leq M \int_0^\infty e^{-(s_1-c)t} dt = \frac{M}{s_1 - c}.$$

Thus, since $s_1 > c$ as $s \to \infty$,

$$F(s) \to 0 \text{ and } |sF(s)| < \frac{M|s|}{|s_1 - c|}, \quad \text{which is finite.}$$

Other important applications of (5.4) are to what are sometimes called the "initial-value theorem" and "final-value theorem." We observe that

Theorem 5.2. Initial-Value Theorem. *If $f(t)$ is continuous, $f'(t)$ piecewise continuous on every finite interval $0 \leq t \leq T$, f, f' of exponential order, then*

$$\lim_{s \to \infty} sF(s) = f(0^+).$$

Proof. We know from (5.4) that

$$\mathcal{L}\{f'\} = sF(s) - f(0^+).$$

Since f' is of exponential order by our previous result, $\mathcal{L}\{f'\} \to 0$ as $s \to \infty$. Hence,

$$\lim_{s \to \infty} sF(s) = f(0^+). \qquad \square$$

Theorem 5.3. Final-Value Theorem. *Suppose f is the same as in the last theorem. Then*

$$\lim_{s \to 0} sF(s) = \lim_{t \to \infty} f(t).$$

Proof. Again by (5.4),

$$\mathcal{L}\{f'\} = sF(s) - f(0^+).$$

Allow $s \to 0$ on both sides and formally take the limit:

$$\lim_{s \to 0} \int_0^\infty f'(t)e^{-st}\,dt = \int_0^\infty f'(t)\,dt = [f(t)]_0^\infty. \tag{5.6}$$

It is clear that for (5.6) to hold $\int_0^\infty f'(t)\,dt$ must exist, so $\lim_{t \to \infty} f(t)$ must exist. Then

$$\lim_{t \to \infty} f(t) - f(0^+) = \lim_{s \to 0} sF(s) - f(0^+),$$

and the desired result follows. $\qquad\qquad\qquad\qquad\qquad\qquad\square$

We see that a condition for this to hold is that $\mathcal{L}\{f'\}$ exists for $\mathfrak{Real}(s) \geq 0$, so that $sF(s)$ has singularities only in the left half-plane, $\mathfrak{Real}(s) < 0$.

5.2.1 The Convolution

One common way of representing the result of inverting a Laplace transform is by the convolution. It arises because of the following observation.

Theorem 5.4. *If $F(s) = \mathcal{L}\{f\}$ and $G(s) = \mathcal{L}\{g\}$, both exist for $\mathfrak{Real}(s) > c$, then*

$$H(s) = F(s)G(s) = \mathcal{L}\{h\}, \quad \mathfrak{Real}(s) > c,$$

$$h(t) = \int_0^t f(t - \tau)g(\tau)\,d\tau = \int_0^t f(\tau)g(t - \tau)\,d\tau. \tag{5.7}$$

The function h is called the "convolution of f and g" and is sometimes written as

$$h(t) = (f * g)(t).$$

The theory of convolutions has a life of its own, apart from transforms, but it is clearly useful to us here because

$$\mathcal{L}^{-1}(F(s)G(s)) = h(t), \tag{5.8}$$

and when the two inverses, of $F(s)$ and $G(s)$, are known, we have a ready representation of the inverse of $H(s)$.

There is generalization of the convolution, which is as follows.

Theorem 5.5. *(Efros) [Antimirov, et al.]. If $G(s)$ and $q(s)$ are analytic functions in $\mathfrak{Real}(s) > c$, where*

$$F(s) = \mathcal{L}\{f\}, \quad \text{and} \quad G(s)e^{-\tau q(s)} \equiv \mathcal{L}\{g(t, \tau)\},$$

then

$$\int_0^\infty f(\tau)g(t, \tau)\,d\tau = \mathcal{L}^{-1}(G(s)F[q(s)]). \tag{5.9}$$

The special case where $q(s) = s$ reduces to the standard convolution.

Proof. Take the Laplace transform of the left side of (5.9) to write

$$\mathcal{L}\left\{\int_0^\infty f(\tau)g(t,\tau)d\tau\right\} = \int_0^\infty e^{-st}dt \int_0^\infty f(\tau)g(t,\tau)d\tau$$

$$= \int_0^\infty f(\tau)d\tau \int_0^\infty e^{-st}g(t,\tau)dt,$$

when that interchange of the order of integration is allowable. The last integral is $G(s)e^{-\tau q(s)}$, by definition. Hence,

$$\mathcal{L}^{-1}\left\{\int_0^\infty f(\tau)G(s)e^{-\tau q(s)}d\tau\right\} = \mathcal{L}^{-1}\left(G(s)F[q(s)]\right)$$

$$= \int_0^\infty f(\tau)g(t,\tau)d\tau,$$

as desired.

Notice that if $q(s) = s$, then

$$\mathcal{L}^{-1}\left\{\int_0^\infty f(\tau)G(s)e^{-\tau s}d\tau\right\} = \mathcal{L}^{-1}\left\{G(s)F(s)\right\}. \tag{5.10}$$

If we suppose that $g(t) = 0$, for $t < 0$, it follows that

$$\mathcal{L}\left\{g(t-\tau)\right\} = \int_0^\infty e^{-st}g(t-\tau)dt = e^{-s\tau}G(s), \quad \tau \geq 0.$$

Hence,

$$\mathcal{L}^{-1}\left\{\int_0^\infty f(\tau)G(s)e^{-\tau s}d\tau\right\} = \int_0^\infty f(\tau)g(t-\tau)H(t-\tau)d\tau$$

$$= \int_0^t f(\tau)g(t-\tau)d\tau. \tag{5.11}$$

From (5.10) and (5.11), the convolution result follows.   □

5.2.2 Completion of the Γ Contour for Laplace Inversion

We have seen that the inverse of a Laplace transform is given by the integral,

$$f(t) = \frac{1}{2\pi i}\int_\Gamma \mathcal{L}\{f\}e^{st}ds, \tag{5.12}$$

where the path Γ is to the right of any singularities of $\mathcal{L}\{f\}$. The following theorems are useful when doing these inversions.

Theorem 5.6. *Consider the Laplace inversion integral, (5.12). If $|\mathcal{L}\{f\}| \leq A|s|^{-k}$, $k > 1$, for $|s| \to \infty$, then*

$$\int_{C_R} \mathcal{L}\{f\}e^{st}ds \to 0, \quad \text{for } R \to \infty, \text{ if either}$$

$t < 0$ and C_R *is the semicircle in the right half of the s* $-$ plane,

or

$t > 0$ and C_R *is the semicircle in the left half s* $-$ plane.

Proof. The proof proceeds as follows: First, for $t < 0$,

$$\left| \int_{C_R} \mathcal{L}\{f\} e^{st} ds \right| \leq \int_{-\pi/2}^{\pi/2} A R^{1-k} |e^{st}| d\theta.$$

Since $|e^{st}| \leq 1$ for $\mathfrak{Real}\{s\} > 0$,

$$\left| \int_{C_R} \mathcal{L}\{f\} e^{st} ds \right| \leq \pi R^{1-k},$$

which vanishes for $k > 1$, when $R \to \infty$.

If $t > 0$, $|e^{st}| < 1$ for $\mathfrak{Real}\{s\} < 0$, and so

$$\left| \int_{C_R} \mathcal{L}\{f\} e^{st} ds \right| \leq \int_{\pi/2}^{3\pi/2} A R^{1-k} |e^{st}| d\theta \leq \pi R^{1-k},$$

which also vanishes for $R \to \infty$. $\qquad\square$

Warning: It should be noted that, even though the semicircle integrals vanish for the conditions listed above, one cannot, in the $t > 0$ case, simply replace the Γ integral by a closed path integral because branch cuts may cross C_R at some locations.

5.2.3 Transform Problems

Problem 5.1.

(a) If $f(t)$ is a function of period T, show that

$$\mathcal{L}\{f\} = \frac{\int_0^T f(t) e^{-st} dt}{1 - e^{-sT}}, \quad \mathfrak{Real}(s) > 0.$$

(b) Use the result of part (a) to verify that the Laplace transform of the square wave

$$f(t) = \begin{cases} 1, & 0 < t < \pi \\ -1, & \pi < t < 2\pi, \end{cases}$$

where $f(t + 2\pi) = f(t)$ is given by

$$F(s) = \frac{1 - e^{-\pi s\}}}{s(1 + e^{-\pi s})}, \quad \mathfrak{Real}(s) > 0.$$

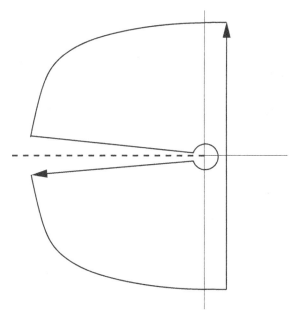

Figure 5.1. Laplace inversion path for Problem 5.5. The dashed line is the branch cut.

Problem 5.2. *Derive the result:*
If

$$f(t) = \sin(2\sqrt{\alpha t}),$$

then

$$F(s) = \frac{\sqrt{\pi \alpha}}{s^{3/2}} e^{-\alpha/s}, \quad \Re\mathrm{eal}(s) > 0.$$

Problem 5.3. *Derive the result:*
If

$$f(t) = \frac{\sin(\alpha t)}{t},$$

then

$$F(s) = \tan^{-1}\left(\frac{\alpha}{s}\right), \quad \Re\mathrm{eal}(s) > 0.$$

Problem 5.4. *Use complex integration to find the inverse Laplace transform of*

$$\frac{1}{s^3 + a^3}, \quad a > 0.$$

Problem 5.5. *Obtain the inverse transform*

$$\mathcal{L}^{-1}\left\{\sqrt{\frac{\pi}{s}} e^{-\alpha\sqrt{s}}\right\} = \frac{1}{\sqrt{t}} e^{-\alpha^2/4t}, \quad \alpha > 0. \tag{5.13}$$

Hint: See Figure 5.1 and adapt Jordan's lemma.

Problem 5.6. *Formally differentiate both sides of (5.13) with respect to α to conclude that*

$$\mathcal{L}^{-1}\left\{\sqrt{\pi}e^{-\alpha\sqrt{s}}\right\} = \frac{\alpha}{2t^{3/2}}e^{-\alpha^2/4t}, \quad \alpha > 0.$$

Problem 5.7. *Verify the following general properties of the Laplace transform, when $F(s) = \mathcal{L}[f]$, and a, b are real constants.*

(a)
$$\mathcal{L}[f(at)e^{-bt}] = \frac{1}{a}F\left(\frac{s+b}{a}\right), \quad (a > 0). \tag{5.14}$$

(b)
$$\mathcal{L}[f(t/a)] = aF(as), \quad (a > 0). \tag{5.15}$$

(c)
$$\mathcal{L}\left[\frac{d^n f(t)}{dt^n}\right] = s^n F(s) - s^{n-1} f(0+)$$
$$- \cdots - f^{(n-1)}(0+), \quad n = 1, 2, \ldots. \tag{5.16}$$

(d)
$$\mathcal{L}[t^n f(t)] = (-1)^n \frac{d^n}{ds^n} F(s), n = 1, 2, \ldots. \tag{5.17}$$

Problem 5.8. *Use the convolution result (5.8) with $g(t) = 1$, to show that*

$$\mathcal{L}^{-1}\left(\frac{F(s)}{s}\right) = \int_0^t f(\tau)d\tau.$$

Problem 5.9. *Use the results of Problems (5.6) and (5.8) to show that*

$$\mathcal{L}^{-1}\left\{\frac{1}{s}e^{-\alpha\sqrt{s}}\right\} = \text{erfc}(\alpha/2\sqrt{t}), \quad \alpha > 0,$$

where erfc *is the complementary error function. The error function is discussed briefly in Section 2.4.*

Problem 5.10.

(a) Show that

$$\mathcal{L}\{J_0(t)\} = \frac{1}{\sqrt{1+s^2}}, \quad \Re\text{eal}(s) > 0.$$

Hints: *There are several ways of doing this problem. One direct way is to expand the Bessel function in its Taylor series, given by Equation (2.19), and to integrate term-by-term. Alternatively, use Equation (2.39), interchange the orders of Integration, and perform a contour integral as in Section 1.4. Still another way would be to solve a suitable initial-value problem for Bessel's equation of order zero.*

(b) Use the convolution result (5.7) to verify that

$$\sin t = \int_0^t J_0(t - \tau)J_0(\tau)\,d\tau.$$

Problem 5.11. *Use the result of Problem 5.5 and Efros's theorem to show that*

$$\mathcal{L}\left\{\int_0^\infty \frac{1}{\sqrt{\pi t}} e^{-\tau^2/4t} f(\tau) d\tau\right\} = \frac{F(\sqrt{s})}{\sqrt{s}},$$

where $F = \mathfrak{L}\{f\}$.

Problem 5.12. *Let $F(s) = \mathcal{L}\{f(t)\}$; define*

$$\tau = \frac{\int_0^\infty t f(t) dt}{\int_0^\infty f(t) dt},$$

and

$$\sigma^2 = \frac{\int_0^\infty (t - \tau)^2 f(t) dt}{\int_0^\infty f(t) dt}.$$

Show that

$$\tau = -\frac{F'(0)}{F(0)},$$

and

$$\sigma^2 = \frac{F''(0)}{F(0)} - \tau^2.$$

5.3 Worked Examples

In this section, we examine a number of initial-value problems for partial differential equations but before that, we construct the solution of an ordinary differential equation.

5.3.1 An Ordinary Differential Equation

Example 5.1. A certain oscillator obeys the following differential equation:

$$\frac{d^2 y}{dt^2} + 3\frac{dy}{dt} + 2y = \cos t.$$

Suppose the initial conditions are homogeneous: $y(0) = \dot{y}(0) = 0$. Then, using (5.3), this equation becomes

$$(s^2 + 3s + 2)Y(s) = \mathcal{L}\{\cos t\} = \frac{s}{s^2 + 1},$$

where Y is the Laplace transform of $y(t)$. Any Laplace transform table (or use of *Mathematica* or *Maple*) indicates that

$$\mathcal{L}\{\cos(at)\} = \frac{s}{s^2 + a^2}.$$

$$\mathcal{L}\{\sin(at)\} = \frac{a}{s^2 + a^2}.$$

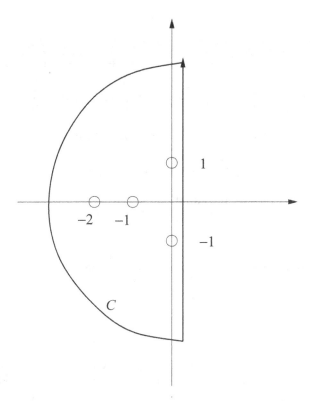

Figure 5.2. Laplace inversion path for Example 5.1. Singularities are located with 'o'.

Doing some algebra on the Y equation gives the transform of the solution as

$$Y(s) = \frac{s}{(s+1)(s+2)(s^2+1)}.$$

The inverse representation of this by partial fraction techniques is an option. Then, tables of Laplace transforms may be readily employed (Boyce and DiPrima). However, as a precursor to our treatment of partial differential equations, we make use of (5.2).

As a function of the complex variable s, $Y(s)\exp(st)$ has simple poles at $s = -1$, $s = -2$, and $s = \pm i$. Since there are no branch points in the integrand function, the integral in (5.2) is equivalent to the integral around the path C shown in Figure 5.2. Thus,

$$y(t) = \sum_{4 \text{ poles}} \text{Residues}.$$

Therefore,

$$\text{Res} = \frac{N}{D'} = \frac{se^{st}}{(s+2)(s^2+1)+(s+1)(s^2+1)+2s(s+1)(s+2)},$$

and so

$$y = \frac{-e^{-t}}{2} + \frac{-2e^{-2t}}{-5} + \frac{ie^{it}}{2i(i+1)(i+2)} + \frac{-ie^{-it}}{-2i(1-i)(2-i)}.$$

$$y = -\frac{1}{2}e^{-t} + \frac{2}{5}e^{-2t} + \frac{1}{10}[\cos t + 3\sin t].$$

Now, we proceed with a sequence of partial differential equation problems.

5.3.2 Translating Plate in a Fluid

Example 5.2. Consider the equation (see (Rosenhead))

$$\frac{\partial u}{\partial t} = v\frac{\partial^2 u}{\partial y^2}, \quad \text{for} \quad y > 0, t > 0,$$

which is subject to the boundary and initial conditions on $u(y, t)$,

$$u(y, 0) = f(y), \quad y > 0$$
$$u(0, t) = g(t), \quad t > 0.$$

The solution describes the motion of an unbounded fluid due to the motion of an infinite plate parallel to itself. Then, $f(y)$ is the initial fluid velocity distribution before the plate starts to move, $g(t)$ is the time history of the plate motion, and v is the kinematic viscosity of the fluid. Performing the Laplace transform in the usual way, we obtain the differential equation for $U(y, s)$

$$v\frac{\partial^2 U}{\partial y^2} - sU = -f(y). \tag{5.18}$$

The solution may be accomplished for general f by a Green's function technique, but here, to emphasize the essence of the solution technique, we consider the simple case when $f(x) = u_o$, a constant. In that case, the general solution to (6.2) is

$$U = \frac{u_o}{s}\left[1 - e^{-y\sqrt{s/v}}\right] + G(s)e^{-y\sqrt{s/v}}, \quad G(s) = \mathcal{L}(g).$$

The inverse of the first term may be carried out directly from (5.2), but that procedure will be deferred to examples 5.2 and 5.3. Using the result given in Problem 5.9, and recalling that the relationship among the error functions is that $\text{erfc}(z) = 1 - \text{erf}(z)$, the solution is

$$u = u_o\text{erf}\left(\frac{y}{2\sqrt{vt}}\right) + \mathcal{L}^{-1}\left(G(s)e^{-y\sqrt{s/v}}\right). \tag{5.19}$$

The final term may be written as a convolution integral; however, that is not very instructive, so we will explore a specific case. Before that, notice that the first term is the transient in the motion due to the initial velocity distribution. For large y, $u = u_o$, but nearer to the wall, in units of $\sqrt{vt}$, the velocity is smaller, and falls to zero at the surface. The second term is clearly the response of the fluid to the wall temporal forcing.

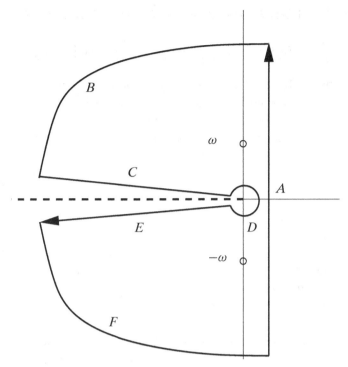

Figure 5.3. Laplace inversion path for Example 5.2.

Exploring the latter, consider the case for which the wall is oscillating. Then, $g(t) = u_1 \cos(\omega t)$, say. Thus, the second term in (5.19) is now written, by (5.2), as

$$u_{\text{FORCED}} = \frac{u_1}{2\pi i} \int_\Gamma \frac{s e^{st}}{s^2 + \omega^2} e^{-y\sqrt{s/\nu}} ds. \qquad (5.20)$$

Because of the square root, the s-plane must be cut, and in this case, since we require that U be analytic in a right half-plane, we place the cut on the negative real axis, starting from the origin. Let $H(s)$ denote the integrand of (5.20), and consider $\oint H ds$ around the closed path shown in Figure 5.3.

Clearly the integral is the sum of six integrals, along the segments labeled A–F. As the circular radius goes to infinity, path A is the path of the inversion, (5.20), and the quarter-circle integrals B and F can be shown, for $t > 0$, to vanish as the radius goes to infinity. Hence, we have

$$\oint H ds = \int_A H ds + \int_C H ds + \int_D H ds + \int_E H ds. \qquad (5.21)$$

By writing $s = \epsilon \exp(i\theta)$ and substituting into $\int_D H ds$, one can easily verify that this integral goes to zero as ϵ goes to zero, so that integral makes no contribution to (5.20). For the integral along C, one must write $s = r \exp(i\pi)$, and for the integral along E, $s = r \exp(-i\pi)$ instead. Putting those two integrals together, and doing the limits $R \to \infty, \epsilon \to 0$, we find that (5.20) reduces to

$$\frac{1}{2\pi i} \oint H ds = u_{\text{FORCED}} + \frac{u_1}{\pi} \int_0^\infty \frac{re^{-rt}}{r^2 + \omega^2} \sin\left(\sqrt{\frac{r}{\nu}} y\right) dr. \qquad (5.22)$$

Clearly, by the Residue theorem, the term on the left of (5.22) is equal to the sum of the residues inside the path. There are two, in this case: simple poles at $\pm i\omega$. Adding those two residues, and replacing the left side of (5.22) with that sum, we find that the forced response is given by

$$u_{\text{FORCED}} = u_1 \cos\left(\omega t - \sqrt{\frac{\omega}{2\nu}} y\right) e^{-y\sqrt{\omega/2\nu}} - \frac{u_1}{\pi} \int_0^\infty \frac{re^{-rt}}{r^2 + \omega^2} \sin\left(\sqrt{\frac{r}{\nu}} y\right) dr.$$

$$(5.23)$$

The first term is the long-time periodic solution for the forced flow, usually referred to as *Stokes's solution* for an oscillating wall (Rosenhead, p. 381). The second term is the transient due to the forcing. That second term integral cannot be written down exactly for arbitrary values of ω. Later, we will investigate how to properly obtain asymptotic evaluations of such an integral at both small and large frequencies. We return to this example at that time. However, as $t \to \infty$, the second term goes to zero, as can be easily seen: Let the integral, the multiplier of u_1/π in the second term in (5.23), be denoted by P. Then

$$|P| \le \int_0^\infty \frac{re^{-rt} dr}{r^2 + \omega^2} \le \frac{1}{2\omega} \int_0^\infty e^{-rt} dr = \frac{1}{2\omega t},$$

which vanishes at $t \to \infty$.

5.3.3 Heat Conduction in a Strip

Example 5.3. Having considered an initial-value problem for a parabolic equation in an unbounded domain, we now turn to an unsteady heat flow problem in a strip, hence

$$\frac{\partial T}{\partial t} = \frac{\partial^2 T}{\partial x^2}, \quad \text{for} \quad 1 > x > 0, t > 0,$$

subject to the boundary and initial conditions on $T(x, t)$,

$$T(x, 0) = f(x), \quad 0 < x < 1,$$

$$T(0, t) = 0, \quad t > 0,$$

$$\text{and} \quad \frac{\partial T}{\partial x} + \alpha(T - g(t)) = 0, \quad \text{at} \quad x = 1.$$

The thermal conductivity and the length have been removed by rescaling in this case, and the quantity α is a Biot number. This problem might typically be approached by a standard Fourier-series technique, but the Laplace transform methodology is robust enough to uncover anomalies in the spectrum of the operator that would not surface in a more ad hoc procedure. For the moment, we consider the case for which the initial temperature, $f \equiv 0$. In such a case, doing the usual Laplace transform operation on the equation leads to the following formula for the transform of the temperature,

$$\mathcal{L}(T) = \frac{\alpha G(s) \sinh(s^{1/2}x)}{s^{1/2} \cosh(s^{1/2}) + \alpha \sinh(s^{1/2})}, \tag{5.24}$$

Apart from singularities in G, it is easy to verify that the function is analytic at the origin – unlike the previous problem for which the origin was a branch point. The only singularities are located at zeroes of the denominator. There is an obvious pole located at the solution of

$$\tanh(\beta) = -\frac{\beta}{\alpha}, \quad s = \beta^2,$$

which has a real solution only for $\alpha < -1$, and leads to a pole on the positive real axis. Whatever the sign of α, however, if $s^{1/2}$ is purely imaginary, then the hyperbolic functions become circular functions, and there are an infinity of zeroes, located at $\{-j_n^2\}$, the solutions of

$$\tan(j_n) = -\frac{j_n}{\alpha}. \tag{5.25}$$

For the simple case of constant $g(t) = T_o$, the residue at these latter points is given by

$$\text{Res}|_{-j_n^2} = \frac{2 T_o \alpha^2 \sin(j_n x)}{j_n[(1+\alpha)\alpha + j_n^2] \cos(j_n)} e^{-j_n^2 t} \equiv a_n(x, t), \tag{5.26}$$

and the residue at the pole located at $s = \beta^2$ is

$$\text{Res}|_{\beta^2} = \frac{2\alpha^2 T_o \sinh(\beta x)}{\beta[\alpha(1+\alpha) - \beta^2] \cosh(\beta)} e^{\beta^2 t} \equiv A(x, t), \quad \text{for} \quad \alpha < -1.$$

Furthermore, for the choice made for g, there is a pole at the origin. Hence, we have the final result that

$$T = \sum_{n=1}^{\infty} a_n(x, t) + \frac{\alpha T_o x}{1 + \alpha} + q A(x, t), \tag{5.27}$$

where $q = 0$, for $\alpha > -1$ and $q = 1$, for $\alpha < -1$. In the case that $\alpha > -1$, there is a steady-state solution, given by the second term, which is dominant for $t \to \infty$. However, in the case that $\alpha < -1$, there is no steady solution, but the temperature grows without bound, given by $A(x, t)$. A naive, separation of variables approach to this problem might not have uncovered the structure that is evident here. Physically, what is happening is quite simple: For $\alpha < -1$, there is massive heat inflow at the right boundary, which causes the temperature of the material to grow without bound.

5.3.4 Telegraph Equation

Example 5.4. Having explored solutions to parabolic problems in both bounded and unbounded domains, we turn now to a hyperbolic problem,

$$\frac{\partial^2 u}{\partial t^2} - \frac{\partial^2 u}{\partial x^2} + \lambda^2 u = 0, \quad \text{in} \quad x > 0, \, t > 0$$

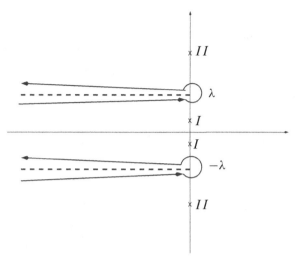

Figure 5.4. Integration path for Example 5.4. Labels I and II indicate the locations of the poles for values of ω smaller than, and larger than λ.

$$\text{with} \quad \frac{\partial u}{\partial t} = 0, \ u = f(x), \quad \text{at} \quad t = 0,$$
$$\text{and} \quad u = g(t), \quad \text{at} \quad x = 0.$$

It should be noted that the prototypical hyperbolic problem, with $\lambda \equiv 0$, presents some difficulties to integral-transform analysis; we return to this point later (see Section 5.3.5). Laplace transforming the equation in the usual way leads to the ordinary differential equation for $U(x, s)$:

$$\frac{\partial^2 U}{\partial x^2} - (\lambda^2 + s^2)U = -sf(x). \tag{5.28}$$

Again, for specificity, we take the special case $f(x) = u_0$, a constant. The solution to (5.22) is then

$$U = \frac{u_o s}{s^2 + \lambda^2} \left[1 - e^{-x(s^2+\lambda^2)^{1/2}} \right] + G(s)e^{-x(s^2+\lambda^2)^{1/2}}. \tag{5.29}$$

We now suppose that u oscillates at the boundary, so that $g(t) = u_1 \cos(\omega t)$. Then, (5.29) takes the special form:

$$U = \frac{u_o s}{s^2 + \lambda^2} \left[1 - e^{-x(s^2+\lambda^2)^{1/2}} \right] + \frac{u_1 s}{s^2 + \omega^2} e^{-x(s^2+\lambda^2)^{1/2}}. \tag{5.30}$$

Clearly, there are branch points at $\pm i\lambda$, so branch cuts are constructed on the lines $\Im\mathrm{mag}(s) = \pm i\lambda$, $\Re\mathrm{eal}(s) \leq 0$. We ignore, for the moment, the inversion of the first term in (5.30) – what turns out to be the "transient," and turn to the second term-the response due to the motion at $x = 0$. Effectively, the solution we will obtain is exact for $u_o \equiv 0$. The inversion is performed by deforming the contour Γ of (5.2) so that it wraps around the two branch cuts as shown in Figure (5.4).

In the deformation of the contour, two poles are crossed, so that the Laplace inversion is given by the sum of the two integrals along the cuts and the residues at $\pm i\omega$. What arises from the cut integrals is a contribution that is spatially oscillatory, but decays in time; it is from the poles that traveling waves emanate. Thus, as $t \to \infty$, all that is in evidence is the traveling-wave portion of the solution. That part of the solution, say u_{forced_1}, is given by

$$u_{\text{forced}_1} = \text{Res}|_{i\omega} + \text{Res}|_{-i\omega}. \tag{5.31}$$

There are two cases to consider: the answer is somewhat different depending on the relative sizes of λ and ω. The poles may be located at location I or at location II, depending on whether λ is larger or smaller than ω, respectively. Crucial to the difference is the fact that

$$b \equiv (s^2 + \lambda^2)^{1/2} = (r_1 r_2)^{1/2} \exp(i(\theta_1 + \theta_2)/2),$$

if we write $r_1 \exp(i\theta_1)$ for $s + i\lambda$, and $r_2 \exp(i\theta_2)$, for $s - i\lambda$. For case *I*, for which $\lambda > \omega$, $\theta_1 + \theta_2 = 0$ for both poles, so that b is purely real, so even though the residues give temporal oscillation, there are no propagating waves but rather exponential spatial decay. Thus,

$$u_{\text{forced}_1} = u_1 \cos(\omega t) e^{-\sqrt{\lambda^2 - \omega^2}\, x}, \qquad \lambda > \omega. \tag{5.32}$$

For case *II*, on the other hand, both θ_1 and θ_2 are $\pm \pi/2$ at the poles, so that $b = \pm i(\omega^2 - \lambda^2)^{1/2}$, and hence, on summing the residues, we have the wavelike result.

$$u_{\text{forced}_1} = u_1 \cos(\omega t - \sqrt{\omega^2 - \lambda^2}\, x), \qquad \omega > \lambda.$$

This result clearly represents a wave moving to the right at speed $(1 - \lambda^2/\omega^2)^{-1/2}$. Notice that, on taking $\lambda \to 0$, we obtain from this solution the solution to the problem posed here for $\lambda \equiv 0$.

Now, we cavalierly deformed the contour from the Laplace inversion path, Γ, to the path shown in the sketch. Doing that assumes that the integral along the semicircle of infinite radius in the left half-plane vanishes. That question deserves more investigation. Careful examination of the integrand of the second term in (5.30) indicates that, for s large and in the left half-plane, the integrand takes the approximate form, for large s

$$\int_{\text{semicircle}} \frac{s}{s^2 + \omega^2} e^{st - x(s^2 + \lambda^2)^{1/2}}\, ds \sim \int_{\text{semicircle}} \frac{e^{st - sx}\, ds}{s} \equiv I_C.$$

Now, on the circle, $s = R \exp(i\theta)$, and so this integral takes the form

$$|I_C| \leq \int_{3\pi/2}^{\pi/2} e^{R(t-x)\cos\theta}\, d\theta.$$

Clearly, this integral goes to zero as R goes to infinity only if $(t - x)\cos\theta < 0$. On the integration path, the cosine is always negative, hence this integral goes to zero *only* if $t > x$. Thus, the assumptions leading to (5.31) require that $t > x$, which is the requirement, then, for (5.32) to be correct. If the time is such that $t < x$, then the same argument means that the contour Γ must be made into a closed contour in the *right* half-plane, which encloses no singularities and, hence, gives $u \equiv 0$.

What remains is to work out the integrals along the branch cuts. Since our primary interest is in the case $\lambda \to 0$, suffice it to say that careful working out of those integrals shows that their sum vanishes for $\lambda \to 0$.

Thus, with all of these things considered, we have the no-dissipation solution by putting $\lambda = 0$, and it is

$$u = 0, \quad t < x,$$
$$u = u_1 \cos(\omega t - \omega x), \quad t > x.$$

5.3.5 A Scattering Problem

Example 5.5. We now turn to a scattering problem for a wave equation, but spatially two-dimensional. The equation for the velocity potential for flow due to acoustic waves in a tube is

$$\frac{\partial^2 \phi}{\partial t^2} - \nabla^2 \phi = 0, \quad \text{in} \quad -\infty < x < \infty, \ 0 < y < 1. \tag{5.33}$$

We take the initial conditions to be homogeneous: $\phi = \partial\phi/\partial t = 0$ at $t = 0$. We suppose that a wave is propagating down the tube, and at $t = 0$ the wall of the tube is altered, so that the walls, for $t > 0$, are given by $y = 1$ and $y = f(x)$, where $|f| \ll 1$. The initial propagating wave, moving to the right, is given by $\phi = \exp(ik(x - t))$, so we write the solution to (5.33) as

$$\phi = e^{ik(x-t)} + \Phi(x, y, t), \tag{5.34}$$

Since the boundary condition at both walls is nonpermeability, we have

$$\frac{\partial \Phi}{\partial y} = 0, \quad \text{at} \quad y = 1 \tag{5.35}$$

and

$$\quad \frac{\partial \Phi}{\partial y} = ikf'(x)e^{ik(x-t)}, \quad \text{at} \quad y = 0. \tag{5.36}$$

Because of the bounded domain and the Neumann boundary conditions, we can write the solution in a form in which one term satisfies the boundary conditions, and the other contains all of the generality. Hence, we write

$$\Phi = \Phi_{bc} + \Phi', \tag{5.37}$$

and we take Φ_{bc} to satisfy the boundary data, and therefore

$$\Phi_{bc} = ikf'(x)e^{ik(x-t)}\left[y - \frac{1}{2}y^2\right]. \tag{5.38}$$

Because Φ obeys the partial differential equation satisfied by ϕ, substitution of this decomposition into that equation leads to

$$\frac{\partial^2 \Phi'}{\partial t^2} - \nabla^2 \Phi' = \left[\left(y - \frac{1}{2}y^2\right)\left(\frac{\partial^2}{\partial x^2} + k^2\right) - 1\right]ikf'(x)e^{ik(x-t)}. \tag{5.39}$$

Because $\partial\Phi'/\partial y$ vanishes at both $y = 0$ and $y = 1$, a Fourier cosine series is suggested for a representation for Φ'. Thus, we write

$$\Phi' = \sum_{n=0}^{\infty} A_n(x,t)\cos(n\pi y). \tag{5.40}$$

The right side of Equation (5.39) must *also* be expanded in such a series, and in so doing, equating Fourier coefficients on both sides of the equation, we obtain the following equations for A_n:

$$A_{0tt} - A_{0xx} = \left(\frac{1}{3}\frac{\partial^2}{\partial x^2} + \frac{k^2}{3} - 1\right) ik f'(x) e^{ik(x-t)},$$

$$A_{ntt} - A_{nxx} + n^2\pi^2 A_n = -\frac{2ik}{n^2\pi^2}\left(\frac{d^2}{dx^2} + k^2\right) f'(x) e^{ik(x-t)}, \; n \geq 1. \tag{5.41}$$

Now, for convenience in what follows, because of the differential operators on the right sides of these equations, it is easier to solve for new variables $\{B_n\}$, and so we get

$$\frac{\partial^2 B_0}{\partial t^2} - \frac{\partial^2 B_0}{\partial x^2} = f'(x) e^{ik(x-t)},$$

$$\frac{\partial^2 B_n}{\partial t^2} - \frac{\partial^2 B_n}{\partial x^2} + n^2\pi^2 B_n = f'(x) e^{ik(x-t)}, \quad n \geq 1$$

$$A_0 = ik\left(\frac{1}{3}\frac{\partial^2}{\partial x^2} + \frac{k^2}{3} - 1\right) B_0, \quad A_n = -\frac{2ik}{n^2\pi^2}\left(\frac{\partial^2}{\partial x^2} + k^2\right) B_n. \tag{5.42}$$

We now specialize the solution to that due to a small step on the wall; f' is then a delta function, $f'(x) = h\delta(x)$, where h is the size of the step, and $h \ll 1$. The solution for this case is essentially the Greens function for the general solution. We develop this briefly as follows. The Laplace transformed equation for B_0 is

$$s^2\mathcal{L}\{B_0\} - (\mathcal{L}\{B_0\})_{xx} = \frac{f'(x)e^{ikx}}{s+ik}. \tag{5.43}$$

The homogeneous solutions are $\exp(\pm sx)$, and so requiring the solution to be bounded for both large and positive and large and negative x, and also requiring continuity at $x = 0$, the solution, since the right side is zero except immediately at $x = 0$, is

$$\mathcal{L}\{B_0\} = Ce^{-|s||x|}. \tag{5.44}$$

How do we determine the constant C for the case given: $f'(x) = h\delta(x)$? We integrate equation (5.43) from just to the left to just to the right of the origin. Then, we get

$$\int_{0-}^{0+} s^2\mathcal{L}\{B_0\}dx - \int_{0-}^{0+} (\mathcal{L}\{B_0\})_{xx}dx = \frac{h}{s+ik}\int_{0-}^{0+} \delta(x)e^{ikx}dx = \frac{h}{s+ik}, \tag{5.45}$$

and the left side integral gives a jump. So we obtain the jump condition at $x = 0$,

$$(\mathcal{L}\{B_0\})_x\Big|_{x=0+} - (\mathcal{L}\{B_0\})_x\Big|_{x=0-} = \frac{h}{s+ik}. \tag{5.46}$$

Substituting solution (5.44) into this equation, we obtain the value of the constant C which is $h/(2|s|(s+ik))$, and so

$$\mathcal{L}(B_0) = \frac{h}{2|s|(s+ik)}\, e^{-|s||x|}.$$

In a similar fashion, the B_n equation can be solved, leading to

$$\mathcal{L}(B_n) = \frac{h}{2(s^2+n^2\pi^2)^{1/2}(s+ik)}e^{-(s^2+n^2\pi^2)^{1/2}|x|}. \tag{5.47}$$

Actually, the inversion of B_0 is tougher, because of the difficulty alluded to at the end of the previous example – the absolute value function, which is not analytic. Here, we write

$$\mathcal{L}(\hat{B}_0(x,t;\epsilon)) = \frac{h}{2(s^2+\epsilon^2)^{1/2}(s+ik)}e^{-(s^2+\epsilon^2)^{1/2}|x|},$$

and then, once the inversion is completed, $B_0 = \hat{B}_0(x,t;0)$. (What this amounts to is letting $n \to 0$ in the eventual result for B_n.) Since the form of $\mathcal{L}(\hat{B}_0)$ and $\mathcal{L}(B_n)$ are similar, they are now written in the generic form,

$$\mathcal{L}(B) = \frac{h}{2(s^2+p^2)^{1/2}(s+ik)}e^{-(s^2+p^2)^{1/2}|x|}. \tag{5.48}$$

In a fashion that is virtually identical to the previous example, branch cuts are located at $\Im\mathrm{mag}(s) = \pm p$, $\Re\mathrm{eal}(s) \leq 0$. Integration paths for $B(x,t)$ are essentially just as in Example 5.4. The integrals along the branch cuts lead to portions in the solution that decay in time. What remains, for $t \to \infty$, is the residue of the pole at $s = -ik$. That residue takes quite a different form depending on the relative sizes of k and p. In a fashion similar to Example 5.4, if $k < p$, the pole lies between the branch points, but if $k > p$, then the pole lies beyond $-p$, and so we have the residues for the two cases:

$$\mathrm{Res}|_{-ik} = \frac{h}{2(p^2-k^2)^{1/2}}e^{-(p^2-k^2)^{1/2}|x|}e^{-ikt}, \, k < p, \tag{5.49}$$

$$\mathrm{Res}|_{-ik} = -\frac{h}{2i(k^2-p^2)^{1/2}}e^{(k^2-p^2)^{1/2}i|x|-ikt}, \, k > p. \tag{5.50}$$

Using these results, we can write down the long-time solution of the problem originally posed. The solution for B_0, on taking the limit of $\hat{B}_0$, as indicated above, corresponds to (5.50), and so

$$B_0 \sim \frac{h}{2ik}e^{ik|x|-ikt}, \quad \Longrightarrow \quad A_0 \sim -\frac{h}{2}e^{ik|x|-ikt}, \quad \text{in} \quad |x| < t$$

which corresponds to the y-independent portion of the reflected wave (in $x < 0$) and the transmitted wave (in $x > 0$). (*Note*: As in the previous problem, examination of the integral on the large semicircle shows that it is appropriate to close the path in the *left* half-plane *only* for $t > |x|$; otherwise, the path is closed to the right, and no singularities are enclosed. Thus, $A_0 = 0$, for $|x| > t$.)

The other terms correspond the reflected and transmitted portions of higher harmonics induced by the wave interaction with the wall. The details will be quite different depending upon the value of n, so let N denote the largest value of n that is smaller than k. Then, we have, using (5.42) again,

$$A_n^w \sim -\frac{kh}{(k^2-n^2\pi^2)^{1/2}}e^{(k^2-n^2\pi^2)^{1/2}i|x|-ikt}, \quad n < N,$$

$$A_n^s \sim -\frac{ikh}{(n^2\pi^2-k^2)^{1/2}}e^{-(n^2\pi^2-k^2)^{1/2}|x|}e^{-ikt}, \quad n > N, \quad \text{for} \quad |x| < t.$$

Then, we have the complete long-time solution by substituting these results into (5.37) and (5.40). The notation ( )^w refers to the reflected/transmitted waves, and ( )^s refers to the scattered waves.

Let's take a specific example. Suppose $\pi < k < 2\pi$. In this case, A_1 is wave-like, but all other modes are damped – and hence constitute the scattered wave field. First off, outside the propagating front, $\Phi \equiv 0$. Thus,

$$\phi = e^{ik(x-t)} \quad \text{in} \quad |x| > t,$$

which represents the incident wave propagating without alteration in these regions. Information about the sudden alteration of wall shape has not yet reached this portion of the fluid in the pipe.

Inside the front, putting together the above results,

$$\phi = e^{ik(x-t)} - \frac{h}{2}e^{ik(|x|-t)} - \frac{kh}{\sqrt{k^2-\pi^2}}e^{i(\sqrt{k^2-\pi^2}|x|-kt)}\cos(\pi y)$$

$$\quad\text{A}\qquad\qquad\text{B}\qquad\qquad\qquad\text{C}$$

$$- ikh\sum_{n=2}^{\infty}\frac{e^{-\sqrt{n^2\pi^2-k^2}|x|}}{\sqrt{n^2\pi^2-k^2}}e^{-ikt}\cos(n\pi y), \quad \text{in} \quad |x| < t.$$

$$\qquad\qquad\text{D}$$

Note that term A is the incident wave. Term B is a wave moving upstream in $x < 0$, with amplitude $h/2$; the same term is a downstream moving wave in $x > 0$, making a total transmitted wave with amplitude $1 - h/2$. Term C is either reflected (in $x < 0$) or transmitted (in $x > 0$), and it has a y-dependent amplitude. Its speed is *faster* than the incident wave, but it still lies wholly within $|x| < t$. Finally, D represents the scattered wavefield. Note that there is no propagation from these terms, only spatial decay, and that there is upstream and downstream symmetry.

5.3.6 Conduction of Heat in a Spherical Shell

Example 5.6. The ability to be able to predict the early stages of a nuclear reactor transient is of great importance. Here we consider unsteady heat conduction in a spherical shell, totally insulated with a spherically symmetric source (Wilkins). It might represent the heating of the pressure shell in a water moderated and cooled reactor.

Owing to the symmetry of the problem, it is described by the partial differential equation for the temperature $T(r, t)$ as

$$\frac{\partial T}{\partial t} = \frac{\partial^2 T}{\partial r^2} + \frac{2}{r}\frac{\partial T}{\partial r} + \frac{f(r)}{r}, \quad t > 0, \quad r_1 < r < r_2,$$

with the initial condition

$$T(r, 0) = 0,$$

and boundary conditions

$$\frac{\partial T}{\partial r}(r_1, t) = \frac{\partial T}{\partial r}(r_2, t) = 0.$$

Here the thermal diffusivity is scaled to unity. If we make the transformation $u(r, t) = r T(r, t)$, the one-dimensional heat conduction problem results are

$$\frac{\partial u}{\partial t} = \frac{\partial^2 u}{\partial r^2} + f(r);$$

the initial condition remains

$$u(r, 0) = 0,$$

but the boundary conditions become

$$r_1\frac{\partial u}{\partial r}(r_1, t) - u(r_1, t) = 0, \tag{5.51}$$

$$r_2\frac{\partial u}{\partial r}(r_2, t) - u(r_2, t) = 0. \tag{5.52}$$

We employ a Laplace transform in t defined as

$$U(r, s) = \int_0^\infty u(r, t)e^{-st}dt.$$

The boundary-value problem for U is

$$\frac{d^2 U}{dr^2} - sU = -\frac{f(r)}{s}$$

with the same boundary conditions in r, (5.51) and (5.52). This boundary-value problem may be solved explicitly to give

$$U(r, s) = \int_{r_1}^{r_2} G(r, \xi; \lambda) f(\xi)d\xi, \tag{5.53}$$

where $G(r, \xi; \lambda)$ is a Green's function with $\lambda = s^{1/2}$,

$$g_1(r; \lambda) = \lambda r_1 \cosh\left(\lambda(r - r_1)\right) + \sinh\left(\lambda(r - r_1)\right),$$

$$g_2(r; \lambda) = \lambda r_2 \cosh\left(\lambda(r_2 - r)\right) - \sinh\left(\lambda(r_2 - r)\right),$$

$$W(\lambda) = \lambda(\lambda^2 r_1 r_2 - 1)\sinh\left(\lambda(r_2 - r_1)\right) + \lambda^2(r_2 - r_1)\cosh\left(\lambda(r_2 - r_1)\right),$$

so that

$$G(r, \xi; s) = \begin{cases} g_1(r;\lambda)g_2(\xi;\lambda)/\lambda^2 W(\lambda), & r \leq \xi \\ g_2(r;\lambda)g_1(\xi;\lambda)/\lambda^2 W(\lambda), & r \geq \xi \end{cases}.$$

Thus, in principle the solution $T(r, t)$ may be found and written as

$$T(r, t) = \mathcal{L}^{-1}\left\{\frac{U(r, s)}{r}\right\}.$$

It turns out that for the particular application in mind it is only the value $T(r_1, t)$ that is of interest. Consider then, from (5.53),

$$U(r_1, s) = \int_{r_1}^{r_2} G(r_1, \xi; \lambda) f(\xi) d\xi$$

$$= \frac{r_1}{\lambda W(\lambda)} \int_{r_1}^{r_2} g_2(\xi; \lambda) f(\xi) d\xi.$$

Call

$$V(\lambda) = \frac{1}{W(\lambda)} \int_{r_1}^{r_2} g_2(\xi; \lambda) f(\xi) d\xi.$$

By a useful relation involving the convolution (see Problem 5.11), we have that for a suitable $v(t)$, such that $\mathcal{L}\{v(t)\} = V(s)$,

$$\mathcal{L}\left\{\frac{1}{\sqrt{\pi t}} \int_0^\infty e^{-\tau^2/4t} v(\tau) d\tau\right\} = \frac{V(\sqrt{s})}{\sqrt{s}} = \frac{V(\lambda)}{\lambda}.$$

Hence,

$$T(r_1, t) = \frac{1}{\sqrt{\pi t}} \int_0^\infty e^{-\tau^2/4t} v(\tau) d\tau.$$

The function needed is $v(t) = \mathcal{L}^{-1}\{V(\lambda)\}$. Though this calculation is unlikely to be done in a simple manner, we will examine a useful approximation to it in Chapter 8.

5.3.7 Boundary Layer Evolution for MHD Flow: Hartmann Layer

Example 5.7. The classic Rayleigh problem for the occurrence of the boundary layer near a wall has been studied in Example 5.2. Here we examine the case where a magnetic field occurs and the resulting velocity field is steady. In the equations of motion there occurs a Lorenz force that leads to magnetic damping (Shercliff).

Here we assume the wall is nonconducting and bounds a semiinfinite region of conducting viscous fluid. The wall moves impulsively at a constant speed $U\mathbf{i}$ parallel to itself. There is an imposed uniform magnetic field $B\mathbf{j}$. Suppose that the resulting flow is two-dimensional and parallel, so that $u(y, t)$ satisfies

$$\rho\frac{\partial u}{\partial t} = \rho v \frac{\partial^2 u}{\partial y^2} - \sigma B^2 u, \ t > 0, \ 0 < x < \infty. \tag{5.54}$$

The density of the fluid is ρ, viscosity ν, and electrical conductivity σ. The appropriate initial and boundary conditions are

$$u(0, t) = U, \quad u \text{ finite as } y \to \infty,$$
$$u(y, 0) = 0.$$

Similar to the analysis of Example 5.2, we employ the Laplace transform and define

$$\mathcal{L}\{u\}(y, s) = \int_0^\infty u(y, t)e^{-st} dt.$$

The result of the Laplace transform on (5.54) is

$$\rho s \mathcal{L}\{u\} = \rho \nu \frac{d^2 \mathcal{L}\{u\}}{dy^2} - \sigma B^2 \mathcal{L}\{u\}.$$

The boundary conditions are transformed to

$$\mathcal{L}\{u\}(0, s) = \frac{U}{s}, \quad \mathcal{L}\{u\} \text{ finite as } y \to \infty.$$

This constant coefficient second-order equation is solved for $\mathcal{L}\{u\}$ as

$$\mathcal{L}\{u\} = \frac{U}{s} e^{-ry},$$

where

$$r = \sqrt{\frac{\sigma B^2}{\rho \nu} + \frac{s}{\nu}}.$$

The solution for $u(y, t)$ is desired as the inverse Laplace transform

$$u(y, t) = \mathcal{L}^{-1} \left\{ \frac{U}{s} e^{-ry} \right\}. \tag{5.55}$$

Using Problem 5.5, we are able to write

$$\mathcal{L}^{-1} \left\{ \sqrt{\pi} e^{-\alpha \sqrt{s}} \right\} = \frac{\alpha}{2t^{3/2}} e^{-\alpha^2/4t}, \quad \alpha > 0$$

and with Problem 5.7,

$$\mathcal{L}[f(at)e^{-bt}] = \frac{1}{a} F\left(\frac{s+b}{a}\right) \quad (a > 0).$$

Take $a = 1$, $b = \sigma B^2/\rho$, $\alpha = y/\sqrt{\nu}$, so that

$$\mathcal{L}^{-1} \left\{ e^{-ry} \right\} = \frac{y}{2\sqrt{\pi \nu}} \frac{1}{t^{3/2}} e^{-y^2/4\nu t} e^{-\sigma B^2 t/\rho}.$$

Consequently, using results from Problem 5.9, we know that

$$\mathcal{L}^{-1} \left(\frac{F(s)}{s}\right) = \int_0^t f(\tau) d\tau$$

and, finally,

$$u(y, t) = \frac{Uy}{2\sqrt{\pi \nu}} \int_0^t \frac{1}{\tau^{3/2}} e^{-y^2/4\nu\tau} e^{-\sigma B^2 \tau/\rho} d\tau \tag{5.56}$$

This integral solution is exact, but it is not immediately apparent from this solution that there is a steady state. In fact, examining $\mathcal{L}\{u\}$ in (5.55), it is clear that there is a branch point beginning at $s = -\sigma B^2/\rho$ and extending to $\Re\text{eal}\{s\} = -\infty$, and there is an origin pole. Deformation of the Laplace inversion contour gives an alternate form of the solution,

$$u = u_s(y) + u_t(t, y), \tag{5.57}$$

where $u_s(y)$ comes from the residue of the origin pole and represents the steady-state solution, and $u_t(t)$ is the transient, corresponding to the integral around the branch cut. The branch-cut integral is very much like some of what has been done above, and so is not repeated here. The steady-state solution is easily found from the origin residue and is

$$u_s(y) = U e^{-(\sigma/\rho v)^{1/2} \, By}. \tag{5.58}$$

It is an interesting exercise to affirm that this result is equal to the value of solution (5.56) with the upper limit on the integral put at infinity.

5.4 Bilateral Laplace Transform

A "two-sided" transform, for variables defined on $(-\infty, \infty)$, can be useful for time-dependent problems for which initial values are not particularly important, but for which there must be decay as $t \to \pm\infty$. One may define the bilateral or "two-sided" Laplace transform as

$$\mathcal{L}_2\{f\} = F_2(s) = \int_{-\infty}^{\infty} e^{-st} f(t)dt, \quad c_1 < \Re\text{eal}(s) < c_2. \tag{5.59}$$

Here, the region of convergence is a vertical strip. The necessity for this restriction is evidenced by the fact that one may define

$$\int_{-\infty}^{\infty} e^{-st} f(t)dt = \int_{-\infty}^{0} e^{-st} f(t)dt + \int_{0}^{\infty} e^{-st} f(t)dt$$

$$= \int_{0}^{\infty} e^{s\tau} f(-\tau)d\tau + \int_{0}^{\infty} e^{-st} f(t)dt.$$

From our knowledge of the unilateral (one-sided) Laplace transform, each of the last two integrals is well defined if $\Re\text{eal}(-s) > c$ and $\Re\text{eal}(s) > c_1$. The intersection of these two regions gives a strip if $-c \equiv c_2 > c_1$.

The strip on which bilateral Laplace transform is defined may degenerate to a line. For instance, for $f(t) = \frac{1}{1+t^2}$,

$$\mathcal{L}_2\{f\} = \int_{-\infty}^{\infty} e^{-st} \frac{1}{1+t^2} dt$$

converges only if $\Re\text{eal}(s) = 0$, on the imaginary axis, in agreement with the Fourier transform, where $s = i\omega$. The bilateral transform is closely related to the Fourier transform with complex argument (Weinberger) but we will not pursue that here. The bilateral transform has played a prominent role in linear systems theory, so it behooves us to give it due diligence in its own right.

We may calculate explicitly that if $f(t) = e^{-a|t|}$, $a > 0$,

$$\int_{-\infty}^{\infty} e^{-st} e^{-a|t|} dt = \int_{-\infty}^{0} e^{-st} e^{at} dt + \int_{0}^{\infty} e^{-st} e^{-at} dt$$

$$= \frac{1}{a - s} \quad (\text{for } a - \Re\text{eal}(s) > 0)$$

$$+ \frac{1}{a + s} \quad (\text{ for } a + \Re\text{eal}(s) > 0)$$

$$= \frac{2a}{a^2 - s^2}, \quad -a < \Re\text{eal}(s) < a.$$

5.4.1 Inverse Bilateral Laplace Transform

The inverse of the bilateral transform is particularly attractive because the "action" all takes place within the strip of convergence. First, we cite a theorem that makes this legitimate.

Theorem 5.7. *((Widder, p. 241)) If $f(t)$ is absolutely integrable in every finite interval and if the integral*

$$F_2(s) = \int_{-\infty}^{\infty} e^{-st} f(t) dt \tag{5.60}$$

converges absolutely on the line $\Re\text{eal}(s) = c$, and if $f(t)$ is of bounded variation in some neighborhood of $t = t_0$, then

$$\lim_{T \to \infty} \frac{1}{2\pi i} \int_{c-iT}^{c+iT} F_2(s) e^{st_0} ds = \frac{f(t_0^+) + f(t_0^-)}{2}.$$

Thus, as with the other transforms we have considered, the inverse will take the average value across a jump. In practice, the actual evaluation of the inverse bilateral transform may depend on the sign of t. Ideally, the inversion, from the Laplace domain to the t-domain is given by the complex integral, where Γ is the const–$\Re\text{eal}(s)$ path in the complex s-plane, with endpoints at $c - i\infty$ and $c + i\infty$; c is chosen so that (5.59) converges absolutely on the line $\Re\text{eal}(s) = c$; and $F_2(s)$ is analytic in in the strip. Then, the singularities of $F_2(s)$ on the edges or outside, right or left determine $f(t)$. For $t > 0$, it appropriate to close the contour on the left, but if $t < 0$, it should be closed on the right.

5.4.2 Worked Examples

A Modified Bessel Function

Example 5.8. Consider the differential equation

$$(ty'(t))' - ty(t) = 0, \quad -\infty < t < \infty, \tag{5.61}$$

which is the modified Bessel equation of order zero. (See Section 2.3.2 for additional details.) We will investigate it for a solution which may be unbounded at infinity.

Perform a bilateral Laplace transform

$$\int_{-\infty}^{\infty} \left[(ty'(t))' - ty(t) \right] e^{-st} dt = 0.$$

No boundary conditions are necessary, simply demanding at this point that all of the indicated integrals converge! Thus, requiring that the boundary terms vanish leads to

$$s^2 \int_{-\infty}^{\infty} ty(t)e^{-st}dt - s \int_{-\infty}^{\infty} y(t)e^{-st}dt - \int_{-\infty}^{\infty} ty(t)e^{-st}dt = 0.$$

Call

$$Y_2(s) = \int_{-\infty}^{\infty} y(t)e^{-st}dt, \tag{5.62}$$

so that

$$Y_2'(s) = \int_{-\infty}^{\infty} -ty(t)e^{-st}dt.$$

A first-order equation for $Y_2(s)$ results,

$$Y_2'(s)(1 - s^2) - sY(s) = 0,$$

which when solved, yields

$$Y_2(s) = \frac{A}{\sqrt{1 - s^2}}. \tag{5.63}$$

This is one solution that must have two possible inverses to supply the linearly independent solutions to the original equation. We examine one that may be determined by making a branch cut on the real s-axis on the interval $[-1, 1]$. Take

$$y(t) = \frac{A}{2\pi i} \int_{c-i\infty}^{c+i\infty} \frac{e^{st} ds}{\sqrt{1 - s^2}}.$$

The vertical line on which the integral is performed lies to right of both singularities according to Figure 5.5. The "strip" where the function is anayltic in this case is $\Re\text{eal}\{s\} > 1$. We make this integral into a closed-path integral for $t < 0$ with a semicircle in $\Re\text{eal}\{s\} > 0$, and then by the Cauchy Theorem, $y \equiv 0$. However, for $t > 0$, closing the contour with a semicircle to the left means that the value of the integral comes from the integral along both sides of the branch cut. On the top $s = x$, and $s = -x$ below. As a result, we have

$$y(t) = \frac{1}{\pi} \int_{-1}^{1} \frac{e^{xt} dx}{\sqrt{1 - x^2}}, \quad t > 0$$

$$= 0, \quad t < 0.$$

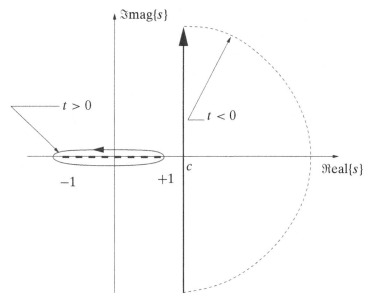

Figure 5.5. Integration paths for the bilateral inverse for $t < 0$ and $t > 0$. (The dashed line marks the branch cut.)

The choice of A, such that $y(0) = 1$ gives $y(t) = I_0(t)$, the modified Bessel function for $t > 0$, while for $t < 0$, the contour is closed to the right where there are no singularities. On the other hand, if we desire the domain of the function to be $t < 0$, choose the vertical line to lie to the left of the cut (that is, $\Re\mathrm{eal}(s) < -1$) and close the contour on the right. The orientation of the contour is reversed. Still, the result is

$$y(t) = \frac{1}{\pi} \int_{-1}^{1} \frac{e^{xt}\, dx}{\sqrt{1 - x^2}}, \quad t < 0$$

$$= 0, \quad t > 0,$$

so that $y(0) = 1$. Indeed, $I_0(t)$ is an even function.

Note that the branch cuts could be chosen differently, being located, for example, from $s = 1$ to $s = +\infty + i0$ on the right and from $s = -1$ to $s = -\infty + i0$ on the left. That leaves a strip $-1 < \Re\mathrm{eal}(s) < 1$ in which Y_2 is analytic. Then, for $t > 0$, for example, the contour may be closed to the left, and what results is an integral along the bottom and top of the left-hand branch cut. Working that out leads to a standard integral representation for $K_0(x)$, the other linearly independent solution of (5.61).

Time-Dependent Boundary Layer with Suction

Example 5.9. Here we investigate the possibility of a two-dimensional exact solution of the Navier–Stokes equations giving a boundary-layer flow from constant suction along an infinite flat wall with a time-dependent free stream. (See (Watson)) The governing equations are scaled to read

$$\frac{\partial u}{\partial t} - 4\frac{\partial u}{\partial y} = f'(t) + 4\frac{\partial^2 u}{\partial y^2}, \quad -\infty < t < \infty, \ y > 0, \qquad (5.64)$$

with the boundary conditions

$$u = 0 \text{ at } y = 0 \text{ and } u \to f(t) \text{ as } y \to \infty.$$

Define a bilateral Laplace transform in t as

$$U_2(y, s) = \int_{-\infty}^{\infty} u(y, t)e^{-st} dt.$$

Perform the transform of (5.64) giving

$$U_{2,yy} + U_{2,y} - \frac{s}{4}U_2 = -\frac{s}{4}F_2(s), \ 0 < y < \infty,$$

so that

$$U_2(0, s) = 0, \quad \text{and} \ U_2 \to F_2(s) \text{ as } y \to \infty.$$

The solution for U_2 may be written as

$$U_2(y, s) = F_2(s)(1 - e^{-ry}), \tag{5.65}$$

with

$$r = \frac{1}{2}(1 + \sqrt{1 + s}),$$

where we take the branch of the square root, which is positive for s real. The inverse Laplace transform is fairly standard. We can also make use of the convolution theorem, which, because the t integrals are defined on $(-\infty, \infty)$, is of the same form as for the Fourier transforms, that is,

$$(f * g)(t) = \int_{-\infty}^{\infty} f(t - \tau)g(\tau)d\tau,$$

and

$$\int_{-\infty}^{\infty} e^{-st}(f * g)(t)dt = F_2(s)G_2(s).$$

Then, in the actual inversion process of (5.65), we obtain

$$u(y, t) = f(t) - \int_{c-i\infty}^{c+i\infty} F_2(s)e^{-ry}e^{st} ds,$$

and it suffices to take $t > 0$. Finally,

$$u(y, t) = f(t) - \frac{ye^{-\frac{1}{2}y}}{4\sqrt{\pi}} \int_0^{\infty} f(t - \tau)\tau^{-3/2}e^{-\tau}e^{-y^2/16\tau} d\tau.$$

REFERENCES

M. Ya. Antimirov, A. A. Kolyshkin, and R. Vaillancourt, *Applied Integral Transforms*, American Mathematical Society, Providence, 1993.

W. E. Boyce and R. C. DiPrima, *Elementary Differential Equations and Boundary Value Problems,* 8th edition, John Wiley & Sons, New York, 2004.

L. Rosenhead, ed., *Laminar Boundary Layers*, Oxford University Press, 1963.

J. A. Shercliff, *A Textbook of Magnetohydrodynamics*, Pergamon, Oxford, 1965.

J. Watson, A Solution of the Navier–Stokes Equations Illustrating the Response of the Laminar Boundary Layer to a Given Change in the External Stream Velocity, *Quart. J. Mech. Appl. Math.* **11**, 302–325 (1958).

H. F. Weinberger, *A First Course in Partial Differential Equations*, Dover, New York, 1965.

D. V. Widder, *The Laplace Transform*, Princeton University Press, 1946.

J. E. Wilkins, Jr., Conduction of Heat in an Insulated Spherical Shell with Arbitrary Spherically Symmetric Sources, *SIAM Rev.* **1**, 149–153 (1959).

EXERCISES

5.1 A two-component system $(x(t), y(t))$ with forcing is governed by the equations

$$\frac{d^2x}{dt^2} + \omega\frac{dy}{dt} = a\cos(\omega_0 t),$$

$$\frac{d^2y}{dt^2} - \omega\frac{dx}{dt} = 0,$$

where a, ω_0 and ω are constants, with initial conditions $x = dx/dt = y = dy/dt = 0$ at $t = 0$. Use Laplace transforms to obtain its solution. What happens as $\omega \to \omega_0$?

5.2 A strip of material is subjected to uniform oscillatory heating. The initial-boundary-value problem is

$$\frac{\partial T}{\partial t} = \frac{\partial^2 T}{\partial x^2} + \cos(\omega t), \quad 0 < x < 1, \quad t > 0.$$

The initial and boundary conditions are

$$T = 0, \quad \text{at} \quad t = 0,$$

$$\frac{\partial T}{\partial x} = 0, \quad \text{at} \quad x = 0, \qquad T = 0, \quad \text{at} \quad x = 1.$$

(a) Show that the Laplace transform of the solution is

$$\mathcal{L}\{T\} = \frac{1}{s^2 + \omega^2}\left[1 - \frac{\cosh(\sqrt{s}\,x)}{\cosh(\sqrt{s})}\right].$$

(b) Discuss the inversion of this transform, showing the proper contour. Formally, write down $T(x, t)$ in terms of sums of all residues and any branch-cut integrals.

(c) Work through the algebra for the temperature at $x = 0$, and show that

$$T(0, t) = \frac{1}{\omega}\left[\left(1 - \frac{\cosh\alpha\cos\alpha}{\cosh^2\alpha - \cos^2\alpha}\right)\sin(\omega t) + \frac{\sinh\alpha\sin\alpha}{\cosh^2\alpha - \cos^2\alpha}\cos(\omega t)\right],$$

where

$$\alpha \equiv \sqrt{\frac{\omega}{2}}.$$

5.3 In this problem, we are going to solve for the fluid flow between parallel plates if the lower plate is at rest, and the upper plate accelerates constantly. Thus, the initial-boundary-value problem for the fluid velocity $u(y, t)$ is

$$\frac{\partial u}{\partial t} = \frac{\partial^2 u}{\partial y^2}, \quad \text{in} \quad t > 0, 0 < y < 1, \tag{5.66}$$

$$u = 0, \quad \text{at} \quad t = 0,$$
$$u = 0, \quad \text{at} \quad y = 0,$$
$$u = at, \quad \text{at} \quad y = 1.$$

Here, a is a constant, and variables have all been made dimensionless for mathematical convenience.

(a) By using Laplace transformation on the equation and boundary conditions, and using the initial condition, show that the Laplace transform of the solution is

$$\mathcal{L}\{u\} = \frac{a \sinh(\sqrt{s}\, y)}{s^2 \sinh(\sqrt{s})}.$$

(b) Show that there is *not* a branch point at $s = 0$. What is the singularity at $s = 0$?

(c) Show that there are singularities at $s = -n^2\pi^2$, with n a positive integer.

(d) By summing all the residues, show that the solution is given by

$$u = ay\left(t + \frac{1}{6}\left(y^2 - 1\right)\right) + 2a \sum_{n=1}^{\infty} \frac{(-1)^{n+1}\sin(n\pi y)}{(n\pi)^3} e^{-n^2\pi^2 t}. \tag{5.67}$$

(Note that the formula for obtaining the residue of a pole other than first order is given by Equation (1.27). Actually, this residue is easier to do with series expansions for sinh.)

(e) The first term in (5.67) is the long-time solution, and the second term is the transient. Show that the first term is a solution to the partial differential equation (5.66).

5.4 In a solid material of width 1 ($0 < x < 1$), there is an alternating electrical current flowing, so that the equation for the temperature in the solid is given in scaled form as

$$\frac{\partial T}{\partial t} - \frac{\partial^2 T}{\partial x^2} = \cos^2(\omega t).$$

The boundary condition and initial condition on the temperature are

$$T = 0, \quad \text{at} \quad x = 0 \text{ and at } x = 1, \quad T = 0, \quad \text{on} \quad t = 0.$$

(a) Laplace transform the equation and boundary conditions, solve the ordinary differential equation for $\mathcal{L}\{T\}$, and apply those boundary conditions to that solution. Show that the Laplace transform of the temperature is then given by

$$\mathcal{L}\{T\} = \frac{s^2 + 2\omega^2}{s(s^2 + 4\omega^2)}\left[1 - \frac{\cosh(\sqrt{s}\, z)}{\cosh(\sqrt{s}/2)}\right],$$

where we use $z \equiv x - 1/2$ simply for convenience in what comes later.

(b) Show that there is no branch point in $\mathcal{L}\{T\}$. Where are the poles?

(c) Using the standard Laplace inversion contour for $t > 0$, show that the solution splits into two parts: A transient that is both an infinite series and a time-periodic solution. For the transient part, get

$$T_{\text{transient}} = -\frac{2}{\pi^2} \sum_{n=0}^{\infty} \frac{2(-1)^n}{(2n+1)} \frac{(2n+1)^2 + 2\omega^2}{(2n+1)^2 + 4\omega^2} \cos\left(\left(n + \frac{1}{2}\right)\pi z\right) e^{-(2n+1)^2\pi^2 t}.$$

The time-periodic piece of the solution is pretty messy. Get as far as you can with it. It is actually convenient to leave it in the form:

$$\mathfrak{Real}\left\{G(z)e^{2i\omega t}\right\},$$

where G is a complicated complex function.

5.5 Analyze the following variant to Example 5.2

$$\frac{\partial u}{\partial t} = \frac{\partial^2 u}{\partial y^2}, \quad \text{for} \quad y > 0, \ t > 0,$$

subject to the boundary and initial conditions on $u(y, t)$,

$$u(y, 0) = f(y), \quad y > 0,$$

$$\frac{\partial u}{\partial y}(0, t) = 0, \quad t > 0.$$

Specialize to the the particular initial condition

$$f(y) = \exp(-y/\sqrt{2})\cos(y/\sqrt{2}).$$

The solution describes the motion of an unbounded fluid originally due to the motion of an infinite plate parallel to itself, when instantly, the plate disappears. Then, $f(y)$ is the initial fluid velocity distribution, from the Stokes solution. Owing to symmetry, no stress is assumed at the boundary. The kinematic viscosity of the fluid is scaled to unity, as is the frequency of oscillation of the plate.

5.6 Using the Laplace transform in t solve the initial-boundary-value problem

$$\frac{\partial u}{\partial t} = \nu \frac{\partial^2 u}{\partial x^2} + Axe^{-\gamma x}, \quad 0 < x, t < \infty, \ \nu, \gamma > 0,$$

$$u(0, t) = 0, \quad u \to 0 \text{ as } x \to \infty,$$

$$u(x, 0) = 0.$$

Show that the solution may be written as

$$u(x, t) = A \int_0^t d\tau \int_0^\infty \xi e^{-\gamma\xi} G(x, \xi, t - \tau)d\xi,$$

where

$$G(x, \xi, t) = \frac{1}{\sqrt{4\pi \nu t}} \cdot \left[e^{-(x-\xi)^2/4\nu t} - e^{-(x+\xi)^2/4\nu t}\right].$$

The function $G(x, \xi, t)$ thus obtained is the "time-dependent Green's function."

5.7 Fluid is flowing downward through a porous plane at $y = 0$. At $t = 0$, the plane starts instantaneously to move at a constant acceleration to the right. The equation, boundary conditions and initial conditions (in a convenient dimensionless form) that describe the fluid motion are

$$\frac{\partial u}{\partial t} - 2\frac{\partial u}{\partial y} = \frac{\partial^2 u}{\partial y^2}, \quad \text{in } y > 0$$

$$u \equiv 0, \quad \text{at } t = 0,$$

$$u = t, \quad \text{at } y = 0,$$

$$u \text{ bounded for } y \to \infty$$

(a) Using the Laplace transformation, show that the Laplace transform of the solution is given by

$$\mathcal{L}\{u\} = \frac{1}{s^2}e^{-(1+\sqrt{1+s})y} \tag{5.68}$$

(b) Indicate by a sketch what path you would use in order to apply the residue theorem to the inversion of (5.68). Indicate in the *same* sketch any branch cuts that need to be inserted.

(c) What is the meaning of the origin pole? Work out that part of the solution due to the origin pole.

5.8 A semi-infinite material is heated by an internal heat source. A reference temperature has been removed, so before the heating begins, the temperature is zero. (The equation has been put in a dimensionless form, so the thermal conductivity is scaled out.)

The mathematical problem is

$$\frac{\partial T}{\partial t} - \frac{\partial^2 T}{\partial y^2} = Q\sin(\omega t)e^{-y}, \quad \text{in } y > 0,$$

$$T = 0, \quad \text{at } t = 0,$$

$$T = 0, \quad \text{at } y = 0.$$

Here, Q is a constant.

(a) Use a Laplace transformation in time, and show that the transform of the solution is

$$\mathcal{L}\{T\} = \frac{Q\omega}{(s^2 + \omega^2)(s - 1)}\left[e^{-y} - e^{-\sqrt{s}y}\right].$$

(b) Using the Laplace inversion integral and the residue theorem, obtain the solution for $T(y, t)$. Note that you will have to cut the plane, so that the inversion contour is like that shown in Figure 5.3. The solution splits into the long-time behavior and a transient. As in Example 5.2, the transient part of the solution will involve an integral that cannot be done in the closed form.

5.9 In a solid occupying $x > 0$, there is an alternating electrical current flowing in a layer near the surface, generating heat. The equation that describes the temperature variation in such a situation is

$$\frac{\partial T}{\partial t} - \frac{\partial^2 T}{\partial x^2} = e^{-x} \cos^2(\omega t).$$

The boundary condition and initial condition on the temperature are

$$T = 0, \quad \text{at} \quad x = 0, \quad T = 0, \quad \text{on} \quad t = 0.$$

(a) Show that the Laplace transform of the temperature is given by

$$\mathcal{L}\{T\} = \frac{s^2 + 2\omega^2}{s(s^2 + 4\omega^2)(s - 1)} \left[e^{-x} - e^{-\sqrt{s}x} \right].$$

(b) Give a discussion of all of the singularities of $\mathcal{L}\{T\}$, classifying each as pole, branch point, and so forth

(c) The Laplace inversion typically consists of residues from isolated singular points and integrals along branch cuts. Indicate which elements are a part of the inversion of this function, identifying any poles and any branch-cut integral paths – use a sketch. For the residues, indicate which correspond to (i) A time-independent portion of the solution, (ii) a time-periodic part of the solution, and (iii) a transient portion of the solution.

(d) If the plane is cut from the origin along the negative real axis, show that the integrals along both sides of the branch cut combine to give the following contribution to the solution:

$$T_{bc} = \frac{1}{\pi} \int_0^\infty \frac{(r^2 + 2\omega^2) \sin(\sqrt{r}x)}{r(r + 1)(r^2 + 4\omega^2)} e^{-rt} dr.$$

(e) Give the solution for $T(x, t)$ that remains after times that are long enough that any transients have disappeared.

5.10 (a) In a problem of flow near a porous wall with curvature, we encounter the problem

$$\frac{\partial u}{\partial t} = \frac{\partial^2 u}{\partial y^2} - v_0 \frac{\partial u}{\partial y} - a^2 u + f(y, t), \quad y > 0, \quad t > 0,$$

with the initial and boundary conditions

$$u(0, t) = u(y, 0) = 0. \tag{5.69}$$

Assuming that the solutions also satisfy

$$\int_0^\infty e^{-v_0 y} (u(y, t))^2 dy < \infty, \tag{5.70}$$

by taking a Laplace transform in t, show that the solution may be written as

$$u(y, t) = \int_0^t d\tau \int_0^\infty f(\eta, \tau) G_0(y, \eta, t - \tau) d\eta,$$

where

$$G_0(y, \eta, t) = e^{v_0(y-\eta)/2} \frac{e^{-(v_0^2/4+a^2)t}}{\sqrt{4\pi t}} \left[e^{-(y-\eta)^2/4t} - e^{-(y+\eta)^2/4t} \right].$$

(b) Suppose, instead of (5.69) the side conditions are

$$u_y(0, t) = u(y, 0) = 0,$$

while (5.70) still holds. Show, again by means of the Laplace transform in t, that now the solution is given by

$$u(y, t) = \int_0^t d\tau \int_0^\infty f(\eta, \tau) G_1(y, \eta, t - \tau) d\eta,$$

where

$$G_1(y, \eta, t) = e^{v_0(y-\eta)/2} \frac{e^{-(v_0^2/4+a^2)t}}{\sqrt{4\pi t}} \left[e^{-(y-\eta)^2/4t} + e^{-(y+\eta)^2/4t} \right].$$

The functions $G_0(y, \eta, t)$ and $G_1(y, \eta, t)$ are time-dependent Green's functions for their respective problems.

5.11 An infinite strip of material has a width h, and is insulated at the upper wall. At the lower wall, an oscillatory temperature load is imposed at $t = 0$. A reference temperature has been removed, so before the heating begins, the temperature is zero.

The mathematical problem is

$$\frac{\partial T}{\partial t} = \frac{\partial^2 T}{\partial y^2}, \quad T = 0, \quad \text{at } t = 0,$$

$$\frac{\partial T}{\partial y} = 0, \quad \text{at } y = h, \quad T = T_o \sin(\omega t), \quad \text{at } y = 0.$$

By performing a Laplace transform in t, determine an expression for the temperature $T(y, t)$. In particular, find the oscillatory part that remains after a long time.

5.12 A cylindrical container rotates rapidly about its axis, at angular speed Ω; the radius and height of the container are a and h respectively, and z is the axial coordinate. The lower container disk, at $z = 0$, is deformed slightly, but in a axisymmetric fashion, and that generates *inertial waves*. If w is the axial velocity perturbation introduced by the boundary deformation, and all velocity components are zero in the rotating frame at $t = 0$, the the initial- boundary-value problem is

$$\frac{\partial^2}{\partial t^2} \nabla^2 w + 4\Omega^2 \frac{\partial^2 w}{\partial z^2} = 0,$$

with initial conditions,

$$w = \frac{\partial w}{\partial t} = 0 \quad \text{at} \quad t = 0,$$

and boundary conditions

$$\frac{\partial w}{\partial r} = 0 \quad \text{at} \quad r = a, \quad w = 0 \quad \text{at} \quad z = h, \quad w = f(r) \quad \text{at} \quad z = 0.$$

Note that, since the fluid is incompressible, f must obey the integral constraint,

$$\int_0^a rf(r)dr = 0.$$

(a) By taking Laplace transforms, and using the cylindrical Laplacian,

$$\nabla^2 = \frac{\partial^2}{\partial r^2} + \frac{1}{r}\frac{\partial}{\partial r} + \frac{\partial^2}{\partial z^2},$$

show that that the transform of w, say $W(r, z, s)$, obeys

$$\frac{\partial^2 W}{\partial r^2} + \frac{1}{r}\frac{\partial W}{\partial r} + \frac{s^2 + 4\Omega^2}{s^2}\frac{\partial^2 W}{\partial z^2} = 0.$$

(b) Expand the solution as a Fourier-Bessel series, so

$$W = \sum_{n=1}^{\infty} A_n(z)J_0(\alpha_n r).$$

Why is $\alpha_n = j_{1,n}/a$, where $j_{1,n}$ is the nth zero of $J_1(z)$? Obtain the ordinary differential equation for $A_n(z)$.

(c) Let f_n denote the Fourier-Bessel series coefficient for the function $f(r)$. Then, show that the solution for A_n is

$$A_n(z) = \frac{f_n}{s}\frac{\sinh(\mu_n(h-z))}{\sinh(\mu_n h)},$$

where

$$\mu_n = \frac{s}{\sqrt{s^2 + 4\Omega^2}}\frac{j_{1,n}}{a}.$$

(d) Examine the Laplace inversion of the solution, by looking carefully at $\mathcal{L}^{-1}\{A_n\} \equiv a_n(z, t)$. Note that the integration path may be deformed to wrap around the the branch cuts that run to $(-\infty \pm 2\Omega i)$ from $\pm 2\Omega i$. In looking carefully at those integrals–which correspond to the transient response, $a_n^{(\text{T})}$, to the disk motion–pay careful attention to the small, circular-path integrals around the branch points themselves; do they contribute? If we write, in general, $a_n = a_n^{(\text{T})} + a_n'$, show that

$$a_n'(z, t) = f_n\left(1 - \frac{z}{h}\right) + f_n\sum_{m=1}^{\infty} R_{n,m}(z, t),$$

where $R_{n,m}$ arise from residues at poles located by solutions of

$$s^2 = -\frac{4\Omega^2 a^2 m^2 \pi^2}{j_{1,n}^2 h^2 + a^2 m^2 \pi^2}.$$

Give the detailed form of $R_{n,m}$. These are the inertial waves. Notice that the temporal frequencies are always less that 2Ω.

6

Fourier Transform Methods

6.1 Preamble

We now turn to problems most easily solved with the Fourier transformation, which is appropriate for functions defined on a doubly infinite domain, say $x \in (-\infty, \infty)$. The occurrence of this transform was suggested in Chapter 3, when we considered singular Sturm–Liouville problems and their expansions. Historically, such transformations have been used to a great advantage in wave propagation problems, and for generating solutions to elliptic equations particularly in unbounded, rectangular geometries.

6.2 The Fourier Transform and Its Inverse

The transform and its inversion formula may be derived from the Fourier Integral theorem, which is intimately connected with Fourier series.

Theorem 6.1. Fourier's integral theorem. *(Sneddon). If $f(x)$ satisfies, on every finite interval (a, b), the conditions for the convergence of its Fourier series, and if $\int_{-\infty}^{\infty} |f(x)| \, dx < \infty$, then for $-\infty < x < \infty$,*

$$\frac{1}{\pi} \int_0^{\infty} dk \int_{-\infty}^{\infty} f(\xi) \cos k(\xi - x) d\xi = \frac{1}{2} \left[f(x^+) + f(x^-) \right]. \tag{6.1}$$

If f is continuous at at the point x, then $f(x^+) = f(x^-) = f(x)$. Another form in which the integral may be written depends on the fact that

$$\int_0^A \cos k(\xi - x) dk = \frac{1}{2} \int_{-A}^A e^{ik(x-\xi)} dk,$$

so that in the limit where $A \to \infty$, where $f(x)$ is continuous

$$f(x) = \frac{1}{2\pi} \int_{-\infty}^{\infty} dk \int_{-\infty}^{\infty} f(\xi) e^{ik(x-\xi)} d\xi. \tag{6.2}$$

This result could also have been derived from the theory of the Laplace transform, but that process is not repeated here. (See (Antimirov et al.) for example.)

The Fourier transform is defined by the inner integrand in (6.2),

$$F(k) = \mathcal{F}[f] \equiv \int_{-\infty}^{\infty} f(x) e^{-ikx} dx. \tag{6.3}$$

As for notation, we will sometimes use an uppercase symbol for the transform, and other times use the $\mathcal{F}$ for the transform operator. Here, k is taken to be a real variable and one sufficient condition is that $|f|$ must decay to zero more rapidly than $1/|x|$ in order that the integral exist. If the transform is known, then the corresponding x function may be found from the outer integral in (6.2)

$$f(x) = \frac{1}{2\pi} \int_{-\infty}^{\infty} F(k) e^{ikx} dk. \tag{6.4}$$

Alternate definitions for $F(k)$ are sometimes encountered. (See, for example (Antimirov et al.) and (Bracewell).) Those variations involve either replacement of i with $(-i)$ in the transform, or the appearance of the 2π factor in other ways, and so the reader is warned to pay careful attention to the definition in a given work.

6.2.1 Problems

Problem 6.1. *(a) Show that the Fourier transform of the function*

$$f_1(x) = \begin{cases} e^{-ax}, & x > 0 \\ 0, & x < 0 \end{cases}, \quad a > 0,$$

is given by

$$F_1(k) = \frac{1}{a + ik}.$$

(b) Show that the Fourier transform of

$$f_2(x) = e^{-a|x|}, \quad a > 0,$$

is

$$F_2(k) = \frac{2a}{a^2 + k^2}.$$

Problem 6.2. *Show that*

$$\mathcal{F}\{e^{-ax^2}\} = \sqrt{\frac{\pi}{a}} e^{-k^2/4a}, \quad a > 0.$$

Hint: Use Cauchy's theorem to change the integral from $\Im\mathrm{mag}(z) = 0$ *to the parallel line* $\Im\mathrm{mag}(z) = k/2a$.

Problem 6.3. *Show that*

$$\mathcal{F}\{\mathrm{sech}(x)\} = \pi \,\mathrm{sech}\frac{k\pi}{2}.$$

Method: Observe that $\mathrm{sech}(z)$ *has poles only on the imaginary axis, where* $z = (2n + 1)\pi i/2, \quad n = 0, \pm 1, \pm 2, \ldots.$ *Use a rectangular contour with sides parallel to*

*the coordinate axes to enclose one of these poles. For instance, let the base of the rect-
angle be along the* $y = 0$, *its sides along* $x = \pm R$, *and its top along* $y = \pi$. *Employ the
residue theorem and allow* $R \to \infty$.

A function f satisfying the condition that $\int_{-\infty}^{\infty} |f(x)|\, dx < \infty$ is said to be *absolutely
integrable* on $(-\infty, \infty)$. However, a function violating this condition may still have
a Fourier transform. Problem 6.4 illustrates this possibility.

Problem 6.4. *Show that if a is real,*

$$\mathcal{F}\{\cos(ax^2)\} = \sqrt{\frac{\pi}{4\,|a|}} \cos\left(\frac{k^2}{4\,|a|} - \frac{\pi}{4}\right).$$

Hint: Even though $\int_{-\infty}^{\infty} |\cos(ax^2)|dx = \infty$, *the Fourier transform may be found
by integrating* e^{iaz^2} *around the sector whose sides are* $\Im\mathrm{mag}(z) = 0$, $0 \le \Re\mathrm{eal}(z) \le
A$; $z = Ae^{i\theta}$, $0 \le \theta \le \pi/4$; $z = Ae^{i\pi/4}$. *Let* $A \to \infty$.

Problem 6.5. *Verify the following general properties of the Fourier transform, when*
$F(k) = \mathcal{F}[f]$ *and* a, b *are real constants:*

(a) $\mathcal{F}[f(ax)e^{-ibx}] = \frac{1}{a} F\left(\dfrac{k+b}{a}\right)$, $(a > 0)$

(b) $\mathcal{F}[f(x/a + b)] = ae^{-iabk} F(ak)$, $(a > 0)$

(c) $\mathcal{F}\left[\dfrac{df(x)}{dx}\right] = ikF(k)$

(d) $\mathcal{F}[xf(x)] = i\dfrac{d}{dk} F(k)$

Problem 6.6. *Derive the results given below if* α, β *are complex constants.*
(a) If $f(x) = e^{-\alpha|x|} \sin \beta x$ *with* $\Re\mathrm{eal}(\alpha) > |\Im\mathrm{mag}(\beta)|$, *then*

$$F(k) = \frac{-4i\alpha\beta k}{(\alpha^2 + \beta^2 + k^2)^2 - 4\beta^2 k^2}.$$

(b) If $f(x) = \tan^{-1}(2/x^2)$, *then*

$$F(k) = \frac{e^{-|k|} \sin k}{k}.$$

6.2.2 The Convolution

For functions f and g defined on $(-\infty, \infty)$, their convolution is defined as

$$h(x) = \int_{-\infty}^{\infty} f(x - \xi)g(\xi)d\xi \equiv f * g.$$

The integral exists if both f and g are absolutely integrable. Notice that $f * g =
g * f$.

Fourier Transforms and Convolutions

Now let $F = \mathcal{F}[f]$ and $G = \mathcal{F}[g]$, then the inverse transform of the product is

$$
\frac{1}{2\pi} \int_{-\infty}^{\infty} e^{ikx} F(k)G(k)dk = \frac{1}{2\pi} \int_{-\infty}^{\infty} e^{ikx} F(k)dk \int_{-\infty}^{\infty} e^{-ik\xi} g(\xi)d\xi
$$

$$
= \frac{1}{2\pi} \int_{-\infty}^{\infty} g(\xi)d\xi \int_{-\infty}^{\infty} e^{ik(x-\xi)} F(k)dk
$$

$$
= \int_{-\infty}^{\infty} g(\xi) f(x - \xi)d\xi. \tag{6.5}
$$

This shows that

$$
\mathcal{F}^{-1}[FG] = f * g.
$$

More directly,

$$
\mathcal{F}[h] = \mathcal{F}[f]\mathcal{F}[g].
$$

6.2.3 Special Properties of Fourier Transforms

When $\int_{-\infty}^{\infty} |f(x)|^2 dx < \infty$, f is said to be square integrable on $(-\infty, \infty)$. For example, the function $f(x) = (1 + x^2)^{-1/2}$ *is square integrable but not absolutely integrable on* $-\infty < x < \infty$.

The function $g(x) = |x|^{-1/2}(1 + x^2)^{-1}$ *is absolutely integrable, but not square integrable on* $-\infty < x < \infty$ *because of the singularity at* $x = 0$.

When f is both absolutely integrable and square integrable, it and its Fourier transforms share a number of important properties (Titchmarsh), too numerous to mention all here. One property to note is

Theorem 6.2. Plancherel's theorem. *If* $F(k) = \mathcal{F}[f]$, *then*

$$
\int_{-\infty}^{\infty} |f(x)|^2 dx = \frac{1}{2\pi} \int_{-\infty}^{\infty} |F(k)|^2 dk.
$$

Another property is a generalization of this.

Theorem 6.3. Parseval's theorem. *If both* f *and* g *satisfy the conditions of Plancherel's theorem, where* $G(k) = \mathcal{F}[g]$, *then*

$$
\int_{-\infty}^{\infty} f(x)\overline{g(x)}dx = \frac{1}{2\pi} \int_{-\infty}^{\infty} F(k)\overline{G(k)}dk. \tag{6.6}
$$

We admit the possibility of complex valued functions and their transforms.

We show how to derive (6.6). Given (6.5), replace g with $\bar{g}$ and set $x = 0$. Then, since

$$G(k) = \int_{-\infty}^{\infty} g(x)e^{-ikx}dx,$$

$$\overline{G(k)} = \int_{-\infty}^{\infty} \overline{g(x)}e^{ikx}dx, \quad (x \to -x)$$

$$= \int_{-\infty}^{\infty} \overline{g(-x)}e^{-ikx}dx;$$

consequently, setting $x = 0$ in (6.5),

$$\frac{1}{2\pi}\int_{-\infty}^{\infty} F(k)\overline{G(k)}dk = \int_{-\infty}^{\infty} \overline{g(-\xi)}f(-\xi)d\xi$$

$$= \int_{-\infty}^{\infty} \overline{g(\xi)}f(\xi)d\xi.$$

6.2.4 Cosine and Sine Transforms

If $f(x)$ is defined only for $x > 0$, we can extend it backward as an even or an odd function. Extending f as even, using (6.1),

$$f(x) = \frac{1}{\pi}\int_0^{\infty} dk \int_{-\infty}^{\infty} f(\xi)\cos k(\xi - x)d\xi,$$

and the evenness of the extension

$$f(x) = \frac{2}{\pi}\int_0^{\infty} dk \int_0^{\infty} f(\xi)\cos k\xi \cos kx\, d\xi$$

$$= \sqrt{\frac{2}{\pi}}\int_0^{\infty} dk \cos kx \sqrt{\frac{2}{\pi}}\int_0^{\infty} f(\xi)\cos k\xi\, d\xi$$

$$= \sqrt{\frac{2}{\pi}}\int_0^{\infty} F_c(k)\cos(kx)dk,$$

where

$$F_c(k) = \sqrt{\frac{2}{\pi}}\int_0^{\infty} f(\xi)\cos k\xi\, d\xi$$

is the cosine transform. In a similar manner, the sine transform is defined as

$$F_s(k) = \sqrt{\frac{2}{\pi}}\int_0^{\infty} f(\xi)\sin k\xi\, d\xi.$$

Its inverse is

$$f(x) = \sqrt{\frac{2}{\pi}}\int_0^{\infty} F_s(k)\sin(kx)dk.$$

6.2.5 Problems

Problem 6.7. *Let $f(t)$ be an absolutely integrable function defined on $(-\infty, \infty)$ and*

$$h(t) = \int_{-\infty}^{t} e^{-2(t-\tau)} f(\tau - \alpha)d\tau,$$

where α is a real constant. Use the convolution theorem to determine $\mathcal{F}[h]$ in terms of $\mathcal{F}[f]$.

Problem 6.8. *(a) The autocorrelation of a function $f(x)$ is defined as*

$$r(x) = \int_{-\infty}^{\infty} \overline{f(\xi)} f(x + \xi)d\xi,$$

which is employed for analyzing noisy signals.

Show that the Fourier transform of $r(x)$, in terms of the Fourier transform of $f(x)$ is

$$R(k) = |F(k)|^2,$$

which is called the energy density spectrum of $f(x)$, because $|F(k)|^2$ gives the energy in the frequency components between k and $k + dk$.

(b) Calculate the energy density spectrum of the function

$$f(x) = e^{-(\alpha + i\beta)x} H(x), \quad \alpha, \beta \text{ real}, \alpha > 0,$$

where $H(x)$ is the Heaviside function. Sketch the energy density spectrum just calculated.

6.3 Worked Examples

In many cases, the real integral in (6.4) can be performed by complex variable methods. We consider now some sample problems for which such a technique is useful. Before turning to partial differential equations, we construct the solution of ordinary differential equations.

6.3.1 Example 1: The Ekman Layer

Example 6.10. Consider the coupled boundary value problem, inspired by the classic Ekman layer problem for flow near a wall (Pedlosky)

$$\frac{d^2u}{dx^2} = -\mu v,$$

$$\frac{d^2v}{dx^2} = \mu u,$$

$0 < x < \infty$, where μ is constant, $u(0) = a$, $v(0) = b$, and as $u, v \to 0$ as $x \to \infty$. Since the differential equations have only even derivatives, as do the boundary conditions, the sine transform is appropriate.

Introducing the Fourier sine transform throughout with

$$\mathcal{F}_s(u) = \sqrt{\frac{2}{\pi}} \int_0^\infty u(x) \sin kx dx$$

$$\mathcal{F}_s(v) = \sqrt{\frac{2}{\pi}} \int_0^\infty v(x) \sin kx dx,$$

$$\mathcal{F}_s[u''] = -\mu \mathcal{F}_s[v], \tag{6.7}$$

$$\mathcal{F}_s[v''] = \mu \mathcal{F}_s[u] \tag{6.8}$$

From the differential equations solutions, we have $u'' \to 0$, and $u' \to 0$ as $x \to \infty$, and using integration-by-parts twice we have

$$\mathcal{F}_s[u''] = \sqrt{\frac{2}{\pi}} \int_0^\infty u''(x) \sin kx dx$$

$$= \sqrt{\frac{2}{\pi}} ku(0) - k^2 \mathcal{F}_s(u) = \sqrt{\frac{2}{\pi}} ka - k^2 \mathcal{F}_s(u),$$

after applying the boundary conditions. Performing the same procedure for $\mathcal{F}_s[v'']$, we obtain, after applying the boundary conditions,

$$\mathcal{F}_s[v''] = \sqrt{\frac{2}{\pi}} kb - k^2 \mathcal{F}_s(v).$$

The transformed system may be written in matrix form:

$$\begin{pmatrix} k^2 & -\mu \\ \mu & k^2 \end{pmatrix} \begin{pmatrix} \mathcal{F}_s(u) \\ \mathcal{F}_s(v) \end{pmatrix} = \sqrt{\frac{2}{\pi}} k \begin{pmatrix} a \\ b \end{pmatrix}. \tag{6.9}$$

and solved by any of a number of means, leading to

$$\begin{pmatrix} \mathcal{F}_s(u) \\ \mathcal{F}_s(v) \end{pmatrix} = \frac{\sqrt{\frac{2}{\pi}} k}{k^4 + \mu^2} \begin{pmatrix} ak^2 + b\mu \\ -a\mu + bk^2 \end{pmatrix}.$$

The solution for the two components is

$$\begin{pmatrix} u(x) \\ v(x) \end{pmatrix} = \mathcal{F}_s^{-1} \left\{ \frac{\sqrt{\frac{2}{\pi}} k}{k^4 + \mu^2} \begin{pmatrix} ak^2 + b\mu \\ -a\mu + bk^2 \end{pmatrix} \right\}$$

$$= \frac{2}{\pi} \int_0^\infty \begin{pmatrix} ak^3 + b\mu k \\ -a\mu k + bk^3 \end{pmatrix} \frac{\sin kx}{k^4 + \mu^2} dk.$$

The integrands are even functions of k, and may be written as

$$\begin{pmatrix} u(x) \\ v(x) \end{pmatrix} = \frac{1}{\pi} \int_{-\infty}^\infty \begin{pmatrix} ak^3 + b\mu k \\ -a\mu k + bk^3 \end{pmatrix} \frac{\sin kx}{k^4 + \mu^2} dk$$

$$= \frac{1}{\pi} \Im \text{mag} \left\{ \int_{-\infty}^\infty \begin{pmatrix} ak^3 + b\mu k \\ -a\mu k + bk^3 \end{pmatrix} \frac{e^{ikx}}{k^4 + \mu^2} dk \right\}.$$

Since $x > 0$, close the contour in the upper-half-plane that encloses the poles in the integrands at $k_1 = \mu^{1/2} e^{\pi i/4}$ and $k_2 = \mu^{1/2} e^{3\pi i/4}$. Employing the residue theorem, we have

$$u(x) = ae^{-\sqrt{\frac{\mu}{2}}x} \cos\left(\sqrt{\frac{\mu}{2}}x\right) + be^{-\sqrt{\frac{\mu}{2}}x} \sin\left(\sqrt{\frac{\mu}{2}}x\right),$$

$$v(x) = -ae^{-\sqrt{\frac{\mu}{2}}x} \sin\left(\sqrt{\frac{\mu}{2}}x\right) + be^{-\sqrt{\frac{\mu}{2}}x} \cos\left(\sqrt{\frac{\mu}{2}}x\right).$$

Now, we proceed with a sequence of partial differential equation problems.

6.3.2 Example 2: Heat Conduction in a Strip

Example 6.2. We consider the heat conduction problem in a strip (Özişik), as stated below,

$$\nabla^2 T = 0, \quad \text{in} \quad -\infty < x < \infty, \quad 0 < y < 1,$$

$$T = 0, \quad \text{on} \quad y = 0 \quad \text{and} \quad T = f(x), \quad \text{on} \quad y = 1,$$

and f decays for large $|x|$ suitably fast. Doing the Fourier transform of the equation, integration by parts shows that for f and therefore T vanishing for $|x| \to \infty$,

$$\mathcal{F}\left[\frac{\partial T}{\partial x}\right] = ik\mathcal{F}[T]. \tag{6.10}$$

Hence, the differential equation and boundary conditions become

$$\frac{\partial^2 \mathcal{F}[T]}{\partial y^2} - k^2 \mathcal{F}[T] = 0, \quad \mathcal{F}[T] = 0, \quad \text{at} \quad y = 0, \quad \mathcal{F}[T] = F(k), \quad \text{at} \quad y = 1.$$

Solution of this constant-coefficient equation is

$$\mathcal{F}[T] = F(k)\frac{\sinh(ky)}{\sinh k}. \tag{6.11}$$

If we suppose that the there is a delta function in the temperature at the wall, then the transform of f is just $F = 1$. Exploring that case further, we can invert (6.11) in a complex k plane by replacing the $-\infty$ to ∞ integral in real k by a closed, contour integral in the k-plane, following the path of Example 1.1 Section 1.4, though taking the contour to avoid the singularities. Then, for the case $x > 0$, the function $\mathcal{F}[T]$ has pole singularities only, located at $k = in\pi$ inside the upper-half-plane contour. A sequence of contours of increasing radii, say $R_n = (n + \frac{1}{2})\pi$, encloses all the poles. It may also be shown that on each such semicircle

$$\left|\frac{\sinh ky}{\sinh k}\right| \le 4e^{(y-1)R_n|\cos\theta|}, \quad 0 < \theta < \pi,$$

and that the integral over the semicircle tends to 0 as $R_n \to \infty$. Hence, we can obtain, by means of the residue theorem, the solution

$$T(x, y) = -\sum_{n=1}^{\infty}(-1)^n e^{-n\pi|x|} \sin(n\pi y) \equiv A(x, y). \tag{6.12}$$

Note that we have also done the $x < 0$ case and combined the results into a formula valid for all x. The series is convergent and may be summed as a pair of geometric series. The result is

$$T(x, y) = \frac{\sin(\pi y)}{2\cosh(\pi x) + 2\cos(\pi y)}.$$

It can be shown – as we do in the next example – that the general solution to this problem for arbitrary surface forcing, $f(x)$, is given as what amounts to a convolution integral,

$$T(x, y) = \int_{-\infty}^{\infty} f(\xi)A(x - \xi, y)d\xi.$$

6.3.3 Example 3: Heat Conduction in a Half-Plane

Example 6.3. We now turn to the problem of heat conduction in a half-plane, so we want to solve in this case

$$\nabla^2 T = 0, \quad \text{in} \quad y > 0.$$

On the boundary, we specify the temperature, $T = f(x)$. The same Fourier transform process leads this time to the solution for the transform bounded in y,

$$\mathcal{F}[T] = F(k)e^{-|k|y}. \tag{6.13}$$

Putting this result into the inversion formula, we have

$$T = \frac{1}{2\pi}\int_{-\infty}^{\infty} F(k)e^{ikx-|k|y}dk = \frac{1}{2\pi}\int_{-\infty}^{\infty}\int_{-\infty}^{\infty} f(\xi)e^{ik(x-\xi)-|k|y}dkd\xi,$$

where we have used the definition of the transform F to write the solution as a general double integral. However, note that the k integration can be done universally, and then a final ξ integration performed. That is, we can write

$$T = \int_{-\infty}^{\infty} f(\xi)G(x - \xi)d\xi, \tag{6.14}$$

and

$$G(x) = \frac{1}{2\pi}\int_{-\infty}^{\infty} e^{ikx-|k|y}dk. \tag{6.15}$$

Clearly this is a convolution form for the solution, and is written in terms of Green's function, G, which is actually the solution of the problem posed above with $f(x) = \delta(x)$, since then $F = 1$ in (6.13). Note that the integral in (6.15) can be done by standard methods in this case, without resort to the complex plane,

$$G(x) = \frac{1}{2\pi} \int_{-\infty}^{\infty} \cos(kx)e^{-|k|y}dk = \frac{1}{\pi}\Re\text{eal}\left\{ \int_{0}^{\infty} e^{ikx-ky}dk \right\} = \frac{1}{\pi}\frac{y}{x^2 + y^2}.$$

Hence, the solution (6.14) is simply

$$T = \frac{y}{\pi} \int_{-\infty}^{\infty} \frac{f(\xi)d\xi}{(x - \xi)^2 + y^2}.$$

Actually, the same procedure for arbitrary f used in this example can be used in Example 6.2, and the result is that (6.14) is the form of solution for Example 6.2 as well, provided we take $G(x)$ to be given by (6.12).

6.3.4 Example 4: Sound Waves

Example 6.4. We consider a hyperbolic problem here, namely, the sound waves that result from the small-amplitude oscillation, normal to itself, of a portions of a wall. We seek the time-periodic solution only, after any transients have settled out, so the vertical velocity, say v, can be written as $v = \Re\text{eal}(V\exp(i\omega t))$, which results in the Helmholtz equation,

$$\nabla^2 V + \lambda^2 V = 0, \quad \text{in} \quad y > 0,$$

and the boundary condition at the wall is

$$V = f(x), \quad \text{at} \quad y = 0.$$

We have written here λ for a/ω, if a is the sound speed. For example, if a portion of the wall oscillates, and the remainder is at rest, then f would be, $f = 0$, $|x| > L$, $f = 1$, $|x| < L$. We take the Fourier transform in x, and that results in the ordinary differential equation for $\Theta \equiv \mathcal{F}[V]$,

$$\frac{\partial^2\Theta}{\partial y^2} + (\lambda^2 - k^2)\Theta = 0.$$

The solution takes two apparently different forms, depending upon the relative sizes of k and λ. Since the solution must be bounded for $y \to \infty$, if F is the Fourier transform of f, then the solutions are easily seen to be

$$\Theta = F(k)e^{-\sqrt{k^2-\lambda^2}y}, \quad \text{for} \quad |k| > \lambda. \tag{6.16}$$

For $|k| < \lambda$, both solutions to the differential equation are permissible, since they oscillate at infinity, not growing or decaying. Hence, we have a dilemma: which solution to keep, suppose we choose to keep the solution given in (6.16). In that case, the square root in the exponential can be given by

$$(k^2 - \lambda^2)^{1/2} = \sqrt{r_1 r_2}e^{i(\theta_1+\theta_2)/2},$$

where we have written $k - \lambda = r_1 e^{i\theta_1}$ and $k + \lambda = r_2 e^{i\theta_2}$ in the usual way. Suppose we cut the plane in a way shown in Figure 6.1. Then, in the region between

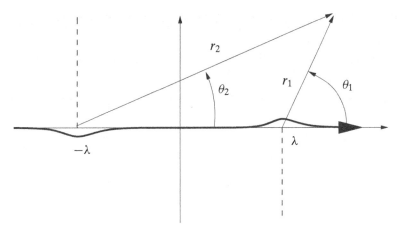

Figure 6.1. Inversion path for Example 6.4. (Dashed lines show the branch cuts.)

the branch points, $(k^2 - \lambda^2)^{1/2}$ is given by $|k^2 - \lambda^2|^{1/2} \times (+i)$. Denote that quantity, for now, simply as Ki. Then, when we come to do the Fourier inversion, the integrand will contain a term like

$$e^{i\omega t - iKy},$$

which corresponds to a wave propagating *outward* – which is what we want, using a *radiation condition*. Had we chosen to cut the plane in another way – with the right-hand cut upward, and the left one downward, then function (6.16) would give a term like $\exp(i\omega t + iKy)$ in the Fourier inversion integral, which corresponds to a wave propagating *inward* form infinity – a physical impossibility!! Therefore, we conclude that we can use function (6.16) for all k, *provided* that the plane is cut as shown. It appears that the choice of branch cuts is crucial in the satisfaction of the radiation condition!

Hence, the solution is given by (6.16) for all k in the cut plane, and using the inverse from (6.4), we have the solution in the form

$$V = \int_{-\infty}^{\infty} f(\xi) G(x - \xi, y) d\xi, \quad \text{and}$$

$$G(x, y) = \frac{1}{2\pi} \int_{-\infty}^{\infty} e^{ikx - \sqrt{k^2 - \lambda^2} y} dk. \tag{6.17}$$

This integral remains difficult to evaluate; we will explore its behavior at large x and y, subsequently. Returning to the physical result by taking the real part of $V \exp(i\omega t)$ as required, we obtain, the Green's function result,

$$v = \int_{-\infty}^{\infty} f(\xi) \hat{G}(x - \xi, y, t) d\xi,$$

$$\hat{G}(x, y, t) = \frac{1}{\pi} \cos(\omega t) \int_{\lambda}^{\infty} \cos(kx) e^{-\sqrt{k^2 - \lambda^2} y} dk$$

$$+ \frac{1}{\pi} \int_{-\lambda}^{\lambda} \cos(\omega t + kx - \sqrt{\lambda^2 - k^2} y) dk. \tag{6.18}$$

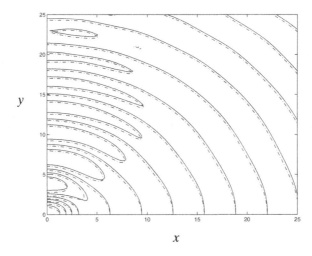

Figure 6.2. Contours of constant-V values for the wavelike part of (6.18), for $\lambda = 1$, at values of $\omega t = \pi$, $- - -$, and $\omega t = \pi + 0.2$, —.

Clearly the second term represents the wavelike portion of the solution. It turns out, for later purposes, that (6.17) is sometimes more useful than this result.

The second term in equation (6.18) is, as noted, wavelike. If we plot contours of constant values of the integral at successive times, as shown in Figure 6.2, we can see the propagation of the waves in the solution. (The integral has been done numerically.)

6.3.5 Example 5: Diffusion in a Force Field

Example 6.5. The theory of random walks leads to a description of the process of diffusion. The basic equation that results is the heat equation. However, if a force field is present, a more general parabolic equation is applicable. We examine the case where there is a one-dimensional harmonic force field. The resulting equation for the probability density $u(x, t)$ of such an elastically bound particle is given by (Lin and Segel)

$$\frac{\partial u}{\partial t} = D\frac{\partial^2 u}{\partial x^2} + \gamma\frac{\partial}{\partial x}(xu), \quad -\infty < x < \infty, \; t > 0. \tag{6.19}$$

We assume an initial condition

$$u(x, 0) = f(x), \quad -\infty < x < \infty.$$

In the usual case we want an expression for u in the form

$$u(x, t) = \int_{-\infty}^{\infty} G(x - \xi; t) f(\xi)d\xi.$$

Since the equation does not have constant coefficients, we cannot be sure that G depends only on $x - \xi$ in space. However, we may take a Fourier transform in x to lower the order of the differential equation.

Define

$$\hat{u}(k,t) = \int_{-\infty}^{\infty} u(x,t)e^{-ikx}dx.$$

Perform this on each term in (6.19), and use integration by parts, to obtain

$$\frac{\partial \hat{u}}{\partial t} = -k^2 D\hat{u} + ik\gamma \int_{-\infty}^{\infty} xu(x,t)e^{-ikx}dx. \qquad (6.20)$$

The last term in (6.20) is related to the derivative of $\hat{u}$ with respect to k, so we obtain

$$\frac{\partial \hat{u}}{\partial t} = -k^2 D\hat{u} - k\gamma \frac{\partial \hat{u}}{\partial k}. \qquad (6.21)$$

This is a first-order linear equation which is probably most easily solved by the method of characteristics. The general method works well for any quasilinear first-order equation, (Garabedian), and here it gives the ordinary differential equations

$$\frac{dt}{1} = \frac{dk}{k\gamma} = -\frac{d\hat{u}}{k^2 D\hat{u}}.$$

Solving two of these equations, we have

$$ke^{-\gamma t} = c_1 \quad \text{and} \quad \hat{u}e^{k^2 D/2\gamma} = c_2$$

as two first integrals of these equations. A general solution would be

$$\Phi(c_1, c_2) = \Phi(ke^{-\gamma t}, \hat{u}e^{k^2 D/2\gamma}) = 0,$$

where Φ is arbitrary. However, because there are only two arguments in Φ, we can simply write

$$\hat{u}e^{k^2 D/2\gamma} = \Phi(ke^{-\gamma t}),$$

with no loss of generality. The transform of the initial condition gives $\hat{u}(k,0) = \hat{f}(k)$. Consequently,

$$\Phi(k) = \hat{f}(k)e^{k^2 D/2\gamma}$$

and

$$\hat{u}(k,t) = e^{-k^2 D/2\gamma} \Phi(ke^{-\gamma t}) = e^{-k^2 D(1-e^{-2\gamma t})/2\gamma} \hat{f}(ke^{-\gamma t}).$$

We then find the inverse transforms

$$\mathcal{F}^{-1}\{e^{-k^2(1-e^{-2\gamma t})D/2\gamma}\} = \sqrt{\frac{\gamma}{2\pi D(1-e^{-2\gamma t})}}e^{-\gamma x^2/(2D(1-e^{-2\gamma t}))}$$

$$\equiv G(x,t) \qquad \text{and}$$

$$\mathcal{F}^{-1}\{\hat{f}(ke^{-\gamma t})\} = e^{\gamma t}f(xe^{\gamma t}).$$

By the convolution theorem, the solution to the initial-value problem is

$$u(x,t) = \int_{-\infty}^{\infty} G(x-\xi, t)e^{\gamma t}f(\xi e^{\gamma t})d\xi.$$

6.3.6 Example 6: An Integro-Differential Equation

Example 6.6. Consider the initial-boundary-value problem,

$$\frac{\partial \eta}{\partial t} + \frac{\partial}{\partial x}\left\{\fint_{-\infty}^{\infty} \frac{\eta(\xi, t)d\xi}{x - \xi}\right\} = 0, \tag{6.22}$$

$$\eta = e^{-x^2/4}, \text{ at } t = 0, \tag{6.23}$$

$$\eta \to 0 \text{ as } x \to \pm\infty. \tag{6.24}$$

The integral is taken as a principal value as defined in Chapter 1.

We analyze the problem by taking the Fourier transform of the integro-differential equation (6.22) under the conditions (6.24) to obtain

$$H_t + ik\mathcal{F}\left\{\fint_{-\infty}^{\infty} \frac{\eta(\xi, t)d\xi}{x - \xi}\right\} = 0, \tag{6.25}$$

where H is the Fourier transform of η and $\mathcal{F}$ denotes the Fourier transform of the linear operator.

For the latter term, under the necessary conditions for the convergence of the integrals (Porter and Stirling), the order of integrations may be reversed. The inner integral is now the principal-value integral,

$$I \equiv \fint_{-\infty}^{\infty} \frac{e^{-ikx}dx}{x - \xi}.$$

This integral may be done for $k > 0$ by using a semicircular integration path that passes underneath the pole at $x = \xi$ and closes in the lower half x-plane. In that case, Cauchy's theorem indicates that

$$I + i\pi \operatorname{Res}(x = \xi) = 0, \quad k > 0.$$

If $k < 0$ on the other hand, then the contour is deformed over the pole, and the semicircle closes in the upper-half x-plane, leading to

$$I - i\pi \operatorname{Res}(x = \xi) = 0.$$

So, we conclude that the integral in question is

$$I = -\operatorname{sgn}(k)i\pi \operatorname{Res}(x = \xi) = -i\pi \operatorname{sgn}(k)e^{-ik\xi}.$$

Then, the outer integral in the second term of (6.25) is

$$\int_{-\infty}^{\infty} (-i\pi \operatorname{sgn}(k))\, \eta(x, \xi)e^{-ik\xi}d\xi = -i\pi \operatorname{sgn}(k)H.$$

Then, (6.25) becomes

$$H_t + i\pi |k| H = 0. \tag{6.26}$$

To solve (6.26), first integrating to obtain its general solution,

$$H = Ce^{-i\pi|k|t},$$

and from the initial condition (6.23),

$$C = \mathcal{F}\{e^{-x^2/4}\} = 2\sqrt{\pi}e^{-k^2}.$$

Hence, the solution for the transform of η is

$$H = 2\sqrt{\pi}e^{-k^2-i\pi|k|t}.$$

The Fourier inversion can be turned into an integral from 0 to ∞ as

$$\eta = \frac{1}{\sqrt{\pi}}\int_{-\infty}^{\infty} e^{-k^2-i\pi|k|t}e^{ikx}dk = \frac{1}{\sqrt{\pi}}\int_{-\infty}^{\infty} e^{-k^2-i\pi|k|t}\cos(kx)dk$$

$$= \frac{2}{\sqrt{\pi}}\int_0^{\infty} e^{-k^2}\cos(\pi kt)\cos(kx)dk = \frac{1}{\sqrt{\pi}}\left[F(x-\pi t) + F(x+\pi t)\right],$$

where we have used a trigonometric identity, and

$$F(x) \equiv \int_0^{\infty} \cos(kx)e^{-k^2}dk,$$

which is done by standard means and is

$$F(x) = \frac{\sqrt{\pi}}{2}e^{-x^2/4}.$$

Hence, the solution is wavelike, with half of the energy propagating to the left and half propagating to the right.

6.3.7 Example 7: Thermal Wake in a Small-Prandtl-Number Fluid

Example 6.7. We consider the three-dimensional thermal wake, so we have the steady three-dimensional heat equation with a point source in a uniform stream. This spatially unbounded problem is

$$U\frac{\partial T^*}{\partial x^*} - \kappa \nabla^{*2} T^* = q\delta(x^*)\delta(y^*)\delta(z^*), \quad T^* \to T_\infty, \quad x^* \to -\infty.$$

For algebraic convenience, we rescale the physical variables, so $\mathbf{x} \equiv U\mathbf{x}^*/\kappa$ and $T^* = T_\infty + (q\kappa/U^2)T$. Then, the rescaled problem is

$$\frac{\partial T}{\partial x} - \nabla^2 T = \delta(x)\delta(y)\delta(z), \quad T \to 0, \quad x \to -\infty. \tag{6.27}$$

The approach here is to do Fourier transformation in both the y and z directions, so that we define

$$F(x, m, n) \equiv \int_{-\infty}^{\infty}\int_{-\infty}^{\infty} T(x, y, z)e^{-imy-inz}dydz.$$

On transforming (6.27), we obtain the following equation for F,

$$\frac{\partial F}{\partial x} - \frac{\partial^2 F}{\partial x^2} + k^2 F = q\delta(x), \quad k^2 \equiv m^2 + n^2. \tag{6.28}$$

Hence, F is actually a Green's function (see Chapter 4). The homogeneous solutions are of the form:

$$\exp\left(\left(-\frac{1}{2} \pm \sqrt{\frac{1}{4} + k^2}\right) x\right).$$

On working out the jump condition as discussed in Section 4.2.1, we find the solution of (6.28) to be

$$F = \begin{cases} \dfrac{e^{(1/2+\sqrt{1/4+k^2})x}}{\sqrt{1+4k^2}}, & x < 0 \\[3mm] \dfrac{e^{-(\sqrt{1/4+k^2}-1/2)x}}{\sqrt{1+4k^2}}, & x > 0. \end{cases} \tag{6.29}$$

Now, the inversion may be performed, so

$$T(x, y, z) = \frac{1}{4\pi^2} \int_{-\infty}^{\infty} \int_{-\infty}^{\infty} F(x, m, n)e^{imy+inz} dmdn.$$

Now, we notice that m and n enter F only through the magnitude of the wavenumber vector, k. Hence, we can convert the integrals over m and n to integrals over polar wave-number coordinates. So,

$$(m, n) = k(\cos\phi, \sin\phi).$$

In addition, it is natural to write the results in a circular polar spatial system, that is,

$$(y, z) = R(\cos\theta, \sin\theta).$$

Since the integration element transforms so that $dmdn = kdkd\phi$, we have

$$T = \frac{1}{4\pi^2} \int_0^{\infty} \int_0^{2\pi} \frac{e^{(1/2-\sqrt{1/4+k^2})x}}{\sqrt{1+4k^2}} e^{ikR\cos(\phi-\theta)} kd\phi dk, \quad x > 0.$$

The inner integral, over ϕ, is well-known and may be found in Chapter 2. It is

$$\int_0^{2\pi} e^{ikR\cos(\phi-\theta)} d\phi = 2\pi J_0(kR),$$

where $J_0(z)$ is the Bessel function of first kind and zero order. Thus, we have the solution in terms of a single integral, namely,

$$T = \frac{1}{2\pi} \int_0^{\infty} \frac{J_0(kR)e^{(-\sqrt{1/4+k^2}+1/2)x} kdk}{\sqrt{1+4k^2}} dk, \quad x > 0. \tag{6.30}$$

The form of (6.30) is the inversion of the Hankel transform, which may be used directly on partial differential equations with Laplacians, if there is axial symmetry in the problem. In fact, the Hankel transform of order ν is defined by

$$\mathcal{H}\{f\} = F_\nu(k) \equiv \int_0^{\infty} f(r)J_\nu(kR)RdR, \tag{6.31}$$

$$f(R) = \int_0^{\infty} F_\nu(k)J_\nu(kR)kdk. \tag{6.32}$$

Specifically, in a separation-of-variables procedure on the Laplace equation, the operator that arises is

$$L\phi \equiv \frac{1}{R}\frac{d}{dR}\left(R\frac{d\phi}{dR}\right) - \frac{v^2}{R^2}\phi.$$

For instance, written in cylindrical coordinates (6.27) becomes

$$\frac{\partial T}{\partial x} - \left(\frac{\partial^2 T}{\partial x^2} + \frac{\partial^2 T}{\partial R^2} + \frac{1}{R}\frac{\partial T}{\partial R}\right) = \frac{\delta(x)\delta(R)}{R}, \quad T \to 0, \quad x \to -\infty.$$

The properties of Bessel functions – specifically using the Bessel equation – make it possible to show, by successive integration by parts, that

$$\mathcal{H}\{L\phi\} = -k^2\mathcal{H}\{\phi\}.$$

Indeed for further elucidation of these properties, see either (Antimirov et al.) and (Sneddon). Not surprisingly, extensive tables exist of Hankel transforms (e.g., (Bateman, Vol. II)). From that source, we may report that (6.30) has the closed form expression

$$T(R, x) = \frac{1}{4\pi\sqrt{R^2 + x^2}}e^{-\frac{1}{2}\sqrt{R^2+x^2}+\frac{1}{2}x}, \quad x > 0. \tag{6.33}$$

If in this solution, R is much smaller than x – that is, near the axis of symmetry, $\sqrt{R^2 + x^2} \approx x + (1/2)(R^2/x)$, leading then to

$$T \sim \frac{1}{4\pi x}e^{-\frac{R^2}{4x}}, \quad x \to \infty,$$

which is exponentially small outside the "wake."

6.3.8 Example 8: Fundamental Solution for Stokes Flow

Example 6.8. Viscous flow at zero Reynolds number is called "Stokes flow." If flow about a three-dimensional body is desired it is often necessary to perform an integral over the body to determine its drag or other properties. For this purpose, the fundamental solution ($\mathbf{G}, \mathbf{h}$), the *stokeslet*, consisting of a matrix and a vector, is of utmost importance. Dyadic notation is used throughout this discussion. It may be simply defined in nondimensional Cartesian coordinates by the system

$$\nabla^2\mathbf{G} - \nabla\mathbf{h} = -\delta(x)\delta(y)\delta(z)\mathbf{I},$$

$$\nabla \cdot \mathbf{G} = \mathbf{0}.$$

Here $\mathbf{I}$ is the identity. The derivation of this is illustrated by performing multiple Fourier transforms defining

$$\hat{\mathbf{G}}(k_1, k_2, k_3) \equiv \iiint_{-\infty}^{\infty}\mathbf{G}(x, y, z)e^{-ik_1x-ik_2y-ik_3z}dxdydz,$$

and similarly for **h**. The result is simply

$$- \|\mathbf{k}\|^2 \hat{\mathbf{G}} + i\mathbf{k}\hat{\mathbf{h}} = -\mathbf{I}, \tag{6.34}$$

where $\|\mathbf{k}\|^2 = (k_1^2 + k_2^2 + k_3^2)$, and

$$\mathbf{k} \cdot \hat{\mathbf{G}} = \mathbf{0}. \tag{6.35}$$

Using (6.35) in (6.34), we see that

$$\hat{\mathbf{h}} = \frac{i\mathbf{k}}{\|\mathbf{k}\|^2}.$$

Using this in (6.34), we may solve for $\hat{\mathbf{G}}$ to obtain

$$\hat{\mathbf{G}} = \frac{\mathbf{I} - \mathbf{k}\mathbf{k}/\|\mathbf{k}\|^2}{\|\mathbf{k}\|^2}.$$

To find the desired functions we perform inverse Fourier transforms. For instance,

$$\frac{1}{(2\pi)^3} \iiint_{-\infty}^{\infty} \frac{1}{\|\mathbf{k}\|^2} e^{ik_1 x + ik_2 y + ik_3 z} dk_1 dk_2 dk_3 = \frac{1}{4\pi \|\mathbf{x}\|},$$

$$\frac{1}{(2\pi)^3} \iiint_{-\infty}^{\infty} \frac{i\mathbf{k}}{\|\mathbf{k}\|^2} e^{ik_1 x + ik_2 y + ik_3 z} dk_1 dk_2 dk_3 = \frac{\mathbf{x}}{4\pi \|\mathbf{x}\|^3},$$

and

$$\frac{1}{(2\pi)^3} \iiint_{-\infty}^{\infty} \frac{\mathbf{k}\mathbf{k}}{\|\mathbf{k}\|^4} e^{ik_1 x + ik_2 y + ik_3 z} dk_1 dk_2 dk_3 = \frac{\mathbf{I} - \mathbf{x}\mathbf{x}/\|\mathbf{x}\|^2}{8\pi \|\mathbf{x}\|}$$

are well-known inverses determining radial functions (Folland; Ladyzhenskaya). These lead to the representation for the stokeslet at the origin

$$\mathbf{G} = \frac{\mathbf{I} + \mathbf{x}\mathbf{x}/\|\mathbf{x}\|^2}{8\pi \|\mathbf{x}\|},$$

$$\mathbf{h} = \frac{\mathbf{x}}{4\pi \|\mathbf{x}\|^3}.$$

The Fourier transform technique is particularly useful in calculating the stokeslet in the presence of a plane boundary (Blake).

6.4 Mellin Transforms

Consider a function $f(x)$ defined on $[0, \infty)$. Its *Mellin transform* is defined as

$$\mathcal{F}(s) = \int_0^{\infty} x^{s-1} f(x) dx, \tag{6.36}$$

where s is in general, a complex parameter. The transform arises naturally in physical problems in polar coordinates. Mathematically, it is easily derived using the bilateral Laplace transform. Its inverse is most easily justified by the Fourier integral theorem (6.2). For instance, suppose $g(t)$ has a bilateral Laplace transform for $a < \Re\text{eal}(s) < b$ as in Section 5.4. Then making the change of variables $x = e^{-t}$, we have

$$G_2(s) = \int_{-\infty}^{\infty} e^{-st} g(t) dt = \int_{0}^{\infty} x^s g(-\log x) \frac{dx}{x}.$$

Identify $f(x) = g(-\log x)$ in which case $\mathcal{F}(s) = G_2(s)$. Also of particular interest is the form that the inverse takes. That is, since by Theorem 5.7

$$g(t) = \lim_{T \to \infty} \frac{1}{2\pi i} \int_{c-iT}^{c+iT} G_2(s) e^{st} ds,$$

we have

$$f(x) = \lim_{T \to \infty} \frac{1}{2\pi i} \int_{c-iT}^{c+iT} \mathcal{F}(s) x^{-s} ds. \tag{6.37}$$

This inverse may also connect with the Fourier transform as follows. Suppose $s = c + ik$, then, using $x = e^{-u}$ in (6.36), the inverse (6.37) becomes

$$\frac{1}{2\pi i} \int_{c-i\infty}^{c+i\infty} \mathcal{F}(s) x^{-s} ds = \frac{1}{2\pi} \int_{-\infty}^{\infty} e^{l(c+ik)} dk \int_{-\infty}^{\infty} e^{-u(c+ik)} f(e^{-u}) du$$

$$= \frac{e^{ct}}{2\pi} \int_{-\infty}^{\infty} dk \int_{-\infty}^{\infty} e^{ik(t-u)} e^{-cu} f(e^{-u}) du.$$

By the Fourier integral theorem, the last double integral is actually

$$e^{ct} e^{-ct} f(e^{-t}) = f(x).$$

Example 6.9. Let

$$f(x) = \begin{cases} 1, & 0 < x < 1 \\ 0, & x > 1 \end{cases},$$

then, its Mellin transform is

$$\mathcal{F}(s) = \int_{0}^{1} x^{s-1} dx = \frac{1}{s}, \quad \Re\text{eal}(s) > 0.$$

And for the inverse, we know that

$$f(x) = \frac{1}{2\pi i} \int_{c-i\infty}^{c+i\infty} \frac{x^{-s}}{s} ds, \quad c > 0.$$

Moreover, since f is discontinuous at $x = 1$, by the Fourier integral theorem (6.1), the inverse must have the value

$$\frac{f(1^+) + f(1^-)}{2} = \frac{1}{2}, \quad \text{at } x = 1.$$

More generally, if

$$f(x) = \begin{cases} x^a, & 0 < x < 1 \\ 0, & x > 1 \end{cases}, \quad \Re\mathrm{eal}(a) > 0,$$

then

$$\mathcal{F}(s) = \frac{1}{s+a}, \quad \Re\mathrm{eal}(s) > -\Re\mathrm{eal}(a).$$

Example 6.10. Another important Mellin transform is for $f(x) = e^{-x}$, whose Mellin transform is simply

$$\int_0^\infty x^{s-1} e^{-x} dx = \Gamma(s), \quad \Re\mathrm{eal}(s) > 0.$$

The inverse is thereby

$$e^{-x} = \frac{1}{2\pi i} \int_{c-i\infty}^{c+i\infty} x^{-s} \Gamma(s) ds, \quad c > 0.$$

Example 6.11. Next consider $f(x) = 1/(x+1)^p$. Then, its Mellin transform is given by

$$\int_0^\infty \frac{x^{s-1} dx}{(x+1)^p}, \tag{6.38}$$

and for this integral to converge, we must assume that $0 < \Re\mathrm{eal}(s) < \Re\mathrm{eal}(p)$, which is a strip in the complex s-plane. Proceeding with the analysis of (6.38), we observe that by referring to Example 2.1 in Chapter 2,

$$\frac{\Gamma(p)\Gamma(q)}{\Gamma(p+q)} = \int_0^\infty \frac{u^{q-1} du}{(1+u)^{p+q}} = \mathrm{B}(p,q),$$

and, hence,

$$M\{(x+1)^{-p}\} = \mathrm{B}(p-s,s), \quad 0 < \Re\mathrm{eal}(s) < \Re\mathrm{eal}(p).$$

6.4.1 Properties

Certain properties of Mellin transforms are useful to know. Suppose that $M[f(x)] = \mathcal{F}(s)$.

Property (i):

$$M[f(ax)] = a^{-s} \mathcal{F}(s), \ a > 0.$$

Property (ii):

$$\frac{d}{ds} \mathcal{F}(s) = M[\ln x f(x)].$$

This is true since within the strip of convergence,

$$\frac{d}{ds}\mathcal{F}(s) = \frac{d}{ds}\int_0^\infty x^{s-1}f(x)dx = \frac{d}{ds}\int_0^\infty e^{(s-1)\ln x}f(x)dx$$

$$= \int_0^\infty \ln x e^{(s-1)\ln x}f(x)dx = \int_0^\infty x^{s-1}\ln x f(x)dx$$

Property (iii):

$$M[xf'(x)] = -s\mathcal{F}(s).$$

We see that

$$M[xf'(x)] = \int_0^\infty x^{s-1}xf'(x)dx = \int_0^\infty x^s f'(x)dx$$

$$= [x^s f(x)]_0^\infty - s\int_0^\infty x^{s-1}f(x)dx.$$

Now, if $M[f(x)]$ exists, we must have

$$\lim_{x\downarrow 0}[x^s f(x)] = 0 \text{ and } \lim_{x\to\infty}[x^s f(x)] = 0.$$

Hence, the desired result follows.

Property (iv): In a similar way we may derive the identity

$$M\left[\int_0^x f(t)dt\right] = -s^{-1}\mathcal{F}(s+1),$$

assuming the left side exists. However, the existence of $\mathcal{F}(s)$ does not imply this. For instance, let

$$f(x) = \frac{x}{x+1}.$$

Then,

$$M[f] = \int_0^\infty \frac{x^{s-1}x}{x+1}dx = \int_0^\infty \frac{x^s dx}{x+1}$$

$$= B(s+1, -s), \quad -1 < \Re\mathrm{eal}(s) < 0.$$

However,

$$\int_0^x f(t)dt = \int_0^x \frac{tdt}{t+1} = x - \ln(x+1)$$

and

$$M[x - \ln(x+1)] = M[x] - \frac{\pi}{s\sin(\pi s)},$$

but $M[x] = \int_0^\infty x^s dx$ is infinite!

Parseval's Formula

With our knowledge of Parseval's formula for Fourier transforms, it is easy to envision that something analogous should hold for Mellin transforms (Cochran). Thus, if $\mathcal{F}(s) = M[f]$ and $\mathcal{G}(s) = M[g]$, then

$$M[f(x)g(x)] = \int_0^\infty x^{s-1} f(x)g(x)dx = \frac{1}{2\pi i}\int_{c-i\infty}^{c+i\infty} \mathcal{F}(s-\tau)\mathcal{G}(\tau)d\tau. \qquad (6.39)$$

The verification of this identity depends on existence of a vertical strip in the τ-plane, in which both $\mathcal{F}(s-\tau)$ and $\mathcal{G}(\tau)$ exist. Then, for $\Re\text{eal}(\tau) = c$,

$$\frac{1}{2\pi i}\int_{c-i\infty}^{c+i\infty} \mathcal{F}(s-\tau)\mathcal{G}(\tau)d\tau = \frac{1}{2\pi i}\int_{c-i\infty}^{c+i\infty} \mathcal{F}(s-\tau)d\tau \int_0^\infty x^{\tau-1}g(x)dx. \qquad (6.40)$$

Suppose that an interchange of the order of integration is permissible,

$$\frac{1}{2\pi i}\int_{c-i\infty}^{c+i\infty} \mathcal{F}(s-\tau)d\tau \int_0^\infty x^{\tau-1}g(x)dx = \frac{1}{2\pi i}\int_0^\infty g(x)dx \int_{c-i\infty}^{c+i\infty} x^{\tau-1}\mathcal{F}(s-\tau)d\tau.$$

Now, making use of the inversion formula (6.37), it follows that

$$\frac{1}{2\pi i}\int_0^\infty g(x)dx \int_{c-i\infty}^{c+i\infty} x^{\tau-1}\mathcal{F}(s-\tau)d\tau = \int_0^\infty g(x)dx \left[\frac{1}{2\pi i}\int_{c'-i\infty}^{c'+i\infty} x^{s-1-u}\mathcal{F}(u)du\right]$$

$$= \int_0^\infty g(x)x^{s-1}f(x)dx,$$

from which, with (6.40), (6.39) follows. An interesting application of (6.39) is to the case where $s = 1$ and $g = \bar{f}$. Then, it is possible to show that

$$\int_0^\infty |f(x)|^2 dx = \frac{1}{2\pi}\int_{-\infty}^\infty \left|\mathcal{F}\left(\frac{1}{2}+ik\right)\right|^2 dk, \qquad (6.41)$$

where the integral in k is along the real line for suitable f.

Problem 6.9. *(a) Complete the derivation of (6.41).*
(b) Verify this identity when

$$f(x) = \frac{1}{1+ix}.$$

Problem 6.10. *Let*

$$f(x) = \begin{cases} x^a, & 0 < x < 1 \\ x^{-a}, & x > 1 \end{cases}.$$

Show that

$$M[f] = \frac{2a}{s^2 - a^2}, \quad -\Re\text{eal}(a) < \Re\text{eal}(s) < \Re\text{eal}(a).$$

Problem 6.11. *Show that*

$$M[x(xf')'] = s^2\mathcal{F}(s).$$

6.5 Worked Examples

6.5.1 Stability of Flow Near a Stagnation Point

Example 6.12. For the problem of the linearized stability of flow near a plane stagnation point, there may be derived the disturbance equation for the streamwise velocity $u(r, z)$ at wave number zero. (The full stability problem is derived in (Brattkus and Davis).)

$$\frac{\partial^2 u}{\partial z^2} + z\frac{\partial u}{\partial z} - r\frac{\partial u}{\partial r} + u = \lambda u, \quad r > 0, \quad z > 0, \tag{6.42}$$

where

$$u(r, 0) = 0, \quad \int_0^\infty e^{z^2/2}|u(r, z)|^2 dz < \infty, \quad u(0, z) \text{ finite.} \tag{6.43}$$

We take a Mellin transform in r to obtain an ordinary boundary value problem, and analyze the spectrum of this problem. We then make use of the theory of Hermite functions to conclude that all eigenvalues satisfy $\Re\text{eal}(\lambda) < 0$. Define

$$\mathcal{U}(z, s) = \int_0^\infty r^{s-1} u(r, z) dr.$$

Based on Property (iii) of Section 6.4.1, we expect that a Mellin transform of (6.42) will be useful. Performing this calculation leads to

$$\frac{d^2\mathcal{U}}{dz^2} + z\frac{d\mathcal{U}}{dz} + s\mathcal{U} + \mathcal{U} = \lambda\mathcal{U}. \tag{6.44}$$

However, this depends on the existence of

$$\int_0^\infty r^{s-1} r\frac{\partial u}{\partial r}(r, z) dr = [r^s u]_0^\infty - \int_0^\infty sr^{s-1} u dr.$$

The boundary condition of the original problem at $r = 0$ requires that $\Re\text{eal}(s) > 0$, while allowing finiteness of u as $r \to \infty$, demands $\Re\text{eal}(s) < 1$. In the equation (6.44), for $\mathcal{U}$, let $\mathcal{Y}(z) = e^{z^2/4}\mathcal{U}$. Then,

$$\mathcal{Y}'' + \left(-\frac{z^2}{4} + s - \lambda + \frac{1}{2}\right)\mathcal{Y} = 0.$$

The solutions are parabolic cylinder functions $\mathcal{Y}_n = e^{-\frac{1}{4}z^2} He_n(z)$, in which $He_n(z)$ are Hermite polynomials (Abramowitz and Stegun). In order to satisfy the boundary condition (6.43) at $z = 0$, choose

$$\mathcal{Y}_n(z) = e^{-\frac{1}{4}z^2}\left[He_n(z) - He_n(-z)\right]. \tag{6.45}$$

However, since $He_n(z) = (-1)^n He_n(-z)$ the eigenfunctions are

$$\mathcal{Y}_n(z) = e^{-\frac{1}{4}z^2} He_{2n+1}(z),$$

and form a complete set with weight function $w(z) = e^{-z^2/2}$. That is,

$$\int_0^\infty w(z)\mathcal{Y}_n(z)\mathcal{Y}_m(z)dz = 0, \ m \neq n. \tag{6.46}$$

Furthermore, the spectrum is discrete so that

$$s - \lambda_n + \frac{1}{2} = 2n + \frac{3}{2}, \ n = 0, 1, \dots.$$

From this we are able to conclude that $\Re\mathrm{eal}(\lambda_n) = \Re\mathrm{eal}(s) - 2n - 1 < 0$, since, by necessity, $0 < \Re\mathrm{eal}(s) < 1$.

6.5.2 Stability of Jeffery–Hamel Flows

Example 6.13. The configuration is an expanding two-dimensional flow between planes from a source at the origin, In polar coordinates (r, θ) the planes lie along the rays $\theta = \pm\alpha$, for $r \geq 0$. The basic flow is described by a streamfunction $\Psi(\theta)$ only, and its description has a wonderful theory all its own. It is determined by solution of the differential equation

$$\Psi^{\mathrm{iv}}(\theta) + \Psi''(\theta)(4 + 2R\Psi(\theta)) = 0,$$

with the boundary conditions,

$$\Psi(\alpha) = 1, \Psi(-\alpha) = -1, \Psi'(\alpha) = \Psi'(-\alpha) = 0, \tag{6.47}$$

where R is the Reynolds number. For our purposes, we may say it determines coefficients in a linearized disturbance equation for the stream function $\psi(r, \theta, t)$

$$\frac{\partial \zeta}{\partial t} + \frac{\Psi'(\theta)}{r}\frac{\partial \zeta}{\partial r} + \frac{2\Psi''(\theta)}{r^4}\frac{\partial \psi}{\partial \theta} + \frac{\Psi'''(\theta)}{r^3}\frac{\partial \psi}{\partial r} = R^{-1}\Delta\zeta, \tag{6.48}$$

where $\zeta = -\Delta\psi$ and ψ satisfies the boundary conditions (6.47) in θ. Here, $\Delta \equiv \nabla^2$. For this flow, time-independent disturbances may lead to spatial instability typified by growth in r. This was first investigated by, W. R. Dean. (See (Drazin) for this exposition.) Our purpose is to see how, by means of Mellin transforms, the completeness of the modes may be determined. Traditionally, it is assumed that $\partial/\partial t = 0$, so that putting

$$\psi(r, \theta) = r^\lambda \phi(\theta), \tag{6.49}$$

into (6.48) leads to

$$\phi^{\mathrm{iv}} + \left[\lambda^2 + (\lambda - 2)^2\right]\phi'' + \lambda^2(\lambda - 2)^2\phi \tag{6.50}$$

$$= R\left[(\lambda - 2)^2\Psi'(\phi'' + \lambda^2\phi) - 2\Psi''\phi' - \lambda\Psi'''\phi\right],$$

with boundary conditions

$$\phi(\alpha) = \phi'(\alpha) = \phi(-\alpha) = \phi'(-\alpha) = 0. \tag{6.51}$$

A number of special solutions of (6.50) have been found. We specialize to the case of *Stokes flow*, where $R = 0$. From our knowledge of Mellin transforms, to justify (6.49), it suffices to set $\lambda = s - 1$. To carry out the spatial spectral analysis in general we would find the steady Green function for (6.48), but, in this case, we find the Green's function $G(r, \theta; r_0, \theta_0)$ satisfying

$$\Delta\Delta G = \frac{\delta(r - r_0)\delta(\theta - \theta_0)}{r}, \tag{6.52}$$

with boundary conditions (6.51). Carrying out the Mellin transform of (6.52) thus gives

$$\mathcal{G}^{\text{iv}} + \left[\lambda^2 + (\lambda - 2)^2\right]\mathcal{G}'' + \lambda^2(\lambda - 2)^2\mathcal{G} = r_0^{\lambda+3}\delta(\theta - \theta_0),$$

where $\mathcal{G}(\theta) = M[G]$. In order to explicitly construct $\mathcal{G}$, we make use of the two basic solutions

$$f_1(\theta) = \cos(\lambda\theta) - \cos((\lambda - 2)\theta) \text{ and } f_2(\theta) = (\lambda - 2)\sin(\lambda\theta) - \lambda\sin((\lambda - 2)\theta).$$

They satisfy boundary conditions $f_n(0) = f_n'(0) = 0, n = 1, 2$. In the manner discussed in Section 4.2.2, we let

$$u_n(\theta) = f_n(\theta + \alpha) \quad \text{and} \quad v_n(\theta) = f_n(\theta - \alpha), n = 1, 2.$$

The concomitants $P(u_j, v_i)$, $i, j = 1, 2$, which are constants, are determined by

$$P(u, v) = u'''v - uv''' - (u''v' - u'v'') + (\lambda^2 + (\lambda - 2)^2)(u'v - uv').$$

and may be evaluated at any convenient point, $-\alpha \leq \theta \leq \alpha$. We omit the detailed expressions, but the Green's function reads

$$\mathcal{G}(\theta, \theta_0; \alpha, \lambda)r_0^{-\lambda-3} = \begin{cases} -\mathbf{U}^T(\theta)\mathbf{P}^{-1}\mathbf{V}(\theta_0), & \theta < \theta_0, \\ -\mathbf{V}^T(\theta)\left(\mathbf{P}^T\right)^{-1}\mathbf{U}(\theta_0), & \theta > \theta_0, \end{cases} \tag{6.53}$$

where

$$\mathbf{U}(\theta) = \begin{pmatrix} u_1(\theta) \\ u_2(\theta) \end{pmatrix}, \quad \mathbf{V}(\theta) = \begin{pmatrix} v_1(\theta) \\ v_2(\theta) \end{pmatrix}.$$

An expansion in terms of the eigenfunctions of (6.50)–(6.51), with $R = 0$ is thereby possible, locating the eigenvalues at the points where $\det \mathbf{P}(\lambda) = 0$.

REFERENCES

M. Abramowitz and I. A. Stegun, eds., *Handbook of Mathematical Functions*, Dover, New York, 1972.

M. Ya. Antimirov, A. A. Kolyshkin, and R. Vaillancourt, *Applied Integral Transforms*, American Mathematical Society, Providence, RI, 1993.

G. A. Articolo, *Partial Differential Equations and Boundary Value Problems with Maple*, Academic Press, San Diego, 1998.

H. Bateman, *Tables of Integral Transforms*, Vols. I, II, McGraw-Hill, New York, 1954.

J. R. Blake, A Note on the Image System for a Stokeslet in a No-Slip Boundary, *Proc. Camb. Phil. Soc.*, **70**, 303–310 (1971).

R. N. Bracewell, Numerical Transforms, *Science* **248**, 697–704 (1990).

K. Brattkus and S. H. Davis, The Linear Stability of Plane Stagnation-Point Flow against General Disturbances, *Q. J. Mech. Appl. Math.* **44**, 135–146 (1991).

J. A. Cochran, *Applied Mathematics: Principles, Techniques and Applications,* Wadsworth, Belmont, CA, 1982.

J. Cooper, *Introduction to Partial Differential Equations with MATLAB*, Birkhäuser, Boston, 1998.

P. G. Drazin, *Introduction to Hydrodynamic Stability*, Cambridge University Press, Cambridge, UK, 2002.

G. B. Folland, *Fourier Analysis and Its Applications*, Brooks/Cole, Pacific Grove, CA, 1992.

P. R. Garabedian, *Partial Differential Equations,* Wiley, New York, 1964.

O. A. Ladyzhenskaya, *The Mathematical Theory of Viscous Incompressible Flow*, 2nd English edition, Gordon and Breach, New York, 1969.

C.-C. Lin and L. A. Segel, *Mathematics Applied to Deterministic Problems in the Natural Sciences*, SIAM, Philadelphia, 1988.

A. Mandelis, *Diffusion-Wave Fields*, Springer-Verlag, New York, 2001.

M. N. Özişik, *Boundary Value Problems in Heat Conduction*, Dover, New York, 1968.

J. Pedlosky, *Geophysical Fluid Dynamics,* Springer-Verlag, New York, 1979.

D. Porter and D. S. G. Stirling, *Integral Equations*, Cambridge University Press, Cambridge, UK, 1990.

Prandtl, L. *Essentials of Fluid Dynamics*, Blackie & Sons Limited, London, 1952.

I. N. Sneddon, *Fourier Transforms*, McGraw-Hill, New York, 1951.

E. C. Titchmarsh, *Introduction to the Theory of Fourier Integrals*, 2nd edition, Clarendon, Oxford, 1948.

G. N. Watson, *Theory of Bessel Functions*, 2nd edition, Cambridge University Press, Cambridge, UK, 1944.

EXERCISES

Some of the following exercises may be facilitated by the use of a computer algebra system (CAS). However, there is no substitute for a personal mastery of the techniques. Nevertheless, once the concepts are understood, the authors suggest that the reader refer to texts devoted to such expositions, namely, (Articolo) and (Cooper).

6.1 Solve the following ordinary differential equation with a Fourier transform

$$y'' - 4y = e^{-|x|},$$

and y goes to zero for $|x| \to \infty$. (You should end up summing the residues of two poles in the upper-half-plane.)

6.2 Obtain the solution to problem

$$\frac{d^2u}{dx^2} = -\mu v,$$

$$\frac{d^2v}{dx^2} = \mu u,$$

$0 < x < \infty$, where μ is constant, $u'(0) = a$, $v'(0) = b$, and as $u, v \to 0$, as $x \to \infty$. In this case, make use of the cosine transform and proceed as in Example 6.3.1.

This problem corresponds to the Ekman layer for flow near a flat surface.

6.3 Consider the initial boundary-value problem

$$\frac{\partial u}{\partial t} = v\frac{\partial^2 u}{\partial x^2}, \quad -\infty < x < \infty, \quad t > 0,$$

where $v > 0$,

$$u(x, 0) = e^{-x^2}.$$

Solve by taking a Fourier transform in x, and show that the solution is

$$u(x, t) = \frac{1}{\sqrt{1 + 4t/v}} e^{-x^2/(1+4t/v)}.$$

Note that this solution exists backward in time up until $t = -v/4$.

6.4 Consider the solution to the thermal-wave problem (Mandelis)

$$\frac{\partial T}{\partial t} = v\frac{\partial^2 T}{\partial x^2} + F_0 \beta e^{-\beta x} \cos(\omega t), \quad 0 < x < \infty,$$

with the boundary condition

$$\frac{\partial T}{\partial x}(0, t) = 0, \quad \text{for all } t.$$

Thus there is zero incident flux on the surface of a semiinfinite solid, and ω is the real frequency.

Solve this problem by taking a cosine transform in x, assuming that the solution is bounded as $x \to \infty$. Note that one may let the time dependency be $e^{i\omega t}$, look for a solution to the time-dependent equation that has the period $2\pi/\omega$, and then take the real part of the solution $T(x, t)$.

6.5 Heat conduction in a material strip $0 < y < h$ is described by the equation and boundary conditions

$$\frac{\partial^2 T}{\partial x^2} + \frac{\partial^2 T}{\partial y^2} = 0,$$

$$T_y = 0, \quad \text{at } y = 0,$$

$$\frac{\partial T}{\partial y} = -B(T - f(x)), \quad \text{at } y = h.$$

Compare this problem to Example 6.2.

(a) Show that the Fourier transform of the solution is

$$\mathcal{F}\{T\} = \frac{BF(k)\cosh(ky)}{B\cosh(kh) + k\sinh(kh)},$$

where $F(k)$ is the Fourier transform of $f(x)$.

(b) By interchanging the order of the integration, following the ideas discussed in Example 6.3, show that the solution is given by

$$T(x, y) = \int_{-\infty}^{\infty} f(\xi)G(x - \xi, y)\, d\xi,$$

where the Green's function is given by

$$G(x, y) = \frac{B}{2\pi} \int_{-\infty}^{\infty} \frac{e^{ikx} \cosh(ky)}{B \cosh(kh) + k \sinh(kh)} dk.$$

(c) Locate all poles, and show the solution for G is given by

$$G(x, y) = B \sum_{n=1}^{\infty} \frac{Bh \, e^{-r_n|x|}}{(Bh + 1)Bh + r_n^2} \frac{\cos(r_n y/h)}{\sin r_n},$$

where the quantities $\{r_n\}$ are solutions of the transcendental equation

$$\tan r = \frac{Bh}{r}.$$

6.6 Consider heat conduction in a strip of height 1, shown in Figure 6.3. The equation for the temperature is

$$\frac{\partial^2 T}{\partial x^2} + \frac{\partial^2 T}{\partial y^2} = 0.$$

and the boundary conditions are

$$T = 0, \quad \text{at} \quad y = 0, \qquad T_y = -\alpha(T - f(x)), \quad \text{at} \quad y = 1.$$

(a) Show that the Fourier transform of the solution is

$$\mathcal{F}(T) = \frac{F(k) \sinh(ky)}{\sinh k + \frac{k}{\alpha} \cosh k},$$

where $F(k)$ is the transform of $f(x)$.

(b) Show that there are poles on the imaginary axis at $k = i\phi$, where ϕ is a solution of the equation

$$\tan \phi = -\frac{\phi}{\alpha}.$$

(c) Sum the residues to obtain the solution in $x > 0$ if the forcing is a delta function (so that $F = 1$).

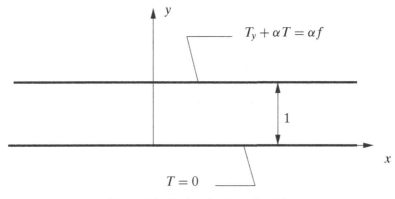

Figure 6.3. Region for Exercise 6.6.

6.7 Consider the following elliptic boundary-value problem:

$$\frac{\partial^2 u}{\partial x^2} + \frac{\partial^2 u}{\partial y^2} = u, \quad \text{in the strip} \quad -\infty < x < \infty, \ 0 < y < 1,$$

with boundary conditions

$$u = 0, \quad \text{at} \quad y = 0, \quad u = \begin{cases} 1, & |x| < L \\ 0, & |x| > L \end{cases}, \quad \text{at} \quad y = 1.$$

Solve the problem by Fourier transformation.

(a) Denote by $\mathcal{F}\{..\}$ the Fourier transform, then show that the boundary condition at $y = 1$ transforms to

$$\mathcal{F}\{u\} = \frac{2}{k} \sin(kL), \quad \text{at} \quad y = 1.$$

(b) Do a Fourier transform of the equation, and solving with the boundary conditions given, show that

$$\mathcal{F}\{u\} = \frac{2\sin(kL)}{k} \frac{\sinh(\sqrt{1+k^2}\, y)}{\sinh(\sqrt{1+k^2})}.$$

(c) Use the Fourier inversion integral, and show that the solution can be put into the form

$$u = A(x+L, y) - A(x-L, y),$$

where

$$A(x, y) = \frac{1}{2\pi i} \int_{-\infty}^{\infty} e^{ikx} \frac{\sinh(\sqrt{1+k^2}\,y)}{\sinh(\sqrt{1+k^2})} \frac{dk}{k}.$$

(*Note*: This splitting of the solution is necessary in order to obtain an integral, A, for which the semicircle integral of the associated contour integral vanishes for $R \to \infty$.)

(d) Use the residue theorem to work out the function $A(x, y)$ and show finally that

$$u(x, y) = \sum_{n=1}^{\infty} \frac{(-1)^n n\pi}{n^2\pi^2 + 1} g_n(x, y)$$

$$g_n(x, y) \equiv \sin(n\pi y)\left[e^{-\sqrt{1+n^2\pi^2}|x-L|} - e^{-\sqrt{1+n^2\pi^2}|x+L|}\right]. \quad (6.54)$$

6.8 Mountain and valley winds in stratified air may be modeled as in (Prandtl). (See Figure 6.4.) Consider laminar flow along a constant sloping surface for the temperature perturbation Θ and velocity w parallel to the wall satisfies the system of differential equations

$$\nu \frac{d^2 w}{dn^2} + g\sin(\alpha)\beta\Theta = 0 \quad (6.55)$$

and

$$a\frac{d^2\Theta}{dn^2} = wB\sin(\alpha).$$

Here α is the angle of inclination of the wall and n is the distance normal to the wall. The other constants are: g, the acceleration due to gravity; B, the ambient temperature gradient; ν, the kinematic viscosity; a, the thermal diffusivity; and β, the expansion coefficient. Suitable boundary conditions are

$$w(0) = 0, \ \Theta(0) = C; \ w, \Theta \to 0 \text{ as } n \to \infty.$$

Use the Fourier sine transform to show that the solution of the system is

$$\Theta = Ce^{-n/l}\cos\left(\frac{n}{l}\right),$$

$$w = C\left(\frac{g\beta a}{\nu B}\right)^{1/2} e^{-n/l}\sin\left(\frac{n}{l}\right),$$

where

$$l = \left(\frac{4\nu a}{g\beta B\sin^2\alpha}\right)^{1/4}$$

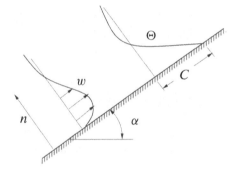

Figure 6.4. Buoyancy layer geometry.

6.9 Some partial differential equations that are not separable may still be solved by transform methods.

(a) Show that the biharmonic equation

$$\nabla^4 u = \frac{\partial^4 u}{\partial x^4} + 2\frac{\partial^4 u}{\partial x^2 \partial y^2} + \frac{\partial^4 u}{\partial y^4} = 0$$

is not separable.

(b) Consider the boundary-value problem:

$$\nabla^4 u = 0, \quad -\infty < x < \infty, \ y > 0,$$

$$u(x, 0) = \varphi(x), \quad \frac{\partial u}{\partial y}(x, 0) = \psi(x), \quad -\infty < x < \infty.$$

Use Fourier transforms in x to establish the formula for the solution

$$u(x, y) = \frac{1}{\pi} \int_{-\infty}^{\infty} \frac{2y^3 \varphi(\xi) d\xi}{((x - \xi)^2 + y^2)^2} + \frac{1}{\pi} \int_{-\infty}^{\infty} \frac{y^2 \psi(\xi) d\xi}{(x - \xi)^2 + y^2}.$$

6.10 The **Hilbert transform** of a function $f(x)$ defined on $(-\infty, \infty)$ is

$$F_{\text{Hi}}(\xi) = \frac{1}{\pi} \int_{-\infty}^{\infty} \frac{f(x) dx}{x - \xi}.$$

Its inverse is defined as

$$f(x) = \frac{1}{\pi} \int_{-\infty}^{\infty} \frac{F_{\text{Hi}}(\xi) d\xi}{x - \xi}.$$

Recognizing that these expressions are convolutions, use the Fourier transform to verify the relationship of this pair. *Hint*: Observe the method of solution to Example 6.6.

6.11 Radially symmetric heating. Use the Hankel transform of order zero (6.31), to solve the initial-boundary problem

$$\frac{\partial u}{\partial t} = \nu \left(\frac{\partial^2 u}{\partial r^2} + \frac{1}{r} \frac{\partial u}{\partial r} \right), \quad 0 < r, t < \infty,$$

$$u(r, 0) = \varphi(r), \ u(0, t) \text{ finite}, \ u \to 0 \text{ as } r \to \infty.$$

Show that

$$u(r, t) = \frac{1}{2\nu t} \int_0^{\infty} \xi \varphi(\xi) e^{-(\xi^2 + r^2)/4\nu t} I_0 \left(\frac{\xi r}{2\nu t} \right) d\xi, \quad t > 0,$$

where I_0 is the modified Bessel function of order 0. *Hint*: One of the interesting identities with Bessel functions is (Watson, p. 395)

$$\int_0^{\infty} e^{-p^2 t^2} J_\nu(at) J_\nu(bt) t \, dt = \frac{1}{2p^2} e^{-(a^2 + b^2)/4p^2} I_\nu \left(\frac{ab}{2p^2} \right),$$

which is valid if $\Re\text{eal}(\nu) > -1$ and $|\arg(p)| < \frac{1}{4}\pi$.

6.12 Consider the problem of solving the two-dimensional circularly symmetric wave equation in a medium with variable sound speed

$$\frac{\partial^2 \phi}{\partial t^2} = c^2 r^2 \left(\frac{\partial^2 \phi}{\partial r^2} + \frac{1}{r} \frac{\partial \phi}{\partial r} \right), \quad r > 0, \ t > 0,$$

under initial conditions

$$\phi(r, 0) = \begin{cases} p_0, & 0 \leq r \leq 1 \\ 0, & r > 1 \end{cases},$$

$$\frac{\partial \phi}{\partial t}(r, 0) = 0, \quad r > 0.$$

Here, p_0 is a constant. Use a Mellin transform in r to solve this problem.

(a) Show that

$$M[\phi(r, t)] = \frac{p_0}{s} \cosh(cst), \quad \Re\text{eal}(s) > 0.$$

(b) Evaluate the inverse Mellin transform to find that

$$\phi(r, t) = \frac{p_0}{4\pi i} \int_{\sigma - i\infty}^{\sigma + i\infty} \left[\frac{e^{cst} e^{-s \ln r} + e^{-cst} e^{-s \ln r}}{s} \right] ds$$

$$= \frac{p_0}{2} [H(ct - \ln r) + H(-ct - \ln r)].$$

(c) Sketch the regions in the (r, t) quarter-plane over which the signal is zero and nonzero.

6.13 The **Abel transform** of a function $f(x)$ defined on $[0, \infty)$ is given by

$$F_A(x) = \int_x^\infty \frac{2r f(r)}{\sqrt{r^2 - x^2}} dr. \tag{6.56}$$

The inverse transform is

$$f(r) = -\frac{1}{\pi} \int_r^\infty \frac{F_A'(x)}{\sqrt{x^2 - r^2}} dx. \tag{6.57}$$

Verify this by means of the Mellin transform applied to (6.56).

Method:

(a) Set $F_A(x) = g(x)$ and

$$K\left(\frac{x}{r}\right) = \begin{cases} 2/\sqrt{1 - \frac{x^2}{r^2}}, & x < r, \ \frac{x}{r} < 1 \\ \\ 0, & x > r, \ \frac{x}{r} > 1. \end{cases}$$

Show by taking Mellin transforms of both sides of (6.56) with $M[g] = \mathcal{G}(s)$, $M[f] = \mathcal{F}(s)$, $M[K] = \mathcal{K}(s)$, that

$$\mathcal{G}(s) = \mathcal{K}(s)\mathcal{F}(1 + s), \tag{6.58}$$

and

$$\mathcal{K}(s) = 2 \int_0^1 \frac{x^{s-1}}{\sqrt{1 - x^2}} dx.$$

(b) By the change of variables $t = x^2$ show that

$$\mathcal{K}(s) = B\left(\frac{s}{2}, \frac{1}{2}\right) = \frac{\Gamma\left(\frac{s}{2}\right)\Gamma\left(\frac{1}{2}\right)}{\Gamma\left(\frac{s+1}{2}\right)}. \tag{6.59}$$

However, we need $\mathcal{F}(s)$ so that from (6.58)

$$\mathcal{F}(s) = \mathcal{G}(s-1)/\mathcal{K}(s-1).$$

Show from (6.59) that

$$\frac{1}{\mathcal{K}(s-1)} = \frac{(s-1)}{2\pi}\mathrm{B}\left(\frac{s}{2},\frac{1}{2}\right) = \frac{(s-1)\mathcal{K}(s)}{2\pi},$$

and that

$$\mathcal{F}(s) = \frac{1}{2\pi}(s-1)\mathcal{G}(s-1)\mathcal{K}(s).$$

(c) Then taking the inverse and using Property (iii) of Section 6.4.1, show that

$$f(r) = -\frac{1}{2\pi}\int_0^\infty g'(x)\frac{1}{x}K\left(\frac{r}{x}\right)dx$$

$$= -\frac{1}{\pi}\int_r^\infty \frac{g'(x)}{\sqrt{x^2-r^2}}dx.$$

The transform pair (6.56)–(6.57) may also be derived by Laplace transforms. It is intimately related to the Hankel and double Fourier transforms (Bracewell).

6.14 Consider the two-dimensional counterpart of Example 6.7, the thermal wake modeled by

$$U\frac{\partial T}{\partial x} - \kappa\nabla^2 T = q\delta(x)\delta(y), \tag{6.60}$$

where κ is the thermal conductivity of the fluid and q is a measure of the heating due to the object in the stream.

(a) Perform a Fourier transform in y on (6.60), which results in

$$U\frac{\partial\mathcal{F}[T]}{\partial x} - \kappa\left(\frac{\partial^2\mathcal{F}[T]}{\partial x^2} - k^2\mathcal{F}[T]\right) = q\delta(x). \tag{6.61}$$

(b) Solve this ordinary differential equation by taking a Fourier transform in x and inverting to obtain

$$\mathcal{F}[T] = \frac{q}{2\kappa R}e^{px-R|x|}, \quad p \equiv \frac{U}{2\kappa}, \quad \text{and} \quad R \equiv \sqrt{k^2+p^2}.$$

Therefore, using the Fourier inversion,

$$T = \frac{q}{2\pi\kappa}e^{px}\int_{-\infty}^\infty \frac{e^{iky-R|x|}dk}{R}.$$

(c) Show that this integral can be expressed in terms of the modified Bessel function K_0 as

$$T(x,y) = \frac{q}{\pi\kappa}e^{px}K_0(pr), \quad \text{where} \quad r = \sqrt{x^2+y^2}.$$

6.15 The equations for velocity perturbations in the far wake of an object embedded in a incompressible fluid flowing at speed U are

$$U\frac{\partial u}{\partial x} + \rho^{-1}\frac{\partial p}{\partial x} = \nu\nabla^2 u - \frac{D}{\rho}\delta(x)\delta(y),$$

$$U\frac{\partial v}{\partial x} + \rho^{-1}\frac{\partial p}{\partial y} = \nu\nabla^2 v,$$

$$\frac{\partial u}{\partial x} + \frac{\partial v}{\partial y} = 0,$$

where D is the object's drag, ρ is the density, and $\mu = \rho\nu$ is the fluid viscosity. By writing the velocity in terms of the streamfunction, so $u = \partial\psi/\partial y$, and $v = -\partial\psi/\partial x$, show that the equations reduce to the equation pair

$$U\frac{\partial\zeta}{\partial x} - \nu\nabla^2\zeta = \frac{D}{\rho}\delta(x)\delta'(y),$$

$$\zeta = -\nabla^2\psi.$$

By doing a Fourier transform in x and solving the first equation for $\mathcal{F}(\zeta)$ in y, then obtaining the solution to the second equation for $\mathcal{F}(\psi)$. Show that eventually the solution for u comes down to a single integral,

$$\frac{u}{U} = -\frac{D}{4\pi\mu i\lambda}\int_{-\infty}^{\infty}\frac{\sqrt{k^2 + ik\lambda}}{k}e^{ikx - |y|\sqrt{k^2 + ik\lambda}}\,dk, \quad \lambda \equiv \frac{\rho U}{\mu}.$$

We examine the asymptotic behavior of this solution in Section 8.6.5.

Particular Physical Problems

7.1 Preamble

For us, the material presented in Chapters 1 through 6 of this book is a preamble: the material allows us to solve problems that arise in the analysis of physical problems, which so often end in partial differential equations. It is true that the analysis of Fourier transforms, for example, or eigenfunction expansions are, in themselves, interesting mathematical pursuits. However, our reason for studying what has come before this chapter, and the motivation for most applied mathematicians, is pragmatic: We are thereby enabled to approach solutions to those very physical problems we wish to solve.

With the advent and now exploding use of tools, such as RANS, DNS, and LES for solving the Navier–Stokes equations numerically, one might presume that the methodologies of this book are out of date. However, it has been our experience in our years of fluid dynamics research that the cross-fertilization of numerics and analysis, functioning synergistically alongside each other, provides insights into physical problems that are not available from either one standing alone. So, this chapter and Chapter 8 present some relatively simple, but real-world problems, that use more than one method from the previous chapters.

7.2 Lee Waves

We now turn to a problem that is important in atmospheric flows, namely the standing gravity waves downstream of mountains, known as Lee waves. Early work on Lee waves may be found in (Janowitz) and (Miles), for example; an excellent, recent summary of the topic is in (Wurtele, et al). The equation that describes such weak waves is

$$\left(\frac{\partial}{\partial t} + U\frac{\partial}{\partial x}\right)^2 \nabla^2\psi + N^2\frac{\partial^2\psi}{\partial x^2} = 0, \tag{7.1}$$

where ψ is the stream function, and, to account for waves due to some disturbance, say a mountain whose shape is given by $y = f(x)$ (See Figure 7.1); the quantity N

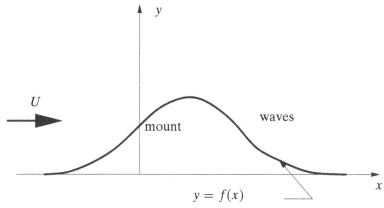

Figure 7.1. Lee wave geometry.

is the Brunt frequency, and U is the speed of the wind past the topography. The boundary condition is

$$\psi = f(x) \quad \text{on} \quad y = 0.$$

In this case, we first do a Fourier transform because ψ will decay to zero for $x^2 + y^2 \to \infty$. If we denote by Ψ the Fourier transform of ψ in x, we have

$$\left(\frac{\partial}{\partial t} + Uik \right)^2 \left(\frac{\partial^2 \Psi}{\partial y^2} - k^2 \Psi \right) - N^2 k^2 \Psi = 0.$$

For looking at the steady wave structure, we could set the time derivatives to zero, but the solution then involves a complicated radiation condition that is not very physical. As an alternative, we approach the steady solution by doing the Laplace transform of Ψ, and examine its long-time behavior, before inverting the Fourier integral. Thus, denoting $\mathcal{L}(\Psi)$ by $\Phi(y, k, s)$, we have the ordinary differential equation and boundary condition for Φ

$$(s + Uik)^2 \left(\frac{\partial^2 \Phi}{\partial y^2} - k^2 \Phi \right) - N^2 k^2 \Phi = 0, \quad \Phi = \frac{F(k)}{s} \quad \text{on} \quad y = 0.$$

Solving for the Green's function – the response due to $f(x) = \delta(x)$,

$$\Phi = \frac{1}{s} e^{-|k| \phi y}, \quad \phi = \frac{[(s + ikU)^2 + N^2]^{1/2}}{s + ikU}.$$

If we now investigate the Laplace inversion, we note a number of features. There are branch points in Φ at $s = i\gamma_1, i\gamma_2$, where $\gamma_1 = -N - kU$ and $\gamma_2 = N - kU$, in addition to an apparent singularity at $-ikU$. Cutting the plane from $i\gamma_1, i\gamma_2$, parallel to the negative real axis, the "cut" integrals give transients. Because we are interested in the steady wave structure, the only portion of the Laplace inversion of importance is the origin pole. The residue is a little tricky, because the two branch points may be both in $\Im \text{mag}(s) > 0$ or in $\Im \text{mag}(s) < 0$ or straddling the origin. For the case in which the origin lies between the two branch points, careful evaluation of the square roots as in Example 5.4 leads to the result that

$$\text{Res}(0) = e^{iy\,\text{sgn}(k)\sqrt{N^2-k^2U^2}/U}, \qquad |k| < \frac{N}{U},$$

where sgn(g) is the sign of g.

However, for $k > N/U$ or $k < -N/U$, the pole lies either below or above both branch points, so evaluation of the square root in such a case gives

$$\text{Res}(0) = e^{-y\sqrt{U^2k^2-N^2}/U}, \qquad |k| > N/U.$$

Having obtained the relevant portions of the Laplace inverse, we know that $\Psi \sim$ Res(0) using these formulae, and so we can now write down the Fourier inversion as

$$\psi = \frac{1}{2\pi}\int_{-\infty}^{-\lambda} e^{ikx-\sqrt{k^2-\lambda^2}\,y}dk + \frac{1}{2\pi}\int_{\lambda}^{\infty} e^{ikx-\sqrt{k^2-\lambda^2}\,y}dk$$

$$+ \frac{1}{2\pi}\int_{-\lambda}^{0} e^{ikx-i\sqrt{\lambda^2-k^2}\,y}dk + \frac{1}{2\pi}\int_{0}^{\lambda} e^{ikx+i\sqrt{\lambda^2-k^2}\,y}dk, \qquad (7.2)$$

where we have used λ for N/U for convenience. The two integrals for $|k| > \lambda$ and the two integrals for $|k| < \lambda$ may be combined, so that (7.2) simplifies to

$$\psi = \frac{1}{\pi}\int_{\lambda}^{\infty} \cos(kx)e^{-\sqrt{k^2-\lambda^2}\,y}dk + \frac{1}{\pi}\int_{0}^{\lambda} \cos(kx + \sqrt{\lambda^2-k^2}\,y)dk. \qquad (7.3)$$

This may be put into a single integral along the real axis in the k-plane, from the origin to ∞ only if the k-plane is properly cut from the branch points located at $\pm\lambda$. Then, we have

$$\psi = \frac{1}{\pi}\Re\text{eal}\left(\int_{0}^{\infty} e^{ikx-(k^2-\lambda^2)^{1/2}y}dk\right),$$

where the plane is cut from λ on the line $k = \lambda + iY$, $Y > 0$ and from $-\lambda$ on the line $-\lambda - iY$, $Y > 0$. Therefore, the integration path passes below the branch point, as shown in the Figure 7.2.

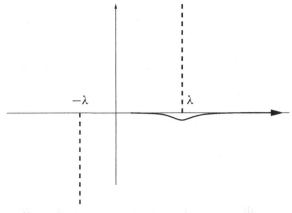

Figure 7.2. The k-plane inversion path for the Lee wave problem of Section 7.2.

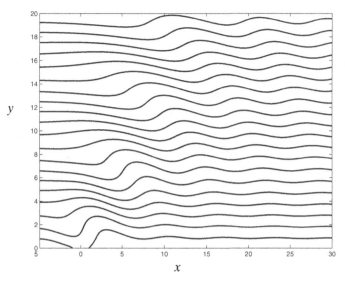

Figure 7.3. Streamline pattern for flow over a rectangular ridge, with $\lambda = 1$.

For flow over a more realistic shape, say a rectangular ridge extending from $x = -1$ to $x = 1$, with a flat top, the transform of the ridge shape, $F(k)$ is given by $F = 2 \sin k / k$. The formula can be written down by inserting $2 \sin k / k$ into the integrands in Equation (7.3). Attached as Figures 7.3, 7.4, and 7.5 are streamline plots for this flow for various values of λ.

7.3 The Far Momentum Wake

We now consider the far-field wake of an object immersed in an incompressible fluid. Writing the velocity vector as that due to a uniform flow with a small

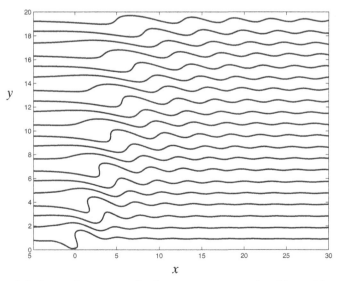

Figure 7.4. Streamline pattern for flow over a rectangular ridge, with $\lambda = 2$.

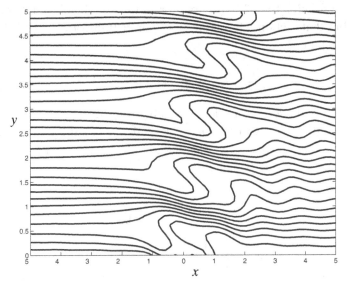

Figure 7.5. Streamline pattern for flow over a rectangular ridge, with $\lambda = 5$.

perturbation, the equation for that perturbation velocity vector is given by

$$\frac{\partial \mathbf{u}}{\partial x} + \nabla p - \nabla^2 \mathbf{u} = -\mathbf{F}, \quad \nabla \cdot \mathbf{u} = 0, \tag{7.4}$$

where for the far-field wake, the effect of the object is to exert a point-force on the fluid, so that

$$\mathbf{F} = \delta(x)\delta(y)\delta(z)\mathbf{i}. \tag{7.5}$$

The dimensional quantities have been scaled out, in a fashion very like that done in Section 6.3.7. The vector $\mathbf{i} = \nabla x$, in the direction of the flow. Taking the gradient of the first equation in (7.4), and using the second, we obtain the Poisson equation for the pressure,

$$\nabla^2 p = -\nabla \cdot \mathbf{F}. \tag{7.6}$$

Then, taking the Laplacian of (7.4) gives

$$\left(\frac{\partial}{\partial x} - \nabla^2\right)\nabla^2 \mathbf{u} = \nabla\left(\nabla \cdot \mathbf{F}\right) - \nabla^2\mathbf{F} = -\nabla_\perp^2\mathbf{F}, \tag{7.7}$$

where $\nabla_\perp^2$ is the "two-dimensional" Laplacian, $\nabla_\perp^2 = \partial^2/\partial y^2 + \partial^2/\partial z^2$ in Cartesian notation. Writing the perturbation velocity vector as $\mathbf{u} = -\mathbf{i}\nabla_\perp^2\phi$, this equation, on using (7.5), becomes

$$\left(\frac{\partial}{\partial x} - \nabla^2\right)\nabla^2\phi = \delta(x)\delta(y)\delta(z). \tag{7.8}$$

Note from Section 6.3.7, in the analysis of the thermal wake, that in writing

$$\nabla^2\phi = T(x, y, z), \tag{7.9}$$

with T as the solution to (6.27), the solution of this problem is simply related to the one solved in that section. On taking the Fourier transforms in y and z as in that section, we obtain

$$\frac{\partial^2 \overline{\phi}}{\partial x^2} - k^2 \overline{\phi} = F \quad \text{and} \quad \overline{u} = -k^2 \overline{\phi}. \tag{7.10}$$

We have here used overlines to denote double transforms of ϕ and u, and also the notation of Section 6.3.7, with $k^2 = m^2 + n^2$, with m and n the y and z Fourier transform wave numbers respectively. F is the double Fourier transform of $T(x, y, z)$. Using the result (6.29) for F, namely

$$F(m, n) = \frac{e^{x/2 - \sqrt{k^2 + 1/4}\,|x|}}{\sqrt{1 + 4k^2}}. \tag{7.11}$$

we may write the Fourier transform of u as

$$\overline{u} = -\frac{k^2}{\sqrt{1 + 4k^2}\left[\frac{1}{2} + \sqrt{k^2 + \frac{1}{4}}\right]} e^{(1/2 + \sqrt{k^2 + 1/4})x}, \quad x < 0, \tag{7.12}$$

$$\overline{u} = -\frac{k^2}{\sqrt{1 + 4k^2}\left[\frac{1}{2} - \sqrt{k^2 + \frac{1}{4}}\right]} e^{(1/2 - \sqrt{k^2 + 1/4})x}, \quad x > 0, \tag{7.13}$$

Then, one may insert these functions into the Fourier inversion integrals, and as in Section 6.3.7, change from Cartesian integration in (m, n) space to polar wavenumber coordinates, and accomplish the angle integration to lead to a final result which is in fact a Hankel inversion. Following that process leads to the following solution in $x > 0$:

$$u = \frac{1}{\pi} \int_0^\infty \frac{k^3 J_0(kR) e^{(1/2 - \sqrt{k^2 + 1/4})x}}{\sqrt{1 + 4k^2}\left[\sqrt{1 + 4k^2} - 1\right]} \, dk, \quad x > 0. \tag{7.14}$$

We give in Section 6.3.7 the exact inversion of the Hankel transform given there. In summary, from that development,

$$T(x, R) = \frac{1}{4\pi^2} \int_{-\infty}^\infty \int_{-\infty}^\infty F(m, n) e^{imy + inz} \, dm \, dn$$

$$= \frac{1}{2\pi} \int_0^\infty \frac{J_0(kR) e^{(1/2 - \sqrt{k^2 + 1/4})x}}{\sqrt{1 + 4k^2}} k \, dk$$

$$= \frac{1}{4\pi \sqrt{R^2 + x^2}} e^{\frac{1}{2}x - \frac{1}{2}\sqrt{R^2 + x^2}} \equiv e^{x/2} \Phi(R, x). \tag{7.15}$$

In this expression, $R = \sqrt{y^2 + z^2}$.

Note that one portion of the integrand in (7.14) may be simplified as

$$\frac{k^3}{\sqrt{1 + 4k^2}\left[\sqrt{1 + 4k^2} - 1\right]} = \frac{k}{4}\left[1 + \frac{1}{\sqrt{1 + 4k^2}}\right].$$

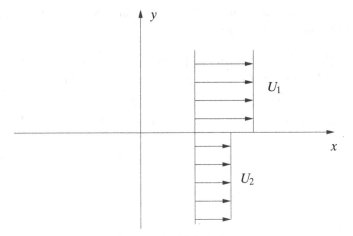

Figure 7.6. Kelvin–Helmholz geometry

So, one integral in (7.14) is exactly the integral arising in the thermal wake (7.15), and the other term is the derivative of Φ. Therefore, we can give an expression for the exact solution for the perturbation velocity in $x > 0$ (and in $x < 0$ as well),

$$u = \frac{e^{x/2}}{2}\left[\Phi - \frac{1}{2}\frac{\partial\Phi}{\partial x}\right], \quad \text{in} \quad x > 0.$$

This result may also be found in (Lagerstrom).

7.4 Kelvin–Helmholtz Instability

We consider here a well-known problem in fluid mechanics: the Kelvin–Helmholtz instability due to a fluid shear (Drazin). We model the situation with two fluid layers, separated by $y = 0$ as shown in Figure 7.6. In the upper layer, $y > 0$, the fluid flows at a speed U_1, and the lower layer, $y < 0$, the speed is U_2. In each region, the perturbation velocity is obtained from a velocity potential, so that $\mathbf{u}_{1,2} = \nabla\phi_{1,2}$. Because the motion is assumed to be irrotational, the velocity potentials are solutions of a Laplace equation, $\nabla^2\phi = 0$. If we use the brackets, [ ], to denote jumps across $y = 0$, i.e., $[Q] \equiv Q_1 - Q_2$, then continuity of pressure implies that

$$\left[\frac{\partial\phi}{\partial t} + U\frac{\partial\phi}{\partial x}\right] = 0 \quad \text{at} \quad y = 0. \tag{7.16}$$

The other condition is that the perturbed interface, located at $y = \eta(x, t)$, is a material interface. That requirement leads to

$$\frac{\partial\phi_1}{\partial y} = \frac{\partial\eta}{\partial t} + U_1\frac{\partial\eta}{\partial x}, \quad \frac{\partial\phi_2}{\partial y} = \frac{\partial\eta}{\partial t} + U_2\frac{\partial\eta}{\partial x} \quad \text{at} \quad y = 0. \tag{7.17}$$

To proceed to a solution, we do a Fourier transform in x and a Laplace transform in time. Denoting this double transform as Φ the Laplace equation solutions are easily written down as

$$\Phi_1 = A_1(k, s)e^{-|k|y}, \qquad \Phi_2 = A_2(k, s)e^{|k|y}.$$

Applying the boundary conditions, after significant algebra, we have the following pair of simultaneous equations for $\{A_1, A_2\}$,

$$
\begin{pmatrix} s + ikU_1 & -s + ikU_2 \\ s + ikU_2 & s + ikU_1 \end{pmatrix} \begin{pmatrix} A_1 \\ A_2 \end{pmatrix} = \begin{pmatrix} [\hat{\phi}]_{y=t=0} \\ i\,\mathrm{sgn}(k)(U_2 - U_1)\hat{\eta}|_{t=0} \end{pmatrix}. \tag{7.18}
$$

In this expression, the hat, ˆ, denotes the Fourier transform of some initial data. The Laplace and Fourier transforms of the surface deflection, η, is H, related to A_1 by

$$
H = \frac{\hat{\eta}|_{t=0}}{s + ikU_1} - \frac{|k|A_1}{s + ikU_1}.
$$

Solving for A_1 from (7.18), we find that, for the case $[\hat{\phi}]_0 \equiv 0$, which we now assume throughout,

$$
H = \frac{2s + ik(U_1 + U_2)}{(s + ikU_1)^2 + (s + ikU_2)^2}\,\hat{\eta}|_{t=0}. \tag{7.19}
$$

The Laplace and Fourier inversions in detail depend on the nature of the initial displacement, η.

Case 1 – Plane-Wave Modes

The response in this case is that due to a single, wavelike initial disturbance, so its Fourier transform, $\hat{\eta}_0 = 2\pi\delta(k - m)$, where m is the wave number of the initial disturbance. Hence, on obtaining the Fourier inversion first, which because of the delta function is easy, we obtain the Laplace transform of the surface deflection as

$$
\mathcal{L}(\eta) = \frac{2s + im(U_1 + U_2)}{(s + imU_1)^2 + (s + imU_2)^2}e^{imx}. \tag{7.20}
$$

What is left, then, is the Laplace inversion of (7.20), and that clearly involves poles located at

$$
s = s_1 = \frac{m}{2}[U] - im\overline{U}, \quad s = s_2 = -\frac{m}{2}[U] - im\overline{U}, \tag{7.21}
$$

where $\overline{U}$ has been written for the average speed, $(U_1 + U_2)/2$. It is a simple matter, then, to sum the residues and that process results in the solution for η given by

$$
\eta = e^{im(x - \overline{U}t)}\cosh(mt[U]/2).
$$

Note that the disturbance is convected with the flow, and that there is always instability regardless of the sign of $[U]$.

Case 2 – Localized Initial Disturbance

Now, we explore a more realistic case: an initially confined disturbance. If we take, say, a Gaussian initial distribution, $\eta = \exp(-\alpha x^2)$, then the Fourier transform, $\hat{\eta}_0$ is also Gaussian, and so (7.19) becomes

$$
H = \left(\frac{\pi}{\alpha}\right)^{1/2} \frac{2s + ik(U_1 + U_2)}{(s + ikU_1)^2 + (s + ikU_2)^2}e^{-\frac{k^2}{4\alpha}}. \tag{7.22}
$$

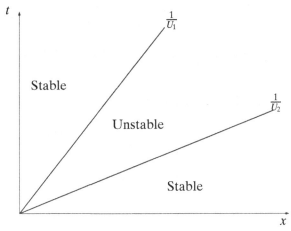

Figure 7.7. Diagram of convective Kelvin–Helmholtz instability.

To complete the solution, the Fourier and Laplace inversions must be performed. The residue calculation for the Laplace inversion is just as before, with the poles given in (7.21) except that the m is now k in those expressions. Thus, the solution comes down to the Fourier inversion,

$$\eta = \frac{1}{\sqrt{4\alpha\pi}} \int_{-\infty}^{\infty} \left(e^{ik(x-\overline{U}t)+k[U]t/2-k^2/4\alpha} + e^{ik(x-\overline{U}t)-k[U]t/2-k^2/4\alpha} \right) dk.$$

Completing the square in the exponents makes the integrals easy, and they may be done without resort to any complex integrations. The result is

$$\eta = e^{-\alpha(x-U_1 t)(x-U_2 t)} \cos\left(\alpha[U]t(x - \overline{U}t)\right). \tag{7.23}$$

We obtain the very interesting result here, which cannot be seen in any sense in the plane–wave analysis, that the flow is convectively unstable, that is, outside the two rays in the x–t plane in Figure 7.7, the flow is stable, that is, as $x, t \to \infty$, $\eta \to 0$. However, within the wedge bounded by those rays, that is for $U_1 t < x < U_2 t$, for $U_1 < U_2$, η is exponentially large as $x \to \infty$. Because this segment of fluid is convected through all of the flow, the entire flow-field becomes unstable, at least in the region $0 < x < \max(U_1, U_2)t$. Figure 7.8 shows this function $\eta(x, t)$, given in Equation (7.23), plotted versus x at two different times.

If the two streams U_1 and U_2 are in opposing directions; say $U_2 = -U_1 > 0$, then the flow is termed "absolutely unstable." This is because at any point x in the flow-field, the instability will occur and remain after some time t. That is, within the wedge $-U_2 t < x < U_2 t$, η is exponentially large as $t \to \infty$. This is illustrated in Figure 7.9.

7.5 The Boundary Layer Signal Problem

Next, we turn to a mathematical model of the boundary-layer signal problem in which a fixed frequency excitation occurs at the wall. This has been the subject of

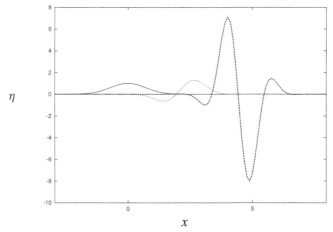

Figure 7.8. Interfacial deflection for Kelvin–Helmholtz instability for $U_1 = 1, U_2 = 2, \alpha = 1$, shown at times $t = 1,—$, $t = 1.5, \ldots$ and $t = 3, — — —$.

several deep studies in, for example (Ashpis), and our purpose here is to simply indicate how the types of transforms we have introduced may be brought to bear on the issue. Consider then the problem for the vorticity, ζ,

$$\frac{\partial \zeta}{\partial t} + U_0 \frac{\partial \zeta}{\partial x} = \frac{1}{R} \nabla^2 \zeta, \quad -\infty < x, t < \infty, \; 0 < y < \infty \tag{7.24}$$

The boundary conditions are

$$\zeta(x, y, t)|_{y=0} = e^{-i\omega_0 t} H(t)\delta'(x) \tag{7.25}$$

and $\zeta \to 0$ as $y \to \infty$. Here U_0, R, and ω_0 are constants. We will assume that ζ is Fourier transformable in x, and, eventually, we will apply the bilateral Laplace transform in t.

Since (7.24) has constant coefficients and is defined on $-\infty < x < \infty$, we may solve it by taking a Fourier transform. Call the Fourier transform

$$\hat{\zeta}(k, y, t) = \int_{-\infty}^{\infty} \zeta(x, y, t) e^{-ikx} dx,$$

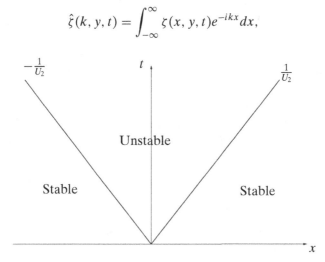

Figure 7.9. Diagram of absolute Kelvin–Helmholtz instability.

so that (7.24) becomes

$$\frac{\partial \hat{\zeta}}{\partial t} + ikU_0 \hat{\zeta} = \frac{1}{R}\left(\frac{\partial^2}{\partial y^2} - k^2\right)\hat{\zeta}. \tag{7.26}$$

The initial conditions become

$$\zeta(x, y, t)|_{y=0} = ike^{-i\omega_0 t}H(t).$$

Next, carry out a bilateral Laplace transform in t on (7.26) with

$$Z(k, y, s) = \int_{-\infty}^{\infty} \hat{\zeta}(k, y, t)e^{-st}\,ds$$

to get

$$sZ + ikU_0 Z = \frac{1}{R}\left(\frac{\partial^2}{\partial y^2} - k^2\right)Z,$$

with the boundary condition

$$Z(k, y, s)|_{y=0} = \frac{-ik}{s + i\omega_0}.$$

Solving for Z, we have

$$Z = \frac{-ik}{s + i\omega_0}e^{-\sqrt{k^2 + R(s+ikU_0)}y}.$$

The expression for ζ is then obtained by inverting both transforms. There are actually two ways of proceeding here. One is to notice that the doubly transformed vorticity, Z, has, in the Laplace domain, a pole at $s = -i\omega_0$, and a branch cut beginning at $s = -ikU_0 - k^2/R$ and extending to $\Re\text{eal}(s) = -\infty$ parallel to the real axis. So, deformation of the Laplace inversion path indicates that the solution is the sum of the pole residue – corresponding to the long-time periodic solution, and an integral around the branch cut – corresponding to the transient. Once those contributions are worked out, then the Fourier inversion may be performed. Alternatively, we may obtain a formal solution using two of the Laplace transform inversion theorems. We use the latter approach in this example. Thus, writing down the two nested inversion integrals,

$$\zeta = \frac{1}{2\pi}\int_{-\infty}^{\infty} e^{ikx}\,dk\,\frac{1}{2\pi i}\int_{\gamma-i\infty}^{\gamma+i\infty} \frac{-ike^{st}}{s + i\omega_0}e^{-\sqrt{k^2 + R(s+ikU_0)}y}\,ds.$$

Evaluating the inverse Laplace transform for $t > 0$, we have, using the convolution,

$$\zeta = \frac{1}{2\pi}\int_{-\infty}^{\infty} -ike^{ikx}\,dk\int_0^t e^{-i\omega_0(t-\tau)}e^{-(k^2/R + ikU_0)\tau}\sqrt{\frac{Ry^2}{4\pi\tau^3}}e^{-Ry^2/(4\tau)}\,d\tau.$$

The inverse Fourier transform is evaluated leaving

$$\zeta = \frac{R^2 y}{16\pi}\int_0^t \frac{(x - U_0\tau)}{\tau^{3/2}}e^{-i\omega_0(t-\tau)}e^{-R(x-U_0\tau)^2/4}e^{-Ry^2/(4\tau)}\,d\tau.$$

7.6 Stability of Plane Couette Flow

The problem of analyzing the stability of plane Couette flow has intrigued research workers for more than a hundred years. (See (Schmidt and Henningson).) Here we show how the techniques used in this book may be brought to bear on the inviscid stability problem, which may be solved fairly explicitly. We employ the standard formulation on the channel,

$$R = \{(x, y, z) | -\infty < x < \infty, -1 \le y \le 1, -\infty < z < \infty\},$$

to solve the initial-boundary value problem for the vertical velocity and vorticity perturbations v and η, respectively,

$$\left(\frac{\partial}{\partial t} + y\frac{\partial}{\partial x}\right)\nabla^2 v = 0, \tag{7.27}$$

$$\left(\frac{\partial}{\partial t} + y\frac{\partial}{\partial x}\right)\eta + \frac{\partial v}{\partial z} = 0. \tag{7.28}$$

The boundary conditions are

$$v = 0 \text{ at } y = \pm 1.$$

The initial conditions are taken as

$$v = v_0(x, y, z) \text{ and } \eta = \eta_0(x, y, z) \text{ at } t = 0.$$

We employ Fourier transforms in x and y setting

$$\hat{v}(k, m; y, t) = \int_{-\infty}^{\infty} \int_{-\infty}^{\infty} e^{-i(kx+mz)} v(x, y, z, t) dx dz$$

and, likewise,

$$\hat{\eta}(k, m; y, t) = \int_{-\infty}^{\infty} \int_{-\infty}^{\infty} e^{-i(kx+mz)} \eta(x, y, z, t) dx dz.$$

The Laplace transform gives

$$\mathcal{V} \equiv \mathcal{L}\{\hat{v}\} = \int_{0}^{\infty} \hat{v} e^{-st} dt,$$

and similarly for $\hat{\eta}$. Carrying out these transforms of (7.27) leads to

$$(iky + s)\left(\frac{d^2}{dy^2} - (k^2 + m^2)\right)\mathcal{V} = \left(\frac{\partial^2}{\partial y^2} - (k^2 + m^2)\right)\hat{v}_0, \tag{7.29}$$

with boundary conditions

$$\mathcal{V} = 0, \quad \text{at} \quad y = \pm 1. \tag{7.30}$$

The transformed version of (7.28) is simply

$$(iky + s)\mathcal{L}\{\hat{\eta}\} + im\mathcal{V} = \hat{\eta}_0. \tag{7.31}$$

If we make changes of notation setting $k^2 + m^2 = \alpha^2$, and $\left(\partial^2/\partial y^2 - \alpha^2\right)\hat{v}_0(y) = \hat{\zeta}_0$, we obtain the nonhomogeneous Rayleigh equation, though not quite in the standard notation,

$$(iky + s)\left(\mathcal{V}''(y) - \alpha^2\mathcal{V}(y)\right) = \hat{\zeta}_0. \tag{7.32}$$

Thus, once $\mathcal{V}$ is determined, the solution for $\mathcal{L}\{\hat{\eta}\}$ follows quite easily from (7.31).

In principle then, we may write down the solution of (7.32) as

$$\mathcal{V} = \int_{-1}^{1} \frac{\hat{\zeta}_0}{ik\xi + s} G(y, \xi; \alpha^2)d\xi,$$

where G is the Green's function for (7.32). The Laplace inverse is straightforward, because there is only a pole,

$$\hat{v} = \int_{-1}^{1} \hat{\zeta}_0 e^{-ik\xi t} G(y, \xi; \alpha^2)d\xi. \tag{7.33}$$

However, the determination of the complete solution depends on the detailed structure of G. By the methods of Section 4.2 this may be put in the form

$$G(y, \xi; \alpha^2) = \begin{cases} \sinh(\alpha(y-1))\sinh(\alpha(\xi+1))/(2\alpha\sinh(2\alpha)), & y > \xi, \\ \sinh(\alpha(\xi-1))\sinh(\alpha(y+1))/(2\alpha\sinh(2\alpha)), & y < \xi \end{cases}$$

$$= \frac{\cosh(\alpha(y+\xi)) - \cosh(\alpha(2 - |y - \xi|))}{4\alpha\sinh(2\alpha)}. \tag{7.34}$$

Still later in Chapter 4 we saw that an eigenfunction expansion of G is possible. This leads to the alternate representation, from Equation (4.41),

$$G(y, \xi; \alpha^2) = -\sum_{n=1}^{\infty} \frac{1}{(\alpha^2 + \frac{n^2\pi^2}{4})} \sin\left[\frac{n\pi}{2}(y+1)\right]\sin\left[\frac{n\pi}{2}(\xi+1)\right]. \tag{7.35}$$

The inverse Fourier transform of (7.33) is more challenging, but in principle it is written as

$$v(x, y, z, t) = \frac{1}{4\pi^2} \int_{-\infty}^{\infty}\int_{-\infty}^{\infty} e^{i(kx+mz)}\hat{v}(k, y, m, t)dkdm.$$

This may be carried out inserting the original expression for $\hat{v}$ in (7.33) setting $k = \alpha\cos\theta$, $m = \alpha\sin\theta$, in polar coordinates to get

$$v(x, y, z, t) = \frac{1}{4\pi^2} \int_{0}^{\infty} \alpha d\alpha \int_{0}^{2\pi} e^{i\alpha(x\cos\theta + z\sin\theta)}d\theta \int_{-1}^{1} \hat{\zeta}_0 e^{-i\alpha\cos\theta\xi t} G(y, \xi; \alpha^2)d\xi.$$

Similarly,

$$\hat{\zeta}_0(\alpha, \theta, \xi) = \int_{-\infty}^{\infty}\int_{-\infty}^{\infty} \zeta(x', \xi, z')e^{-i\alpha\cos\theta x' - i\alpha\sin\theta z'}dx'dz'.$$

Reordering the integrals we evaluate first with respect to θ to obtain a Bessel integral

$$\int_0^{2\pi} e^{i\alpha(x\cos\theta+z\sin\theta)} e^{-i\alpha\cos\theta\xi t} e^{-i\alpha\cos\theta x'-i\alpha\sin\theta z'} d\theta$$

$$= \int_0^{2\pi} e^{i\alpha(x-x'-\xi t)\cos\theta+i\alpha(z-z')\sin\theta} d\theta = 2\pi J_0(\alpha\rho),$$

where $\rho^2 = [(x - x') - t\xi]^2 + (z - z')^2$. The next integral with respect to α is a Hankel transform, and inserting (7.35), we have

$$\int_0^\infty \alpha J_0(\alpha\rho)G(y,\xi;\alpha^2)d\alpha = -\sum_{n=1}^\infty \int_0^\infty \frac{\alpha J_0(\alpha\rho)d\alpha}{(\alpha^2 + \frac{n^2\pi^2}{4})} \sin\left[\frac{n\pi}{2}(y+1)\right] \sin\left[\frac{n\pi}{2}(\xi+1)\right].$$

(A few more details on the Hankel transform may be found in Section 6.3.7.) Due to the nature of this transform, the result is naturally expressible in terms of the modified Bessel function K_0 (see (Bateman)). From this sequence of integrals a final general form is

$$v(x,y,z,t) = -\frac{1}{2\pi} \sum_{n=1}^\infty A_n(x,z,t) \sin\left[\frac{n\pi}{2}(y+1)\right], \tag{7.36}$$

$$\text{where } A_n = \int_{-\infty}^\infty \int_{-1}^1 \int_{-\infty}^\infty \zeta_0(x',\xi,z') \sin\left[\frac{n\pi}{2}(\xi+1)\right] K_0\left(\frac{n\pi\rho}{2}\right) dx'\,d\xi\,dz'.$$

This expression has only fairly recently been derived for the first time (Criminale et al.).

For our purposes, we observe that when the disturbance is a monochromatic impulse of the form

$$\zeta_0(x,y,z) = e^{i(k_0x+m_0z)}\delta(y-y_0), \tag{7.37}$$

it is simpler to employ (7.34) in (7.33) leading to

$$\hat{v} = \int_{-1}^1 \hat{\zeta}_0 e^{-ik\xi t} G(y,\xi;\alpha^2)d\xi$$

$$= \hat{v} = \int_{-1}^1 \delta(k-k_0)\delta(m-m_0)\delta(\xi-y_0)e^{-ik_0\xi t} G(y,\xi;\alpha_0^2)d\xi$$

$$= \delta(k-k_0)\delta(m-m_0)\left[\frac{\cosh(\alpha_0(y+y_0)) - \cosh(\alpha_0(2-|y-y_0|))}{4\alpha_0\sinh(2\alpha_0)}\right]e^{-ik_0y_0t}.$$

This gives, simply,

$$v(x,y,z,t) = e^{i(k_0x+m_0z)-ik_0y_0t}\left[\frac{\cosh(\alpha_0(y+y_0)) - \cosh(\alpha_0(2-|y-y_0|))}{4\alpha_0\sinh(2\alpha_0)}\right].$$

So, the solution represents a wave propagating to the right at speed y_0, and in the terminology of stability theory, the disturbances are "neutrally stable."

7.7 Generalized Transform Techniques

The Laplace transform and Fourier transform methods utilized so far are for problems for which the partial differential equations have constant coefficients. That is not, of course, always the case in many problems of practical interest. We consider in this section a generalized approach to transform method that is effective in problems where the coefficients are polynomials of an order smaller than the order of the differential equation. We discuss this method here because its details are determined problem-by-problem. A thorough discussion of the method may also be found in (Ince).

7.7.1 A Model Problem

Example 7.1. We consider in this section the solution to the ordinary differential equation and boundary conditions,

$$\frac{d^2y}{dx^2} + x\frac{dy}{dx} - y = 0, \quad \text{in} \quad 0 < x < \infty, \quad y(0) = 1, \quad y(\infty) = 0. \quad (7.38)$$

We now suppose that the solutions to (7.38) can be written as

$$y(x) = \int_A^B Q(p)e^{ipx}dp. \quad (7.39)$$

In this expression, the quantity Q is not yet known, and the integration path, in a complex p-plane, and in particular the endpoints, A and B, are not yet known. Substitution into (7.38) of (7.39) leads to

$$\int_A^B \left[-p^2Q + ipxQ - Q\right]e^{ipx}dp = 0. \quad (7.40)$$

The middle term can be rewritten, then integrated by parts, so we have

$$\int_A^B ipxQe^{ipx}dp = \int_A^B Q\frac{\partial}{\partial p}\left(e^{ipx}\right)dp = Qe^{ipx}\Big|_A^B - \int_A^B e^{ipx}\frac{\partial}{\partial p}(pQ)dp. \quad (7.41)$$

Substituting this simplification into (7.40), we obtain the two conditions,

$$\int_A^B \left[-p^2Q - \frac{\partial(pQ)}{\partial p} - Q\right]e^{ipx}dp = 0, \quad Qe^{ipx}\Big|_A^B = 0. \quad (7.42)$$

So, (7.38) is satisfied if the square bracket in the integrand of (7.42) is zero, and the boundary term listed are simultaneously satisfied – a feature of the endpoints A and B. Putting the square bracket to zero gives a first-order ordinary differential equation, whose solution is

$$Q = \frac{C}{p^2}e^{-p^2/2}. \quad (7.43)$$

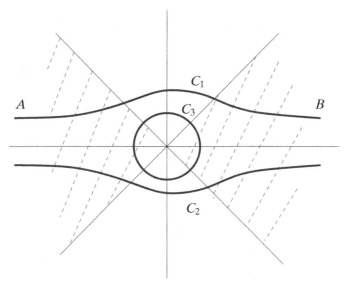

Figure 7.10. Integration paths for Example 7.1.

The constant C is not yet known. We can then substitute the boundary term into (7.42) to obtain the criterion for the choice of A and B, namely,

$$\frac{1}{p^2} e^{-p^2/2} e^{ipx} \bigg|_A^B = 0. \tag{7.44}$$

There are two simple ways for this equation to be satisfied: (1) Choose a closed path for the curve, so A and B are identical. (2) Choose an open path, with A and B lying in the 45 degrees regions shown in the sketch in Figure 7.10.

That can be seen as follows. Note that if we write $p = R\exp(i\theta)$, then

$$\frac{1}{p^2} e^{-p^2/2+ipx} = R^{-2} e^{-2i\theta} e^{-R^2(\cos 2\theta + i \sin 2\theta)/2 + i Rx \cos\theta - xR\sin\theta}.$$

Note that this quantity vanishes for $R \to \infty$ provided that $\cos 2\theta > 0$, which means that A and B lie in the shaded regions shown. Because the problem requires that y be bounded for $x \to \infty$, we can see too that $\sin\theta > 0$, which is the upper-half-plane. Therefore, we choose a path wholly lying in the upper-half-plane and terminating at A and B that satisfy requirement (7.44); hence, the path C_1 shown in Figure 7.10. Note that choosing a path C_2 would give a solution bounded in a lower-half physical plane. Note also that the two paths C_1 and C_2 can be combined into a path C_3 as shown, which is an alternative second solution. That solution is easy to evaluate,

$$y = \oint_{C_2} \frac{e^{ipx-p^2/2} dp}{p^2} = -2\pi x,$$

by the residue theorem. It can easily be verified that x is a solution to (7.38), but this is not relevant to solving this problem.

The former choice, then, gives a solution that is well-behaved in all x. Therefore, we have

$$y(x) = C \int_{C_1} \frac{e^{ipx - p^2/2} dp}{p^2}.$$

All that remains is to determine the constant C. The boundary condition then requires that

$$1 = C \int_{C_1} \frac{e^{-p^2/2} dp}{p^2}.$$

The integral around C_1 can be changed by writing $p = \sqrt{2u}$. Substitution eventually gives

$$1 = -C \frac{e^{3i\pi/2}}{2^{3/2}} \int_{C_o} \frac{e^{-u} du}{(-u)^{3/2}},$$

where C_o starts at $u = \infty e^{i2\pi}$, encircles the origin in a clockwise fashion, and terminates at ∞e^{i0}. This integral can be related to the gamma function, written in terms of a path, C_Γ the encircles the origin, starting at ∞e^{i0} and ending at $\infty e^{2\pi i}$,

$$\frac{1}{\Gamma(z)} = \int_{C_\Gamma} (-t)^{-z} e^{-t} dt.$$

(See Section 2.2.1.)

Therefore, we have

$$1 = -\frac{C\pi}{2^{1/2}} \frac{1}{\Gamma(3/2)} = -\sqrt{2\pi} C.$$

Hence, the complete solution is

$$y(x) = -\frac{1}{\sqrt{2\pi}} \int_{C_1} \frac{e^{ipx - p^2/2} dp}{p^2}. \tag{7.45}$$

The path C_1 must be symmetric about the imaginary p-axis to assure that this integral is purely real. It can be shown, by methods to be discussed in Chapter 8, that,

$$y \sim \frac{1}{x^2} e^{-x^2/2}, \quad \text{for} \quad x \to \infty.$$

7.7.2 A Boundary-Layer Example

Example 7.2. The boundary on a flat plate in a stratified flow is the solution of the following problem for the velocity component parallel to the plate,

$$\epsilon^3 \frac{\partial^4 u}{\partial y^4} - \frac{\partial u}{\partial x} = 0, \quad u(x, 0) = 0, \quad u(x, \infty) = U. \tag{7.46}$$

In this equation, the quantity ϵ is a parameter involving the viscosity and stratification magnitude, and is small if there is a boundary layer. (More details may be

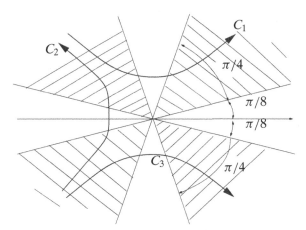

Figure 7.11. Inversion path for Example 7.2.

found in (Foster).) There are some suitable upstream conditions as well. There is a similarity solution to this problem, given by

$$u = UF(\eta), \quad \eta \equiv \frac{y}{(-\epsilon^3 x)^{1/4}}.$$

Substitution into (7.46) leads to the linear equation

$$F^{(iv)} + \frac{1}{4}\eta F' = 0, \quad F(\infty) = 1, \quad F(0) = 0. \tag{7.47}$$

We now write the solution to this equation as

$$F(\eta) = \int_A^B Q(p)e^{ip\eta}dp$$

and substitute into (7.47), which results, with the same integration by parts technique as before, in the differential equation for Q,

$$\frac{dQ}{dp} = 4p^3 Q,$$

plus the usual terms that need to vanish at endpoints. The solution is easy, and, hence, we have

$$F(\eta) = \int_A^B \frac{e^{p^4 + ip\eta}}{p}dp. \tag{7.48}$$

This is a fourth-order problem, so there are four possible choices for paths that are independent and that give the four linearly independent solutions. Of those, the simplest is that resulting from a circular path enclosing the origin which leads to a solution $F_1 = 1$, on working out the residue of the first-order pole. The other solutions use the paths shown in the Figure 7.11.

Noticing that one element of the integrand is $\exp(ip\eta)$, which can be written as

$$\exp(i\Re\text{eal}(p)\eta - \Im\text{mag}(p)\eta),$$

it is immediately evident that on the path C_1, the solution will be bounded for $\eta \to +\infty$ and on the path C_3 the solution will be bounded for $\eta \to -\infty$. On the C_2 path, the solution is bounded neither in $\eta > 0$ or $\eta < 0$. Here, we clearly choose path C_1, and hence the solution appropriate to this problem is

$$F = A_1 + A_2 \int_{C_1} \frac{e^{p^4+ip\eta}}{p} dp, \tag{7.49}$$

where A_1 and A_2 are constants. Because the second term vanishes for large η, matching to the outer flow requires that $A_1 = 1$. "No-slip" on the wall then requires that

$$1 + A_2 \int_{C_1} \frac{e^{p^4}}{p} dp = 0.$$

Evaluation of this integral may be accomplished by making a change of variable, $p = r^{1/4}$, in which case, we have

$$1 + \frac{A_2}{4} \int_{C_r} \frac{e^r}{r} dr = 0.$$

Using the limits of the p-integration, the path C_r begins at $\infty e^{i\pi}$, encircles the origin in the counterclockwise direction, and goes to infinity again along $\infty e^{3\pi i}$. The result will be equal to $2\pi i$ times the residue at the origin. Thus, the integral has value $2\pi i$, and hence we have the constant A_2 determined to be $2i/\pi$. Therefore, we have the solution from (7.49) as

$$F = 1 - \frac{2}{i\pi} \int_{C_1} \frac{e^{p^4+ip\eta}}{p} dp, \tag{7.50}$$

REFERENCES

D. E. Ashpis and E. Reshotko, The Vibrating Ribbon Problem Revisited, *J. Fluid Mech.* **213**, 531 (1990).

H. Bateman, *Tables of Integral Transforms*, Vols. I, II, McGraw-Hill, New York, 1954.

W. O. Criminale, B. Long, and M. Zhu, General Three-Dimensional Disturbances to Inviscid Couette Flow, *Stud. Appl. Math.* **85**, 249–267 (1991).

P. G. Drazin, *Introduction to Hydrodynamic Stability*, Cambridge University Press, Cambridge, UK, 2002.

M. R. Foster, The Boundary Layer on Flat Plate in a Rotating and Stratified Fluid, *J. Eng. Math.* **44**, 117–132 (1980).

P. R. Garabedian, *Partial Differential Equations*, Wiley, New York, 1964.

P. Schmid and S. Henningson, *Stability and transition in shear flows*, New York: Springer, 2001.

E. L. Ince, *Ordinary Differential Equations*, Dover, 1956.

P. Lagerstrom, Laminar Flow Theory, in *Theory of Laminar Flows*, F.K. Moore, ed., Princeton University Press (1964).

John Miles, Lee Waves in a Stratified Flow. Part 1. Thin barrier, *J. Fluid Mech.* **32**, 549 (1968).

G. S. Janowitz, Unbounded Statified Flow Over a Vertical Barrier, *J. Fluid Mech.* **58**, 375 (1973).

M. G. Wurtele, R. D. Sharman, and A. Datta, Atmospheric Lee Waves, *Ann. Rev. Fluid Mech.* **28**, 429 (1996).

EXERCISES

7.1 We examine lee waves in a channel with an upper lid. The equation for steady-state waves is, as seen in Section 7.2,

$$\nabla^2 \psi + \lambda^2 \psi = 0,$$

where $\lambda = N/U$, and ψ is a streamfunction. If there is topography on the lower wall, then the boundary conditions are

$$\psi = 0, \quad \text{on} \quad y = 1, \qquad \psi = f(x), \quad \text{on} \quad y = 0.$$

(a) Do a Fourier transform in x, and show that the Fourier transform of ψ, denoted by Ψ, is

$$\Psi = F(k) \frac{\sinh(a(1-y))}{\sinh(a)}, \quad a \equiv \sqrt{k^2 - \lambda^2}.$$

Here, F is the Fourier transform of $f(x)$.

(b) Verify that there are no branch points in this transform.

(c) Take the disturbance to be a delta function, so that $F = 1$.
Discuss the location of the poles for Ψ.
In your discussion, assume that λ has a numerical value that is between 2π and 3π.

(d) Using the λ range given above, obtain ψ upstream ($x < 0$). The Fourier inversion path should be *below* the real k-axis, underneath any real-axis poles. Your result should be an infinite series of spatially decaying modes.

(e) For that same λ range, obtain ψ downstream ($x > 0$). Your result should consist of an infinite series of spatially decaying modes, and two wavelike components.

7.2 A Kelvin–Helmholz analysis as in Section 7.4, but allowing for the possibility density (ρ) differences in the upper and lower surfaces and ignoring any surface tension at the interface, replaces condition (7.16) with

$$\rho_1 \left[\frac{\partial \phi_1}{\partial t} + U_1 \frac{\partial \phi_1}{\partial x} \right] - \rho_2 \left[\frac{\partial \phi_2}{\partial t} + U_2 \frac{\partial \phi_2}{\partial x} \right] + (\rho_1 - \rho_2)g\eta, \quad \text{at} \quad y = 0,$$

where g is the acceleration of gravity. Equation (7.17) is unchanged. Repeating that analysis, show that the poles in the Laplace domain are now located at

$$s = -ik\overline{U} \pm \sqrt{k^2 [U]^2 + \frac{g(\rho_1 - \rho_2)|k|}{\rho_1 + \rho_2}},$$

if we define the velocity parameters slightly differently, as

$$\overline{U} \equiv \frac{\rho_1 U_1 + \rho_2 U_2}{\rho_1 + \rho_2}, \quad [U] \equiv \frac{\sqrt{\rho_1 \rho_2}}{\rho_1 + \rho_2}(U_1 - U_2).$$

(a) If the fluid velocity is exactly zero, do Case I of Section 7.4. Notice that the character of the solution is different depending on the sign of $\rho_1 - \rho_2$.

(b) If $\rho_1 - \rho_2$ is negative, heavy fluid lies under light fluid, and waves propagate on the inteface but there is no instability. If U is not zero, and $\rho_1 - \rho_2 < 0$, is the flow always unstable?

7.3 Consider the transport-diffusion equation

$$\frac{\partial u}{\partial t} + V\frac{\partial u}{\partial x} = \sigma u + \nu\frac{\partial^2 u}{\partial x^2}, \quad -\infty < x < \infty,$$

where $V > 0$, $\nu > 0$. We will study this equation for the possibility of a convective instability. Pose the initial condition $u(x, 0) = f(x)$ and employ a Fourier transform in x. Show that the solution may be written formally as

$$u(x, t) = \frac{e^{\sigma t}}{2\pi}\int_0^\infty \hat{f}(k)e^{ik(x-Vt)}e^{-\nu k^2 t}dk,$$

where $\hat{f}(k) = \mathcal{F}\{f\}$, the Fourier transform. Examine the special case where $f(x) = e^{-\alpha x^2}$, $\alpha > 0$. Show that the solution is stable if $\sigma < 0$, but that the solution grows along the rays $x - Vt = x_0$, and therefore is convectively unstable if $\sigma > 0$.

7.4 Fluid fills a channel of dimensionless height one, and a small actuator oscillates on the floor of the channel at $x = 0$, and that can be modeled by a delta function. In this problem, we work out the sound wave field so generated. We remove the time with an $\exp(i\omega t)$ dependence, that is, we have written the actual streamfunction, $\psi_{\text{physical}} = \Re\text{eal}\{\psi(x, y)\exp(-i\omega t)\}$. So, the steady-state amplitude is described by the boundary-value problem,

$$\frac{\partial^2 \psi}{\partial x^2} + \frac{\partial^2 \psi}{\partial y^2} + \lambda^2\psi = 0,$$

where λ is a prescribed constant.

$$\psi = 0, \text{ at } y = 1; \quad \psi = \delta(x), \text{ at } y = 0.$$

(a) By doing a Fourier transform on the equation and boundary conditions, show that the Fourier transform of ψ, written as Ψ, has the solution

$$\Psi = \frac{\sinh(\sqrt{k^2 - \lambda^2}(1 - y))}{\sinh(\sqrt{k^2 - \lambda^2})}.$$

(b) Why are there no branch points in the k-plane? Locate all poles. Under what circumstances are there poles on the real axis?

(c) We know that the inversion path for the Fourier transform is the real axis in the k-plane. However, the path must be slightly modified since it cannot cut through poles. We *choose* the path shown in Figure 7.12. Why?? *Hint*: Look at ψ_{physical} for large positive and negative x.

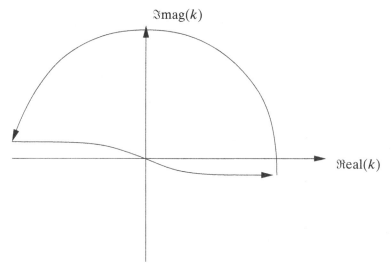

Figure 7.12. Inversion path for Exercise 7.4.

Using the path shown, for *Only* the case for which $\pi < \lambda < 2\pi$, work out ψ, and show that

$$\psi = \frac{\pi}{\sqrt{\lambda^2 - \pi^2}} \sin(\pi y) e^{i\sqrt{\lambda^2 - \pi^2}x}$$

$$+ \pi \sum_{n=2}^{\infty} \frac{n}{\sqrt{n^2\pi^2 - \lambda^2}} \sin(n\pi y) e^{-\sqrt{n^2\pi^2 - \lambda^2}x}, \quad x > 0.$$

(There is a similar expression in $x < 0$.)

7.5 This problem analyzes the growth of a shear layer in a fluid with a spatially varying viscosity. We seek to solve

$$\frac{\partial u}{\partial t} + \frac{\partial u}{\partial x} = (1 + \lambda x)\frac{\partial^2 u}{\partial y^2}, \quad \text{in} \quad |y| < \infty, \quad x > 0, \quad t > 0.$$

The initial and boundary conditions are

$$u = 0, \quad \text{at} \quad t = 0, \quad u = \sin t \delta(y), \quad \text{at} \quad x = 0.$$

(a) Doing a Laplace transform in time and a Fourier transform in y, show that the solution for the double transform is

$$\mathcal{L}\{\mathcal{F}\{u\}\} = \frac{1}{s^2 + 1} e^{-sx - k^2(x + \lambda x^2/2)}.$$

(b) Do the Laplace inversion first, then the Fourier inversion and show that

$$u = \begin{cases} 0, & x > t \\ \dfrac{\sin(t-x)}{2\pi\sqrt{x(1 + \frac{1}{2}\lambda x)}} e^{-\frac{y^2}{4x(1+\lambda x/2)}}, & x < t. \end{cases}$$

(c) Make a plot of u versus x at time $t = \pi$ for $y = 1$ and the parameter $\lambda = 0$.

(d) Repeat for $\lambda = 2$.

7.6 Consider the following initial-boundary-value problem:

$$\frac{\partial u}{\partial t} + \frac{\partial u}{\partial x} = (1 + \lambda x)\frac{\partial^2 u}{\partial y^2} \qquad \text{in } x > 0,\ t > 0,\ -\infty < y < \infty,$$

$$u = 0, \quad \text{at}\quad t = 0,$$

$$u = e^{-\alpha y^2}\cos t, \quad \text{at } x = 0.$$

In addition, the solution must decay for large $|y|$. This is the growth of a shear layer in a fluid with a spatially varying viscosity (Exercise 7.5) when the boundary condition is not local.

(a) By doing a Laplace transform in time and a Fourier transform in y, show that

$$\mathcal{L}\{\mathcal{F}\{u\}\} = \sqrt{\frac{\pi}{\alpha}}\,\frac{se^{-sx}}{s^2 + 1}e^{-k^2\left(\frac{1}{4\alpha}+x\right)}e^{-\lambda k^2 x^2/2}.$$

(b) By doing the Laplace inversion first (which can be done simply, independent of the value of k), and then the Fourier inversion, obtain the **exact** solution to this problem. (Be careful to note that the sign of $t - x$ is critical to the solution.)

7.7 We consider in this problem sound waves reflecting from an uneven wall, as shown in Figure 7.13 The equation that describes the propagation is the equation for the velocity potential,

$$a^2\nabla^2\phi = \frac{\partial^2\phi}{\partial t^2}, \tag{7.51}$$

where a is the sound speed. The boundary condition is that $\partial\phi/\partial n = 0$ on the surface.

(a) We begin with a flat surface, as shown by the solid line. Show that an incident wave given by

$$\phi_i = \cos(\alpha x \cos\theta - \alpha y \sin\theta - a\alpha t)$$

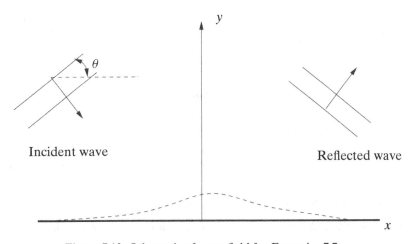

Figure 7.13. Schematic of wave field for Excercise 7.7.

satisfies the equation (7.51), where θ and α are constants. As shown in the sketch, θ is related to the direction of propagation, and k is a wave number. In what follows, it is easiest to work with complex exponentials, so instead we write

$$\phi_i' = e^{i\alpha(x\cos\theta - y\sin\theta - at)}. \tag{7.52}$$

Then, when we are all finished, $\phi = \Re\text{eal}(\phi')$.

The reflected wave will have the same time dependence and will continue to propagate in the x-direction, but it will be moving *outward* instead of *inward*. Thus, we can write

$$\phi_r' = Ae^{i\alpha(x\cos\theta + y\sin\theta - at)}. \tag{7.53}$$

Thus, the solution for a flat wall is

$$\phi_{\text{flat}}' = \phi_i' + \phi_r'.$$

The boundary condition for the flat wall is

$$\frac{\partial \phi_{\text{flat}}'}{\partial y} = 0, \quad \text{at} \quad y = 0.$$

(b) Show using (7.52) and (7.53) in this condition that the reflected amplitude must be $A = 1$. Thus (do not show), the combined incident and reflected waves are given by

$$\phi_{\text{flat}}' = 2\cos(\alpha y \sin\theta)e^{i\alpha(x\cos\theta - at)}.$$

For an uneven surface, given by $y = f(x)$, we write $\phi' = \phi_{\text{flat}}' + \phi_R'$, because the extra stuff will be added reflections. Clearly, ϕ_R' still satisfies equation (7.51), and the boundary condition turns out to be (you are not required to show this!)

$$\frac{\partial \phi_R'}{\partial y} = 2f'(x)i\alpha\cos\theta \, e^{i\alpha(x\cos\theta - at)}, \quad \text{at} \quad y = 0.$$

Now write $\phi_R' = p(x, y)\exp(-i\alpha at)$, and substitute into Equation (7.51). Show this that leads to

$$\nabla^2 p + \alpha^2 p = 0.$$

(c) Show that, on using a Fourier transform in x, with P defined as the Fourier transform of p, we obtain the following differential equation and boundary condition combination:

$$\frac{\partial^2 P}{\partial y^2} - (k^2 - \alpha^2)P = 0, \quad \frac{\partial P}{\partial y} = 2\cos\theta\alpha(\alpha\cos\theta - k)F(k - \alpha\cos\theta),$$

$$\tag{7.54}$$

where F is the Fourier transform of $f(x)$.

(d) Make the bottom wall a unit step function, so that $f = 0$, in $x < 0$ and $f = 1$, in $x > 0$. It can be shown that $F(k) = -i/k$ for such a surface. Use this result to modify (7.54) for this special case. Show that the solution of (7.54) this case is

$$P = \frac{2i\alpha\cos\theta}{\sqrt{k^2-\alpha^2}}e^{-\sqrt{k^2-\alpha^2}y}.$$

There are branch points at $k = \pm\alpha$. Indicate in a sketch how the k plane should be cut to assure that the waves represented by ϕ'_R propagate OUT-WARD. Therefore, show that the solution is

$$\phi'_R = \frac{2i\alpha\cos\theta}{\pi}e^{-i\alpha at}\int_\alpha^\infty \frac{\cos(kx)}{\sqrt{k^2-\alpha^2}}e^{-\sqrt{k^2-\alpha^2}y}dk$$

$$- \frac{2\alpha\cos\theta}{\pi}\int_0^\alpha \frac{e^{i(kx+\sqrt{\alpha^2-k^2}y-\alpha at)}}{\sqrt{\alpha^2-k^2}}dk.$$

We will do some approximate evaluation of this solution in Chapter 8. How does this approach compare with a Laplace transform procedure for this problem in Section 5.35?

7.8 (a) By taking a Fourier transform in x, show that the harmonic function $u(x, y)$ in the half-plane $-\infty < x < \infty$, $y < 0$ which, vanishes at infinity and satis-fies the boundary condition

$$g\frac{\partial u}{\partial y}(x, 0) + s^2 u(x, 0) = sf(x),$$

where s and g are constants is given by

$$u(x, y) = \frac{s}{\pi}\int_0^\infty \int_{-\infty}^\infty \frac{f(\xi)e^{ky}\cos(k(x-\xi))}{s^2+gk}d\xi\,dk.$$

(b) The function $\varphi(x, y, t)$ satisfies

$$\frac{\partial^2\varphi}{\partial x^2}(x, y, t) + \frac{\partial^2\varphi}{\partial y^2}(x, y, t) = 0, \quad -\infty < x < \infty, \ y < 0, \ t > 0,$$

together with the boundary condition

$$g\frac{\partial\varphi}{\partial y}(x, 0, t) + \frac{\partial^2\varphi}{\partial t^2}(x, 0, t) = 0,$$

and the initial conditions

$$\frac{\partial\varphi}{\partial t}(x, 0, 0) = 0, \quad \varphi(x, 0, 0) = \eta(x).$$

Taking a Laplace transform in t, reduce the problem to one formulated in part (a) and deduce, for, $t > 0$,

$$\varphi(x, y, t) = \frac{1}{\pi}\int_0^\infty \int_{-\infty}^\infty \eta(\xi)e^{ky}\cos(t\sqrt{gk})\cos(k(x-\xi))d\xi\,dk.$$

The function just calculated is called a "*linearized water wave potential.*"

7.9 A semiinfinite slab of material is exposed to a periodic thermal variation along the side. The initial-boundary-value problem is

$$\nabla^2 T = \frac{\partial T}{\partial t}, \quad \text{in } -\infty < x < \infty, \ 0 < y < \infty,$$

under

$$T = 0, \quad \text{at } t = 0, \quad T = f(x)\cos t, \quad \text{on } y = 0.$$

To do this problem, we write $T = \Re\text{eal}\{\theta\}$, so the equivalent problem is now

$$\nabla^2 \theta = \theta_t \quad \text{in } -\infty < x < \infty, \ 0 < y < \infty,$$

with

$$\theta = 0, \quad \text{at } t = 0, \quad \theta = f(x)e^{-it}, \quad \text{on } y = 0.$$

(a) Do a Laplace transform in t and a Fourier transform in x, which satisfy initial and boundary conditions, and then show that the Fourier–Laplace transform of the solution is

$$\mathcal{L}\{\mathcal{F}\{\theta\}\} = \frac{F(k)}{s+i} e^{-\sqrt{k^2+s}\,y}.$$

(b) Looking at the Laplace inverse, notice that it consists of two parts: a transient due to an integration around a branch cut that emanates from $s = -k^2$, and the long-time, periodic response due to the residue at $s = -i$. Denote this latter portion of the solution as θ_s. Show that, when finding the residue,

$$\mathcal{F}\{\theta_s\} = F(k)e^{-\sqrt{k^2-i}\,y-it}.$$

Conclude then that the solution for θ_s is given in terms of the Fourier inversion integral,

$$\theta_s = \frac{e^{-it}}{2\pi} \int_{-\infty}^{\infty} F(k)e^{rh(k)}dk, \quad h(k) = ik\cos\phi - \sqrt{k^2-i}\sin\phi,$$

where polar coordinates have now been used in the physical plane.

7.10 Integral Representations of Airy Functions. Airy's equation is defined as $w''(z) - zw(z) = 0$, and its solutions were discussed in Section 2.3.6. Here, we look for a solution of the form:

$$w(z) = \frac{1}{2\pi i} \int_C e^{sz} W(s)ds, \tag{7.55}$$

which is an *inverse* Laplace transform, where C is an infinite contour.

Substitute (7.55) into Airy's equation and integrate by parts. Find that to require boundary terms to vanish at infinity on C, s must tend to infinity along rays $|s| \to \infty$ asymptotic to $|s|e^{2n\pi i/3}$, $n = 0, 1, 2$. Choose in particular the rays, $n = 1, 2$. Let C begin at infinity along $|s|e^{4\pi i/3}$ and end at infinity along $|s|e^{2\pi i/3}$. This defines $Ai(z)$, the Airy function of the first kind, and already discussed in Section 2.3.6 in Chapter 2.

Furthermore, for z $\Re$eal, say $z = x$, show that the contour can be deformed onto the imaginary s-axis, so that

$$Ai(x) = \frac{1}{2\pi i}\int_C e^{sx - s^3/3}ds = \frac{1}{\pi}\int_0^\infty \cos\left(\frac{u^3}{3} + ux\right)du,$$

and show that $Ai(0) = 1/\left(3^{2/3}\Gamma(\frac{2}{3})\right)$.

7.11 Use the method of generalized transforms of Section 7.7 to show that

$$xy''' + y = 0, \quad x > 0$$

has a solution

$$y_1(x) = \int_0^\infty ve^{-x/v}e^{-v^2/2}dv,$$

for the case $y(0) = 1$ and $y \to 0$ as $x \to \infty$. Notice from the equation that if $y_1(x)$ is a solution, so is $y_2(x) = y_1(-x)$, so these functions represent two of the three linearly independent solutions. The third solution can most easily be found by working out the residue at the essential singularity at $v = 0$; show that it can be written as

$$y_3(x) = x^2 \sum_{n=0}^\infty \frac{(-1)^n}{n!(2(n+1))!}\left(\frac{x^2}{2}\right)^n.$$

Thus, the general solution is $y = c_1y_1 + c_2y_2 + c_3y_3$; y_1 is bounded in $x > 0$, y_2 is bounded in $x < 0$, and y_3 is finite near $x = 0$ but unbounded for $|x| \to \infty$. (*Note*: It looks at first that y_3 is a Bessel function, but simplification of the factorial in the denominator indicates otherwise.)

7.12 Complete the solution of the inviscid plane Couette flow problem for the vertical component of the disturbance vorticity $\eta(x, y, z, t)$. Show that from (7.31), its Laplace–Fourier transform is given by

$$\mathcal{L}\{\hat{\eta}\} = \frac{\hat{\eta}_0}{iky + s} - im\int_{-1}^1 \frac{\hat{\zeta}_0(k, m, \xi)}{(ik\xi + s)(iky + s)}G(y, \xi; \alpha^2)d\xi,$$

Perform the inverse Laplace transform, noting that in the integrand there are two s-poles, to obtain

$$\hat{\eta} = \hat{\eta}_0 e^{-ikyt} - im\int_{-1}^1 \hat{\zeta}_0(k, m, \xi)\left[\frac{e^{-ikyt} - e^{-ik\xi t}}{ik(y - \xi)}\right]G(y, \xi; \alpha^2)d\xi.$$

In the special case where the initial impulsive wave is of the form (7.37), show using (7.34), that the part of η due to the impulse is given by

$$\eta_I(x, y, z, t) = m_0 e^{i(k_0 x + m_0 z)}\frac{e^{-ik_0 yt} - e^{-ik_0 y_0 t}}{k_0(y - y_0)}$$

$$\times \left[\frac{\cosh(\alpha_0(y + y_0)) - \cosh(\alpha_0(2 - |y - y_0|))}{4\alpha_0\sinh(2\alpha_0)}\right].$$

Verify that this contribution experiences linear growth in time at the point $y = y_0$, a three-dimensional effect!

8

Asymptotic Expansions of Integrals

8.1 Preamble

Having developed transform methods applicable to a variety of physical problems, we have found the solutions are generally in the form of definite integrals. In any but the simplest cases, those integrals are intractable. For many engineering and scientific applications, what is needed is an approximate value for say heat transfer, momentum defect, or some other significant parameter, in the limit as one parameter or another is very large or very small. It has also been recognized that this process is intimately connected to what are sometimes called "perturbation expansions." This connection will not be thoroughly explored here, but the interested reader is referred to an article by (Segel), as well any of the thoughtful textbooks on that subject by (Bender and Orszag), (Cole and Kevorkian), and (Holmes). Here, we will primarily be interested in the outcome of expressing the solution to a differential equation in integral form. Those issues can sometimes be explored by means of standard asymptotic methods for integrals to which we now turn.

Most such integrals are in one of the following forms:

$$f(x) = \int_a^b g(x, y)dy, \quad f(x) = \int_x^b g(y)dy, \tag{8.1}$$

where a and b are constants, and $g(x, y)$ is well-behaved and integrable for y in (a, b) or (x, b), respectively.

There are more extensive treatments of this subject to be found elsewhere. The reader who desires more grounding in this subject is referred to, in particular, the books by (Erdélyi) and (Bleistein and Handelsman).

8.2 Asymptotic Expansions

Consider the asymptotic behavior of $f(x)$ given by either of (8.1) as, say, $x \to 0$. That behavior can be expressed as an asymptotic series given by

$$f(x) = f_0(x) + f_1(x) + f_2(x) + \cdots, \tag{8.2}$$

where the functions $\{f_n(x)\}$ obey the property that

$$\lim_{x \to 0} \left(\frac{f_{n+1}(x)}{f_n(x)} \right) = 0, \quad \text{all } n. \tag{8.3}$$

This may be written with the ordering symbol, o, as $f_{n+1} = o(f_n)$. If (8.3) is satisfied, then (8.2) is indeed an asymptotic series. For example, if we write

$$\frac{1}{1 + x^3} = 1 - x^3 + x^6 + \cdots,$$

we know from elementary considerations that the series is convergent in $|x| < 1$, and any finite set of terms in the series forms an asymptotic expansion of the function as $x \to 0$, since

$$\lim_{x \to 0} \left(\frac{(-1)^{n+1} x^{3(n+1)}}{(-1)^n x^{3n}} \right) = 0.$$

Series may be asymptotic without being convergent. For example, we might have obtained an asymptotic series in the form:

$$f(x) = x - x^3 + 16x^4 - 1456x^7 + \mathcal{O}(x^8).$$

We have an idea, from the order term, of the size of the next term in the series, but we do not have the general term. The worry might be that the coefficients are growing, so as an (eventual) sum, the infinite series may be divergent. None of that matters from the point of view of asymptotic expansions, since (8.3) assures us that the series is indeed asymptotic to $\mathcal{O}(x^7)$.

With regard to our notation in this chapter, we will sometimes write

$$f(x) \sim \sum_{n=0}^{\infty} f_n(x) \quad \text{for} \quad x \to x_o$$

as an asymptotic representation, even though the series may not be convergent. Cole and Kevorkian always use a finite upper limit, emphasizing the fact that what is represented is *not* a convergent series, but an asymptotic series. A more complete discussion of the ideas and pitfalls of asymptotic series may be found in that book.

We turn now to the asymptotic evaluation of definite integrals, which will lead us to the construction from (8.1) of series in the form (8.2).

8.3 Integration by Parts

The simplest method for obtaining asymptotic series for definite integrals is by means of integration by parts. The procedure is a bit ad hoc in that it sometimes takes experience to dertermine how to choose u and dv in the integration-by-parts procedure in order to generate the requisite expansion.

8.3.1 Worked Examples

Example 8.1. Consider, for example, the integral,

$$f(x) = \int_x^\infty \frac{e^{-t^2} dt}{t}. \tag{8.4}$$

Using integration by parts, in identifying t^{-2} with u and $t \exp(-t^2)$ with dv, we find that (8.4) can be written, without approximation as

$$f(x) = \frac{1}{2x^2} e^{-x^2} - \int_x^\infty \frac{e^{-t^2} dt}{t^3}. \tag{8.5}$$

Integrating by parts again, we find that

$$\int_x^\infty \frac{e^{-t^2} dt}{t} = \frac{1}{2x^2} e^{-x^2} - \frac{1}{2x^4} e^{-x^2} + 2 \int_x^\infty \frac{e^{-t^2} dt}{t^5}. \tag{8.6}$$

This last integral can be shown to be of order $x^{-6} \exp(-x^2)$, so we can write (8.6) as

$$\int_x^\infty \frac{e^{-t^2} dt}{t} = \frac{1}{2x^2} e^{-x^2} - \frac{1}{2x^4} e^{-x^2} + \mathcal{O}(x^{-6} e^{-x^2}).$$

This series is clearly asymptotic for $x \to \infty$.

Example 8.2. Consider the more complicated example,

$$g(x) = \int_1^\infty \frac{dt}{t^{3/2}(x + t)}.$$

By choosing $u = t^{-3/2}$ and $dv = dt/(x + t)$, two successive integrations by parts lead to

$$g = -\log(x + 1) - \frac{3}{2}(x + 1)\log(x + 1) + \frac{3}{2}(x + 1) + \mathcal{O}((x + 1)^2 \log(x + 1)).$$

The process has yielded a series that is asymptotic for $(x + 1) \to 0$, since it is evident that

$$\lim_{x \to -1} \frac{(x + 1)^{n+1} \log(x + 1)}{(x + 1)^n \log(x + 1)} = \lim_{x \to -1} (x + 1) = 0,$$

hence obeying (8.3).

8.4 Laplace-Type Integrals; Watson's Lemma

In this section, we consider a class of definite integrals of the form

$$f(x) = \int_0^\infty \phi(t) e^{-xt} dt, \tag{8.7}$$

where ϕ is well behaved at 0 and of exponential order as $t \to \infty$, so that the integral in fact exists. The theory of such integrals is intimately connected with Laplace transforms, since we may define $f(s) = \mathcal{L}\{\phi\}$. We now develop a general approach as a systematic procedure for dealing with integrals in the form (8.7), as in (Erdélyi).

Suppose that ϕ has an asymptotic expansion near $t = 0$,

$$\phi(t) = \sum_{n=0}^{N} \Phi_n(t) + R_N(t), \tag{8.8}$$

where R_N is a remainder, the error incurred in an N-term approximation, so that $R_N = o(\Phi_N)$ and the functions $\{\Phi_n\}$ are arranged as required by (8.3), so that $\Phi_{n+1} = o(\Phi_n)$, to assure that the series is indeed asymptotic. Then, it is desired to show that

$$\mathcal{L}\{\phi\} \sim \sum_{n=0}^{N} \mathcal{L}\{\Phi_n\} \quad \text{as } x \to \infty. \tag{8.9}$$

Let a be a value of t sufficiently close to zero to make the approximation (8.8) as good as we like. Then, (8.7) becomes

$$f = \sum_{n=0}^{N} \int_{0}^{a} \Phi_n(t)e^{-xt}dt + \int_{0}^{a} R_N(t)e^{-xt}dt + \int_{a}^{\infty} \phi(t)e^{-xt}dt. \tag{8.10}$$

We see that the middle term, because of the ordering indicated in (8.10), vanishes as $x \to \infty$ when compared with the last term in the series. (This follows, by definition, since on $0 < t \le a$, $R_N(t) \le \varepsilon\Phi_N(t)$ for some ε, $0 < \varepsilon < 1$.) Note that the final term in (8.10) can be bounded by

$$\left| \int_{a}^{\infty} \phi(t)e^{-xt}dt \right| \le \int_{a}^{\infty} Me^{ct}e^{-xt}dt = \frac{M}{x-c}e^{-a(x-c)}, \quad \text{for } x > c, \tag{8.11}$$

if $|\phi(t)| \le Me^{ct}$ on $[a, \infty)$. The integral in the sum in (8.10) is

$$\int_{0}^{a} \Phi_n(t)e^{-xt}dt = \int_{0}^{\infty} \Phi_n(t)e^{-xt}dt - \int_{a}^{\infty} \Phi_n(t)e^{-xt}dt. \tag{8.12}$$

The final term in this expression is bounded easily as well as

$$\left| \int_{a}^{\infty} \Phi_n e^{-xt}dt \right| < \int_{a}^{\infty} |\Phi_n|e^{-xt}dt < \frac{M_n}{x-c}e^{-a(x-c)}, \quad \text{for } x > c, \tag{8.13}$$

where $|\Phi_n(t)| \le M_n e^{ct}$, $n = 0, 1, \ldots, N$. Thus, putting this result together with (8.11), and inserting into (8.10), we have

$$f = \sum_{n=0}^{N} f_n(x) + o(f_N) + \mathcal{O}\left(\frac{e^{-a(x-c)}}{x-c}\right), \quad \text{for } x > c, \tag{8.14}$$

where a is some positive constant, and the terms in the asymptotic series for f are given by

$$f_n(x) \equiv \int_{0}^{\infty} \Phi_n(t)e^{-xt}dt. \tag{8.15}$$

It is easy to observe that this series in indeed asymptotic. Recall that the asymptotic series for ϕ had the property that $\Phi_{n+1} = o(\Phi_n)$, that is to say

$$\lim_{t \to 0} \left(\frac{|\Phi_{n+1}(t)|}{|\Phi_n(t)|} \right) = 0. \tag{8.16}$$

Hence, we have that, in summary

$$|f_{n+1}| < \int_0^\infty |\Phi_{n+1}(t)| e^{-xt} dt = o\left(\int_0^\infty |\Phi_n(t)| e^{-xt} dt \right)$$

$$= o(|f_n(x)|), \quad \text{as } x \to \infty, \tag{8.17}$$

because the series $\{f_n\}$ is an asymptotic series.

This result requires only that the function $\phi(t)$ be integrable near $t = 0$. Actually, as a corollary, if the function $\phi(t)$ is analytic in the neighborhood of $t = 0$, (and of exponential order as $t \to \infty$) then the asymptotic series for ϕ is in fact a convergent Taylor series, and in that case, $\Phi_n = \phi^{(n)}(0) t^n / n!$, and substituting into (8.15) gives the special case for (8.14), namely,

$$f(x) \sim \sum_{n=0}^N \frac{\phi^{(n)}(0)}{x^{n+1}} + o(x^{-(N+1)}), \quad \text{as } x \to \infty, \tag{8.18}$$

which is the standard version of Watson's lemma. (See (Bender and Orszag), (Bleistein and Handelsman), (Carrier, Krook, and Pearson).) We now turn to a few examples that illustrate the use of Watson's lemma.

8.4.1 Worked Examples

Example 8.3. Consider the integral,

$$f(x) = \int_0^\infty (t^2 + 1)^3 e^{-xt} dt. \tag{8.19}$$

The function ϕ is clearly analytic in this case, so Watson's lemma (8.18) applies, and since $\phi = 1 + 3t^2 + 3t^4 + t^6$, it is trivial to obtain

$$f \sim \frac{1}{x} + \frac{6}{x^3} + \frac{72}{x^5} + \frac{6!}{x^7}, \quad \text{for } x \to \infty.$$

Now, in this case this asymptotic series ends at four terms, and so this is the exact solution for the integral!

Example 8.4. Here we consider a more complicated example,

$$g(x) = \int_0^\infty \log\left(\frac{\sqrt{1 + t^2}}{t} \right) e^{-xt} dt. \tag{8.20}$$

Clearly ϕ has a branch point at $t = 0$, so we need to use the more general technique by writing ϕ in an asymptotic series valid near $t = 0$. Doing that, we obtain

$$\log\left(\frac{\sqrt{1+t^2}}{t}\right) = \frac{1}{2}\log(1+t^2) - \log t = -\log t + \frac{1}{2}t^2 - \frac{1}{4}t^4 + \mathcal{O}(t^6).$$

Substituting into (8.20), we have

$$g(x) = \int_0^\infty \left(-\log t + \frac{1}{2}t^2 + \frac{1}{4}t^4 + \mathcal{O}(t^6)\right) e^{-xt}dt + R, \qquad (8.21)$$

where R denotes the remainder, estimated from (8.14). Making a change of variable $u = xt$, the integrals can be written down, and, hence,

$$f \sim \frac{\log x}{x} + \frac{Q}{x} + \frac{1}{x^3} - \frac{6}{x^5} + \mathcal{O}\left(\frac{1}{x^8}\right), \qquad \text{for } x \to \infty,$$

$$Q \equiv \int_0^\infty \log u\, e^{-u}du = -0.57727104.$$

Example 8.5. The function

$$f(x) = \int_0^1 t^{1/2}e^{-xt}\sin(\log t)dt,$$

is not obviously a candidate for Watson's lemma because $\sin(\log t)$ has no Taylor series about $t = 0$. Write the integral as

$$f(x) = \Im\text{mag} \int_0^1 t^{1/2}e^{-xt}e^{i\log t}dt = \Im\text{mag}\int_0^1 t^{\frac{1}{2}+i}e^{-xt}dt \equiv \Im\text{mag}(F).$$

We make use of the idea that arises in the Watson's lemma proof, that the upper limit may be taken to infinity in an asymptotic sense (see also exercise 8.30) to conclude that

$$F(x) \sim \int_0^\infty t^{\frac{1}{2}+i}e^{-xt}dt = \frac{\Gamma(\frac{3}{2}+i)}{x^{\frac{3}{2}+i}} = \frac{e^\alpha(\cos\beta + i\sin\beta)}{x^{\frac{3}{2}}x^i},$$

having written $\Gamma = \exp(\alpha + i\beta)$. From a suitable table of values of the Γ-function for $\Gamma(3/2 + i)$ (see (Abramowitz and Stegun)),

$$\alpha = \Re\text{eal}\{\log\Gamma(z)\} = -0.541218868547$$
$$\beta = \Im\text{mag}\{\log\Gamma(z)\} = 0.152140993452. \qquad (8.22)$$

Since $\log\Gamma(z) = \alpha + i\beta$, $e^\alpha \approx 0.582038392$, $\cos\beta = 0.988448866$, $\sin\beta = 0.151554741$ the desired function is found to be

$$f(x) \sim \Im\text{mag}\frac{e^\alpha(\cos\beta + i\sin\beta)}{x^{\frac{3}{2}}}e^{-i\log x}$$

$$= \Im\text{mag}\frac{e^\alpha(\cos\beta + i\sin\beta)}{x^{\frac{3}{2}}}(\cos(\log x) - i\sin(\log x)).$$

Finally,

$$f(x) \sim \frac{e^\alpha}{x^{3/2}} \left(\sin(\beta) \cos(\log x) - \cos(\beta) \sin(\log x) \right)$$

$$f(x) \sim \frac{0.08821068 \cos(\log x) - 0.57531519 \sin(\log x)}{x^{3/2}} \qquad \text{as } x \to \infty.$$

Example 8.6. To anticipate the generalization of the method for Laplace-type integrals to integrals of a more general character, consider the example,

$$f(x) = \int_0^\infty t^2 e^{-(t^3/3 - t)x} dt.$$

As $x \to \infty$, the exponential gets smaller and smaller, at least for $t^3/3 - t > 0$; however for values of t between 0 and $\sqrt{3}$, the coefficient of x in the exponential is positive. The largest contribution to the exponential, and therefore the integral, comes from the neighborhood of $t = 1$, where the function $h(t) = -(t^3/3 - t)$ has its maximum. In that neighborhood, we write $t = 1 + u$, and then the integral takes the form:

$$f(x) = \int_{-1}^\infty (u+1)^2 e^{-(u^3/3 + u^2 - 2/3)x} du.$$

This is of course exact at this point. However, let $u = v/\sqrt{x}$ and substitute,

$$f(x) = \frac{e^{2x/3}}{\sqrt{x}} \int_{-\sqrt{x}}^\infty \left(\frac{v}{\sqrt{x}} + 1 \right)^2 e^{-v^2} e^{-v^3/3\sqrt{x}} dv.$$

Now, as $x \to \infty$, the leading-order behavior would appear to be given by

$$f(x) \sim \frac{e^{2x/3}}{\sqrt{x}} \int_{-\infty}^\infty e^{-v^2} dv = \sqrt{\frac{\pi}{x}} e^{2x/3}, \qquad \text{as } x \to \infty. \qquad (8.23)$$

Example 8.7. Though it is not often needed, an infinite asymptotic expansion may sometimes be calculated. So for the complementary error function

$$\text{erfc}(x) = \frac{2}{\sqrt{\pi}} \int_x^\infty e^{-z^2} dz,$$

let $t = z - x$. Then

$$\text{erfc}(x) = \frac{2}{\sqrt{\pi}} \int_0^\infty e^{-(t+x)^2} dt = \frac{2e^{-x^2}}{\sqrt{\pi}} \int_0^\infty e^{-t^2} e^{-2xt} dt.$$

Here, since $f(t) = e^{-t^2}$ is analytic at $t = 0$, the result (8.18) applies with x replaced by $2x$. Knowing that

$$f(t) = \sum_{n=0}^\infty (-1)^n \frac{t^{2n}}{n!},$$

$$\text{erfc}(x) \sim \frac{2e^{-x^2}}{\sqrt{\pi}} \sum_{n=0}^\infty \frac{(-1)^n}{n!} \frac{(2n)!}{(2x)^{2n+1}}, \qquad \text{as } x \to \infty.$$

Though the series expansion is clearly asymptotic, it can be shown quite easily by the ratio test, that is divergent if $x \neq \infty$.

8.4.2 Application: Early-Time Heat Transfer

In Section 5.6, we determined the solution to the problem of heat conduction in a spherical shell $r_1 \leq r \leq r_2$, with spherically symmetric sources $f(r)/r$. This is a model of the effect of transients in an environment where the early-time behavior is likely to be crucial such as in a nuclear reactor.

The temperature T at the inner surface $r = r_1$, at time t, is given by

$$T(r_1, t) = \frac{1}{\sqrt{\pi t}} \int_0^\infty e^{-\tau^2/4t} v(\tau) d\tau. \tag{8.24}$$

The function v is defined in terms of its Laplace transform by

$$V(\lambda) := \int_0^\infty e^{-\lambda \tau} v(\tau) d\tau = \frac{1}{W(\lambda)} \int_{r_1}^{r_2} g_2(\xi; \lambda) f(\xi) d\xi. \tag{8.25}$$

The functions g_2 and W have the forms:

$$g_2(r; \lambda) = \lambda r_2 \cosh(\lambda(r_2 - r)) - \sinh(\lambda(r_2 - r))$$

and

$$W(\lambda) = \lambda(\lambda^2 r_1 r_2 - 1) \sinh(\lambda(r_2 - r_1)) + \lambda^2(r_2 - r_1) \cosh(\lambda(r_2 - r_1)).$$

By Watson's lemma, we know that for (8.24)

$$T(r_1, t) \sim \frac{v(0)}{\Gamma(1)} + \frac{t^{1/2} v'(0)}{\Gamma\left(\frac{3}{2}\right)} + \frac{t v''(0)}{\Gamma(2)} + \cdots, \quad \text{as } t \to 0^+.$$

The lemma's application to the first integral in (8.25) as $\lambda \to \infty$ leads to

$$V(\lambda) \sim \frac{v(0)}{\lambda} + \frac{v'(0)}{\lambda^2} + \frac{v''(0)}{\lambda^3} + \cdots. \tag{8.26}$$

We make use of the second integral in (8.25), setting

$$\lambda(r - r_2) = k, \quad \lambda(r_2 - r_1) = \beta,$$

which gives exactly

$$V(\lambda) = \frac{1}{\lambda^2} \int_0^\beta \frac{[(\lambda r_2 - \tanh \beta) \cosh k + (1 - \lambda r_2 \tanh \beta) \sinh k]}{(\lambda^2 r_1 r_2 - 1) \tanh \beta + \lambda(r_2 - r_1)}$$

$$\times f(r_1 + k/\lambda) dk. \tag{8.27}$$

Our interest is in the behavior as $\lambda \to \infty$ for which

$$\tanh \beta = 1 - 2e^{-2\beta} + O(e^{-4\beta}).$$

The behavior of $V(\lambda)$ is then

$$V(\lambda) \sim \frac{1}{\lambda^2} \int_0^\infty \frac{e^{-k} f(r_1 + k/\lambda)}{\lambda r_1 + 1} dk \quad \text{(with } k/\lambda = w\text{)}$$

$$= \frac{1}{\lambda (\lambda r_1 + 1)} \int_0^\infty e^{-\lambda w} f(r_1 + w) dw$$

$$\sim \frac{1}{\lambda^2 (\lambda r_1 + 1)} \left[f(r_1) + \frac{f'(r_1)}{\lambda} + \frac{f''(r_1)}{\lambda^2} + \cdots \right],$$

as $\lambda \to \infty$, again by Watson's lemma. Comparing this result with (8.26) gives $v(0) = 0$, $v'(0) = 0$ and

$$v''(0) = \frac{f(r_1)}{r_1}.$$

The desired expansion is thus

$$T(r_1, t) \sim \frac{f(r_1)}{r_1} t \quad \text{as } t \to 0^+.$$

Though the leading order expansion could likely be obtained by other methods, this approach has the advantage that it leads to an expansion of any desired order. This was done by (Wilkins).

The result has physical significance since the total amount of heat produced in the shell in time t is given by

$$\int_{r_1}^{r_2} 4\pi t r f(r) dr.$$

The average temperature rise T_{ave} is determined by

$$\frac{4\pi}{3} (r_2^3 - r_1^3) T_{\text{ave}} = \int_{r_1}^{r_2} 4\pi t r f(r) dr.$$

In the limit as $r_2 \to r_1$, for a thin shell,

$$T_{\text{ave}} = \frac{f(r_1)}{r_1} t.$$

For the special case where the source of heat is due to neutrons or isotropic gamma rays, then $f(r)$ has the form

$$f(r) = \frac{H e^{-\mu r}}{4\pi r},$$

where H is a constant. Then,

$$T(r_1, t) \sim \frac{H e^{-\mu r_1}}{4\pi r_1^2} t.$$

8.5 Generalized Laplace Integrals: Laplace's Method

Equation (8.7) may take a slightly more general form

$$f(x) = \int_\alpha^\beta g(t)e^{xh(t)}dt. \tag{8.28}$$

In this situation, the integral may have significant large-x contributions from two sources: the neighborhoods of α and β, and in the vicinity of any maxima of h that happen to lie on the interval between α and β.

Suppose that there is one point in (α, β), say at $t = t_o$, at which $h' = 0$ and, further, $h''(t_o) < 0$, so that the point is indeed a maximum of h. In that case, we decompose (8.28) into two pieces by writing $t = t_o - v$ in $t < t_o$ and $t = t_o + u$ in $t > t_o$, so that (8.28) becomes

$$f(x) = e^{xh(t_o)} \int_0^{t_o - \alpha} g(t_o - v))e^{x(h(t_o - v) - h(t_o))}dv$$

$$+ e^{xh(t_o)} \int_0^{\beta - t_o} g(t_o + u))e^{x(h(t_o + u) - h(t_o))}du. \tag{8.29}$$

In the first of these integrals, we define a new integration variable by

$$\xi = h(t_o) - h(t_o - v).$$

This expression may be inverted in principle to $v = V(\xi)$. In that case, the first term in (8.29) becomes

$$f_1(x) = e^{xh(t_o)} \int_0^{h(t_o) - h(\alpha)} G_1(\xi)e^{-x\xi}d\xi, \quad G_1(\xi) \equiv \frac{g(t_o - V(\xi))}{h'(t_o - V(\xi))}.$$

The second integral can be done in precisely the same way, with the result that (8.29) becomes

$$f(x) = e^{xh(t_o)} \int_0^{h(t_o) - h(\alpha)} G_1(\xi)e^{-x\xi}d\xi - e^{xh(t_o)} \int_0^{h(t_o) - h(\beta)} G_2(\xi')e^{-x\xi'}d\xi', \tag{8.30}$$

where,

$$G_2(\xi') \equiv \frac{g(t_o + U(\xi'))}{h'(t_o + U(\xi'))},$$

with $u = U(\xi')$ the inverse of $\xi' = h(t_o) - h(t_o + u)$.

Each of these integrals, apart from the exponential multiplier, is an asymptotic integral of the type (8.7), for which Watson's lemma was proved. Hence, we can proceed as usual for such an integral: expand the functions G_1 and G_2 in asymptotic series for $\xi \to 0$, then integrate term-by-term.

To proceed, for now consider the first integral only. From the definition of ξ above, notice that when $\xi = 0$, as is $v = 0$, so we we can write $\xi = -h''(t_o)v^2/2 + h'''(t_o)v^3/6 + \cdots$. This series may be reversed by Lagrange's method (see (Markusevich, Vol. II, chapter 3)), and becomes an expression for the inverse of the function, as

$$V(\xi) = \sqrt{\frac{2\xi}{-h''(t_o)}} - \frac{h'''(t_o)}{3h''(t_o)^2}\xi + \cdots.$$

This asymptotic expansion for V is then substituted into G_1 with the result that

$$G_1(\xi) \sim \frac{g(t_o)}{\sqrt{-2h''(t_o)\xi}} + \frac{1}{h''(t_o)}\left[g'(t_o) - \frac{1}{3}\frac{g(t_o)h'''(t_o)}{h''(t_o)}\right] + \mathcal{O}(\xi^{1/2}).$$

Substitution into the first integral of (8.29) gives integrals in ξ easily evaluated, and eventually, we have

$$f_1(x) \sim g(t_o)\sqrt{\frac{\pi}{-2xh''(t_o)}}e^{xh(t_o)} + \frac{1}{xh''(t_o)}\left[g'(t_o) - \frac{1}{3}\frac{g(t_o)h'''(t_o)}{h''(t_o)}\right]e^{xh(t_o)}$$

$$+ \mathcal{O}\left(x^{-3/2}e^{xh(t_o)}\right), \quad \text{for } x \to \infty.$$

A similar expression may be worked out for f_2, by exactly the same procedure, and the result is

$$f_2(x) \sim g(t_o)\sqrt{\frac{\pi}{-2xh''(t_o)}}e^{xh(t_o)} - \frac{1}{xh''(t_o)}\left[g'(t_o) - \frac{1}{3}\frac{g(t_o)h'''(t_o)}{h''(t_o)}\right]e^{xh(t_o)}$$

$$+ \mathcal{O}\left(x^{-3/2}e^{xh(t_o)}\right), \quad \text{for } x \to \infty.$$

Combining with the f_1 asymptotics above,

$$f \sim g(t_o)\sqrt{\frac{2\pi}{-xh''(t_o)}}e^{xh(t_o)} + \mathcal{O}\left(x^{-3/2}e^{xh(t_o)}\right), \quad \text{for } x \to \infty. \tag{8.31}$$

If the critical point at which $h(t)$ is a maximum had occurred at the end-point of the interval, then only one of the asymptotic expressions would have been needed. This result then uses Watson's lemma and the fundamental idea: asymptotic expansion of the function g near the critical point of the exponential ($h' = 0$) leads to a term-by-term integration which gives the asymptotic expansion for the integral. There are situations in which it may be desirable and possible to obtain many more terms in the asymptotic expansion of the generalized Laplace integral. This has been recently considered in detail in a review article by (Wojdylo).

Example 8.8. In Example 8.6, we illustrated the ideas put forward here in this more general setting. That problem is

$$f(x) = \int_0^\infty t^2 e^{-(t^3/3-t)x}\,dt.$$

After performing the indicated changes of variables near the critical point $t = 1$, we have exactly,

$$f(x) = \int_0^1 (1-v)^2 e^{(v^3/3-v^2-2/3)x}\,dv + \int_0^\infty (u+1)^2 e^{-(u^3/3+u^2-2/3)x}\,du.$$

The choice of a new variable in the first integral's exponent is

$$\xi = v^2 - \frac{v^3}{3}. \tag{8.32}$$

Here, $\xi \sim v^2$ near $v = 0$, so we expect to reverse the series as

$$v = \xi^{1/2} + b_2\xi + b_3\xi^{3/2} + \cdots.$$

In fact, this is possible from the theory of inverses of multivalued functions as in (Markusevich, Vol. II, chapter 3). Such an expansion, substituted in (8.32) gives $b_2 = \frac{1}{6}$, $b_3 = \frac{35}{216}$, and so forth. This type of expansion has the advantage that

$$dv = \left(\frac{1}{2}\xi^{-1/2} + b_2 + \frac{3}{2}b_3\xi^{1/2} + \cdots\right)d\xi,$$

and thereby may be substituted directly into the expansions of the integrand, and for application of Watson's lemma, the upper limit is taken to be ∞. Likewise, in the second integrand take

$$\eta = u^3/3 + u^2,$$

where $\eta \sim u^2$ near $u = 0$. Because of the difference in sign, it turns out that

$$u = \eta^{1/2} - b_2\eta + b_3\eta^{3/2} + \cdots,$$

with the same values of b_2 and b_3 as before. Inserting each of these expansions into their respective integrals there follows:

$$f(x) \sim e^{2x/3}\int_0^\infty \left(1 - (\xi^{1/2} + b_2\xi + \cdots)^2\right)\left(\frac{1}{2}\xi^{-1/2} + b_2 + \cdots\right)e^{-x\xi}\,d\xi$$

$$+ e^{2x/3}\int_0^\infty \left(1 + (\eta^{1/2} - b_2\eta + \cdots)^2\right)\left(\frac{1}{2}\eta^{-1/2} - b_2 + \cdots\right)e^{-x\eta}\,d\eta.$$

Computing the integrals leads to

$$f(x) \sim \frac{\sqrt{\pi}}{x^{1/2}}e^{2x/3}\left(1 + \frac{35}{144x} + O(x^{-3/2})\right),$$

as $x \to \infty$. A computer algebra system, such as *Maple*, is extremely useful for this type of calculation.

8.5.1 Stirling's Formula

We have on several occasions encountered the Γ-function with the integral representation

$$\Gamma(s + 1) = \int_0^\infty z^s e^{-z}\,dz, \quad s > -1.$$

When s is an integer n, $n! = \Gamma(n + 1)$. In order to employ Laplace's method, write the integral as

$$\Gamma(s + 1) = \int_0^\infty e^{s\log z}e^{-z}\,dz.$$

Examine the point in the exponential where $s \log z - z$ has a critical point, that is, where $z = s$, which is not fixed. Hence introduce a new variable of integration $t = z/s$. Thus, equivalently

$$\Gamma(s+1) = s^{s+1} \int_0^\infty e^{s(\log t - t)} dt.$$

The integral is now in a form suitable for Laplace's method (8.28), with $g(t) = 1$, $h(t) = \log t - t$, $t_0 = 1$, $h''(t) = -1/t^2$, and we have Stirling's formula

$$s! = \Gamma(s+1) \sim \sqrt{2\pi} e^{-s} s^{s+1/2}, \quad \text{as} \quad s \to \infty.$$

Using the method of steepest descent to be discussed in Section 8.6, this can be extended to complex values of s.

8.6 Method of Steepest Descent

We consider in this section an extension of the method for generalized Laplace integrals we have just done above.

$$f(z) = \int_\alpha^\beta g(t) e^{zh(t)} dt, \tag{8.33}$$

where z is itself a complex variable and the integration occurs in a complex t plane from α to β. We seek the asymptotic behavior of $f(z)$ for $|z| \to \infty$, and in fact that behavior may be quite different depending on $\arg(z)$.

Suppose for the moment that we deform the integration path from α to β in a way that does not cross a singularity, and furthermore the path has a particular character: it is a path, C_s, along which $\Im\text{mag}(h) = $ constant. Such a path is called a "steepest path" ((Jeffreys and Jeffreys); (Bleistein and Handelsman); (Olver)), which we will describe in what follows. If we write $h = \phi + i\psi$ on that path, then from (8.33) we have

$$f(z) = e^{i\psi z} \int_{C_S} g(t) e^{z\phi(t)} dt. \tag{8.34}$$

Interestingly, this is precisely the sort of integral done by Laplace's method, since the exponential multiplier of z is entirely real. The dominant contribution to the integral for $|z| \to \infty$ is as before from the neighborhood of the maximum of ϕ, at least for $\Re\text{eal}(z) > 0$, which we now assume. (If $\Re\text{eal}(z) < 0$, then we replace h by $(-h)$ and proceed as here.)

The maxima of $\Re\text{eal}(h)$ would occur where $\partial\phi/\partial x = \partial\phi/\partial y = 0$. Since we know that $h'(t) = \partial\phi/\partial x - i\partial\phi/\partial y$, we conclude that the possible maxima are at the critical points of h.

Critical Points: $h'(t) = 0$

However, since h is analytic, $\nabla^2 \phi = 0$, and consequently the critical point is a saddlepoint of ϕ. For this reason, the method is sometimes referred to by that name.

The crux of the method then, is to be able to so deform the path as to obtain C_s just described. This is indeed possible. At the critical point, we also have $\partial \psi / \partial x = \partial \psi / \partial y = 0$, by the Cauchy–Riemann equations, and $\nabla \phi \cdot \nabla \psi = 0$, so the two vector fields are orthogonal. Thus, $\Im \mathrm{mag}(h) = $ constant is compatible with this requirement.

Suppose that the deformed path has resulted in a critical point located at t_o, say. Then, near that location

$$\phi(t) = \phi(t_o) + \frac{1}{2}\phi''(t_o)(t - t_o)^2 + \frac{1}{6}\phi'''(t_o)(t - t_o)^3 + \cdots.$$

We then make a change of variable, remembering that in the case of Watson's lemma we need only write g in an asymptotic series near, in this case, the critical point t_o. In the particular case that g is analytic at t_o, on writing $t - t_o = u$, we have, from (8.28),

$$f(z) = e^{h(t_o)z} \int_{C_S} \left[g(t_o) + g'(t_o)u + \mathcal{O}(u^2) \right] e^{z(h''(t_o)u^2/2 + \mathcal{O}(u^3|z|)} du. \tag{8.35}$$

In order that we are assured that the path C_S is one along which, near the critical point, ϕ is strictly decreasing away from t_o, we write the complex quantities z and $\phi''(t_o)$ in the polar form $z = |z| \exp(i\alpha)$ and $h''(t_o) = |h''(t_o)| \exp(i\beta)$. Near $u = 0$ suppose the path is oriented at an angle γ relative to the positive real axis. Then $u = r \exp(i\gamma)$ on a path of steepest descent through t_o and

$$zh''(t_o)u^2 = r^2 |z| |h''(t_o)| e^{2i\gamma + i\beta + i\alpha}.$$

We desire the exponent to be so that

$$\cos(\alpha + \beta + 2\gamma) = -1 \Rightarrow \gamma = \left(n + \frac{1}{2} \right) \pi - \frac{\beta + \alpha}{2}. \tag{8.36}$$

Doing the integrals in sequence in the fashion suggested by Watson's lemma (see also Exercise 8.22) then yields for (8.35)

$$f(z) \sim \sqrt{\frac{2\pi}{|zh''(t_o)|}} g(t_o) e^{zh(t_o) + i\gamma} + \mathcal{O}(|z|^{-3/2} e^{zh(t_o)}), \quad \text{for } |z| \to \infty. \tag{8.37}$$

There are modifications to this result, should the critical point be nonsimple; these are discussed in more specialized texts as in (Bleistein and Handelsman). We turn now to several examples of the use of the steepest descent technique.

Remark 8.1. *If on the original contour in (8.33), $\Re \mathrm{eal}(h(t)) \leq M$, then no critical point $t = t_o$ at which $\Re \mathrm{eal}(h(t_o)) > M$ can contribute to the asymptotic expansion of the integral along C.*

8.6.1 Application: A Special Function

Consider the integral

$$f(x, s) = 2 \int_0^\infty \cos(st^3) e^{-xt^2} \, dt, \tag{8.38}$$

which can be written more conveniently as

$$f(x, s) = \int_{-\infty}^\infty e^{ist^3 - xt^2} \, dt = \frac{x}{s} \int_{-\infty}^\infty e^{p(iu^3 - u^2)} \, du, \quad p \equiv \frac{x^3}{s^2}. \tag{8.39}$$

The first form in (8.39) leads to the second in the following way. The exponential might be written as $\varphi(t; x, s) = ist^3 - xt^2$. Then, consider the places where

$$\frac{\partial \varphi}{\partial t} = 3ist^2 - 2xt = 0.$$

Aside from $t = 0$, we have $t = -2ix/3s$. This critical point moves with (s, t). Hence set $u = t/(x/s)$, so that for $x > 0$, $t > 0$, u increases with t. Making this substitution gives the second form in (8.39). Clearly, we can obtain the behavior for $|p| \to \infty$ from the second form, but for $p \to 0$, we will see that it is appropriate to use the first.

Behavior as $p \to +\infty$

From (8.39), since $h(u) = iu^3 - u^2$, putting $h' = 0$ in the required way leads to critical points at $u = 0$ and $u = -2i/3$. At $u = 0$, the steepest descent path is along the real axis where $\Re\text{eal}(h) = -u^2$. Thus curves through that point can be steepest paths as required. On the other hand, at $u = -2i/3$, $\Re\text{eal}(h(-2i/3)) = 4/27 > 0$, the maximum of $\Re\text{eal}(h)$ on the original path and so is called an "inadmissible critical point." (See p. 267 of (Bleistein and Handelsman) and Remark 8.1.) The steepest descent paths themselves are given, by taking $\Im\text{mag}(h) = \text{const.}$ by

$$u = \xi + i\eta : \quad \xi(\xi^2 - 3\eta^2 - 2\eta) = \text{const.}$$

In particular, putting the constant to zero for the paths through the origin leads to hyperbolic steepest descent paths sketched in Figure 8.1.

In the neighborhood of the origin the path is parallel to the real axis, so we have

$$f(x, s) = \frac{x}{s\sqrt{p}} \int_C e^{p^{-1/2} iv^3} e^{-v^2} \, dv$$

$$\sim \frac{x}{s\sqrt{p}} \int_{-\infty}^\infty e^{-v^2} \left[1 + \frac{iv^3}{p^{1/2}} - \frac{1}{2} \frac{v^6}{p} + \cdots \right] dv, \quad \text{for } p \to \infty.$$

Doing the v integrals, we have the asymptotic behavior as

$$f(x, s) \sim \sqrt{\frac{\pi}{x}} \left[1 - \frac{15}{16p} + \mathcal{O}(p^{-2}) \right], \quad \text{for } p \to +\infty. \tag{8.40}$$

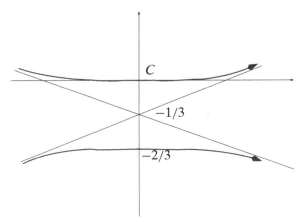

Figure 8.1. Integration paths for steepest descent for Section 8.6.1.

Behavior as $p \to -\infty$

Now examine the limit as $p \to -\infty$. In this case, let $p = -q$, so now $q \to \infty$, and the quantity h is now $h = u^2 - iu^3$. What happens is that paths from the critical point $u = 0$ are now of *steepest ascent* (see (Bleistein and Handelsman)) and are not appropriate. The maximum of $\Re\text{eal}(h)$ along $\Im\text{mag}(h) = \text{const.}$ occurs at $u = -2i/3$. Because of the geometry of the steepest paths shown above, we write $u = -2i/3 + v$ and substitute into (8.39), which results in

$$f(x,s) = \frac{x}{s\sqrt{q}} e^{-8q/27} \int_C e^{-q^{-1/2} i r^3} e^{-r^2} \, dr$$

$$\sim \frac{x}{s\sqrt{q}} e^{-8q/27} \int_{-\infty}^{\infty} e^{-r^2} \left[1 - \frac{ir^3}{q^{1/2}} - \frac{1}{2} \frac{r^6}{q} + \cdots \right] dr, \quad \text{for } q \to \infty,$$

where the new variable is $r = vq^{1/2}$. Again, doing the r integrals in the usual way,

$$f(x,s) \sim \sqrt{\frac{\pi}{-x}} e^{8p/27} \left[1 + \frac{15}{16p} + \mathcal{O}(p^{-2}) \right], \quad \text{for } p \to -\infty.$$

Behavior as $p \to 0$

We now explore the behavior of the integral as $p \to 0$, which corresponds to taking $x \to 0$ and $s = \mathcal{O}(1)$. In this case, from (8.39), the relevant $h = it^3$, which obviously has its only critical point at the origin. The steepest paths are located on lines $\cos(3\theta) = 0$, if $t = r \exp(i\theta)$. Note that, along such a straight line, we can write

$$it^3 = r^3 (i \cos(3\theta) - \sin(3\theta)) = -r^3 \sin(3\theta).$$

On the line $\theta = \pi/6$, clearly $it^3 < 0$, and also on $\theta = 5\pi/6$. However, on the line $\theta = \pi/2, it^3 > 0$. Therefore, it seems that in order to have the origin be a maximum of ϕ on the path, we must choose the deformed integration path to be given as in the sketch (Figure 8.2).

Figure 8.2. Integration path for $p \to 0$.

We must integrate on these two paths, and combine. Finally that leads to

$$f = \frac{2}{s^{1/3}} \Re\text{eal}\left[e^{i\pi/6} \int_0^\infty e^{-v^3} e^{p^{1/3}v^2 \exp(i\pi/3)} dv \right].$$

Doing the usual asymptotic expansion in p, then integrating, finally we find that

$$f(x,s) \sim \frac{1}{\sqrt{3}s^{1/3}}\left[\Gamma\left(\frac{1}{3}\right) - p^{2/3}\Gamma\left(\frac{5}{3}\right) + \mathcal{O}(p^{4/3}) \right], \quad p \to 0.$$

Hence, we have a quite complete picture of how f behaves for both small and large values of p.

8.6.2 Application: The Oscillating Plate

We recall now the results given in Example 5.2, the flow due to the oscillation of a plate. We found that the solution split into two parts: a long-time periodic solution, and a transient which, from Equation (5.23), is given by

$$u_T = -u_1 I(y,t), \quad I(y,t) = \frac{1}{\pi}\Im\text{mag}\left(\int_0^\infty e^{-rt + i(r/v)^{1/2}y} \frac{r\,dr}{r^2 + \omega^2} \right)$$

$$= \frac{1}{\pi}\int_0^\infty e^{-rt}\sin\left(r^{1/2}y/\sqrt{v}\right)\frac{r\,dr}{r^2 + \omega^2}. \tag{8.41}$$

We can use Watson's lemma to obtain the large-t behavior of $I(y,t)$ quite simply, taking y to be order one, and so the large-t behavior is governed by expanding the integrand about $r = 0$, so we get

$$\frac{\sin\left(r^{1/2}y/\sqrt{v}\right)r}{\omega^2\left[(r/\omega)^2 + 1\right]} \sim \frac{r}{\omega^2}\left[1 - \frac{r^2}{\omega^2} + \cdots \right] \times \left[r^{1/2}\tau - \frac{1}{6}r^{3/2}\tau^3 + \mathcal{O}(r^{5/2}\tau^5) \right],$$

where $\tau \equiv y/\sqrt{v}$. Inserting this into the integral and evaluating (see Exercise 8.30), using the fact that $\int_0^\infty \varrho^k \exp(-\varrho)d\varrho = \Gamma(k+1)$ gives the asymptotic series

$$I \sim \frac{3y}{2\omega^2 t\sqrt{v\pi t}}\left[1 - \frac{5y^2}{12vt} + \mathcal{O}\left(\frac{1}{t^2}\right) \right].$$

It is immediately obvious from this result that this series is asymptotic for large t, with y not large, that is, $t \to \infty$, y fixed.

We can also explore the behavior for large time if y is also large, by making a change of variable in the form $r = (y^2/vt^2)v$ (in order to fix the critical point), so that the integral (8.41) becomes

$$I = \frac{1}{\pi}\Im\text{mag}\left(\int_0^\infty \frac{e^{p^2(iv^{1/2}-v)}v\,dv}{v^2 + \omega^2 t^2/p^4} \right), \tag{8.42}$$

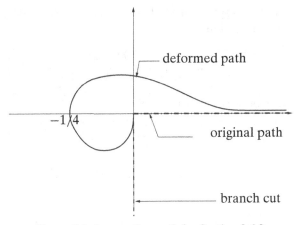

Figure 8.3. Integration path for Section 8.6.2.

where $p = y/\sqrt{vt}$. The critical point is determined by putting the derivative of h to zero, which leads to $\sqrt{v_0} = i/2$. We cut the plane from the origin as shown in Figure 8.3, along the negative imaginary axis. This critical point is then clearly located as shown in the sketch, at $v_0 = -1/4$. In the vicinity of that critical point, we expand h, so

$$iv^{1/2} - v = h(v_0) + \frac{1}{2}h''(v_0)\left(v + \frac{1}{4}\right)^2 + \cdots = -\frac{1}{4} + \left(v + \frac{1}{4}\right)^2 + \cdots. \quad (8.43)$$

So $h'' = 2$ at the critical point, and ultimately we use a path as indicated in the Figure 8.3. Now a brief examination of the nature of the curves of constant imaginary part of h, near the critical point will lead to this.

The steepest descent paths themselves are given by taking $\Im\text{mag}(h) = \text{constant}$ with

$$v = \xi + i\eta: \quad h(v) = iv^{1/2} - v = -\frac{1}{4}.$$

The polar form of the curve is found by setting $v = \rho e^{i\theta}$, with $\Im\text{mag}(h) = 0$ so that

$$\rho = \frac{1}{2(1 - \cos\theta)}, \quad \frac{\pi}{2} < \theta < \frac{3\pi}{2}.$$

Back in Cartesian coordinates, this is the parabola $\xi = -\frac{1}{4} + \eta^2$.

Thus, inserting (8.43) into the exponential and also expanding the remainder of the integrand gives, on writing $v = -1/4 + p^{-1}s$,

$$I = -\frac{4e^{-p^2/4}}{\pi p}\Im\text{mag}\left(\int_P \frac{e^{s^2}(1 - 4s/p) + \cdots}{1 - 8s/p + \cdots}ds\right).$$

The integration path must be deformed to go through the critical point, so we denote this path, for the time being, as P. Note that the critical point must be crossed in a direction so that h decreases away from the critical point. Note also that writing

$s = \exp(i\pi/2)\ell$ gives an argument of the exponential that decreases away from $s = 0$. Making that substitution,

$$I = -\frac{4e^{-p^2/4}}{\pi p}\Im\text{mag}\left(\int_P \frac{ie^{-\ell^2}(1 - 4i\ell/p) + \cdots}{1 - 8i\ell/p + \cdots}d\ell\right).$$

Now, the path must be deformed as shown in Figure 8.3, in order that it passes through the critical point along a line $\Im\text{mag}(h) = \text{const}$. Doing that, and working to first-order in p^{-1} only, we have

$$I \sim -\frac{4e^{-p^2/4}}{\pi p}\Im\text{mag}\left(\int_{-\infty}^{\infty} ie^{-\ell^2}d\ell\right) = -\frac{4e^{-p^2/4}}{\sqrt{\pi}\,p}.$$

The asymptotic form may also be found based on a modified form of Watson's lemma (Exercise 8.22). In fact a single change of variable in (8.42), say $v = -\left(\frac{1}{2} + iu\right)^2$ near $v = -\frac{1}{4}$, will achieve this objective. The result of the usual asymptotic expansion of the integrand, $-\infty < u < \infty$, is

$$I \sim \frac{1}{\sqrt{\pi}\,p}\frac{2e^{-p^2/4}}{[1 + 16(\omega t/p^2)^2]}, \quad \text{for } p \to \infty.$$

Here, it was assumed that the poles of (8.41) do not affect the asymptotic analysis, in that in deforming the contour, they are not crossed. The necessary modifications to the outcome of the method of steepest descents when the integrand has a pole are discussed in detail in an article by (Northover). An estimate is derived of the distance the pole must be from the saddlepoint in order that it may have little effect upon the usual results.

8.6.3 Application: Lee Waves

We now return to the Lee wave problem examined by transform methods earlier, in Section 7.1. We obtained that the result for the perturbation stream function due to the mountain as

$$\psi = \frac{1}{\pi}\Re\text{eal}\left(\int_0^{\infty} e^{ikx - (k^2 - \lambda^2)^{1/2}y}dk\right),$$

and the integration path is taken *below* the branch point at $k = \lambda$ in the complex-k plane; the path is shown in Figure 7.2. For convenience, we make a polar coordinate transformation, so $(x, y) = r(\cos\phi, \sin\phi)$. Then, we have this integral in a steepest-descent form as

$$\psi = \frac{1}{\pi}\Re\text{eal}\left(\int_0^{\infty} e^{h(k)r}dk\right), \quad h = ik\cos\phi - (k^2 - \lambda^2)^{1/2}\sin\phi. \quad (8.44)$$

We now seek the far-field structure of the solution by letting $r \to \infty$. The critical points are located at $h' = 0$ as usual, and clearly they are located at solutions of

$$\frac{i\cos\phi}{\sin\phi} - \frac{|k|}{\sqrt{r_1 r_2}}e^{i\theta - i(\theta_1 + \theta_2)/2} = 0, \quad (8.45)$$

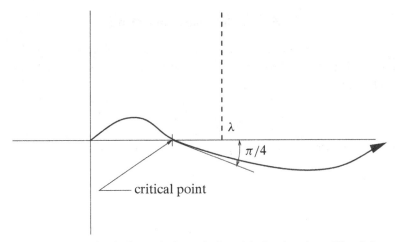

Figure 8.4. For Section 8.6.3, the path through the critical point; dotted line is branch cut.

where we have written, to handle the branch points properly, $k \pm \lambda = r_{1,2} \exp(i\theta_{1,2})$ and $k = r \exp(i\theta)$. Squaring the equation and solving, we obtain solutions $k = \pm \lambda \cos \phi$. However, on insertion into (8.45) we find that, since $\theta_2 = -i\pi$, the only solution of (8.45) is at $k = \lambda \cos \phi$, but only in $\cos \phi > 0$ – that is, downstream of the bump. Writing $k = i\lambda \cos \phi + u$, we find

$$h = i\lambda - \frac{i}{2\lambda \sin^2 \phi} u^2 - \frac{i}{2\lambda \sin^2 \phi} u^3 + \cdots .$$

In order to determine the direction of the path through the critical point, let $u = |u| \exp(i\alpha)$. In that case, the second term in the above equation becomes

$$-\frac{1}{2\lambda \sin^2 \phi} |u|^2 e^{i\pi/2 + 2i\alpha},$$

from which it is immediately evident that if the critical point is to sit atop a maximum of $\Re\text{eal}(h)$ along $\Im\text{mag}(h) = \text{const.}$, then $\alpha = -i\pi/4$. Hence, we put $u = (2\lambda \sin^2 \phi / r)^{1/2} \exp(-i\pi/4) v$, and substituting into (8.44), we have

$$\psi = \frac{1}{\pi} \left(\frac{2\lambda \sin^2 \phi}{r} \right)^{1/2} \Re\text{eal} \left(e^{i\lambda r - i\pi/4} \int_C e^{-v^2} e^{\beta v^3 / r^{1/2}} dv \right),$$

$$\beta \equiv -\frac{\sqrt{2} e^{i\pi/4}}{\lambda^{3/2} \sin \phi}. \tag{8.46}$$

The path C is as sketched in Figure 8.4. Hence, to leading order, working out the integral gives

$$\psi \sim \left(\frac{2\lambda \sin^2 \phi}{\pi r} \right)^{1/2} \cos \left(\lambda r - \frac{\pi}{4} \right) + \mathcal{O}(r^{-3/2}), \quad r \to \infty.$$

8.6.4 Application: Sound Waves

In Example 6.4, we considered the sound field resulting from the oscillation of a portion of a plane normal to itself. That resulted in a solution which can, for our purposes here, be put in the form

$$V(x, y) = \frac{1}{2\pi} \int_C F(k) e^{ikx - \sqrt{k^2 - \lambda^2} y} dk. \tag{8.47}$$

The integration path is shown in Figure 6.1. The critical points are located at zeros of the derivative in the exponential, so

$$ix = \frac{yk}{\sqrt{k^2 - \lambda^2}}. \tag{8.48}$$

Squaring and solving, we find that solutions of (8.48) appear to be at $k = \pm \lambda \cos \theta$, where (r, θ) are polar coordinates related in the usual way to (x, y). However, great care must be exercised here. We write $k - \lambda = r_1 \exp(i\theta_1)$ and $l + \lambda = r_2 \exp(i\theta_2)$, in which case (8.48) can be written as

$$i = \frac{k}{\sqrt{r_1 r_2}} \tan \theta e^{-i(\theta_1 + \theta_2)/2}. \tag{8.49}$$

Anywhere on the line between the two branch points $\theta_1 = \pi$ and $\theta_2 = 0$. Thus, if we consider for the moment the first quadrant, that is, $x, y > 0$, then the only solution of (8.49) is the one for $k < 0$. (If $x < 0$, then we must use the $k > 0$ critical point.) Considering now the $x, y > 0$ case, and substituting in the usual way the value of the critical point, $k = -\lambda \cos \theta$, we find that

$$h(k_{cp}) = -i\lambda r \quad \text{and} \quad h''(k_{cp}) = \frac{i}{\lambda \sin^3 \theta}.$$

Thus, the leading-order behavior is given by

$$V(x, y) \sim \frac{1}{2\pi} F(k_{cp}) \int_C F(k_{cp}) e^{xh(k)} dk,$$
$$h \sim -i\lambda r + \frac{ri}{2\lambda \sin^2 \theta} (k - k_{cp})^2 + \cdots. \tag{8.50}$$

The steepest path must pass through the critical point in such a way that the coefficient in the second derivative term is real and negative. Hence, we write $k - k_{cp} = \exp(i\pi/4) R(2\lambda \sin^2 \theta / r)^{1/2}$. That results in the standard sort of integral for these problems, $\int_{-\infty}^{\infty} \exp(-v^2) dv = \sqrt{\pi}$. Thus, finally, we have

$$V \sim \sqrt{\frac{\lambda}{2\pi r}} \sin \theta F(k_{cp}) e^{-i\lambda r + i\pi/4}. \tag{8.51}$$

The solution is completed once the forcing function is known. Say that a width of the plate, from $-L$ to L, moves and the rest is stationary. Then, in the notation of Section 6.4, $f = 0, |x| > L$, $f = 1, |x| < L$. The Fourier transform is easily found to be

$$F(k) = \frac{2 \sin(kL)}{k}.$$

In this case, from (8.51),

$$V \sim \sqrt{\frac{2}{\lambda \pi r}} \tan \theta \sin(\lambda L \cos \theta) e^{-i\lambda r + i\pi/4}, \quad r \to \infty. \tag{8.52}$$

Combining with the fact that the actual vertical velocity component is given by $v = \Re\text{eal}\{V \exp(i\omega t)\}$, we have the final result that for motion of a plate of width $2L$, the sound in the far field is given by

$$v \sim \sqrt{\frac{2}{\lambda \pi r}} \tan \theta \sin(\lambda L \cos \theta) \cos\left(\frac{\pi}{4} + \omega t - \lambda r\right). \tag{8.53}$$

An incorrect choice of either the critical point or the path direction can lead to disastrous results: either waves that propagate toward the wavemaker rather than away, or incorrect phase information. Alternatively, if there are two plates, with the plate in $-L < x < 0$ moving 180 degrees out of phase with the plate in $0 < x < L$, then the general result (8.51) still applies but with a different F, and in this case, it turns out that

$$v \sim 2\sqrt{\frac{2}{\lambda \pi r}} \tan \theta \cos^2(\lambda L \cos \theta/2) \cos\left(\frac{3\pi}{4} + \omega t - \lambda r\right). \tag{8.54}$$

The results are in conformity with what is known about two-dimensional waves, namely, that the geometric attenuation factor is $r^{-1/2}$. That could be asserted at the outset, without any of this detailed analysis. However, without the steepest-descent technique coupled with the transform analysis, it is extremely difficult to determine the geometric factors in the solution.

8.6.5 Application: Two-dimensional Laminar Wake

In an analysis similar to that performed for the thermal wake in Section 7.3 (See Problem 6.15), it is possible to show that in a two-dimensional case, a perturbation, u, to the horizontal x-direction velocity, U, in which an object is embedded is given by a single Fourier inversion in x,

$$\frac{u}{U} = -\frac{D}{4\pi \mu i \lambda} \int_{-\infty}^{\infty} G(y, k) dk, \tag{8.55}$$

$$G(y, k) \equiv \frac{\sqrt{k^2 + ik\lambda}}{k} e^{ikx - |y|\sqrt{k^2 + ik\lambda}}, \tag{8.56}$$

where D is the drag on the object, μ is the coefficient of viscosity, and the parameter $\lambda = \rho U/\mu$, with ρ the fluid density. This result is restricted to low-speed flow, though any subsonic flow can be simply related to this one via a similarity transformation. Unlike the three-dimensional case, there is no solution for the two-dimensional wake in terms of transcendental and/or special functions. So, we turn to asymptotics.

To obtain the large-r behavior, we take $h = ik - \tan\theta\sqrt{k^2 + ik\lambda}$, with θ again, as above a polar coordinate, $\tan^{-1}(y/x)$. Then, we have

$$h'(k) = i - \frac{k + i\lambda/2}{\sqrt{k^2 + ik\lambda}}\tan\theta, \quad h''(k) = -\lambda^2\frac{\tan\theta}{(k^2 + ik\lambda)^{3/2}}. \tag{8.57}$$

Putting $h' = 0$ leads to two apparent solutions, obtained by squaring the equation, namely,

$$k_{cp} = -\frac{i\lambda}{2} \pm \frac{i\lambda}{2}\cos\theta. \tag{8.58}$$

A careful assessment precisely like that done in Example 4 above leads to the conclusion that the only one of these solutions that is in fact a solution to the original equation $h' = 0$ is given by the solution in (8.58) with the UPPER sign. Note that this means that in the second quadrant, where the cosine is negative, that the critical point will be further from the real k-axis. Careful substitution of the critical point locations from (8.58) and (8.57), and similar substitution into the rest of the integrand, in G, leads to

$$\frac{u}{U} \sim -\frac{D}{4\mu\pi\lambda\tan\theta}\int_{-\infty}^{\infty} e^{\lambda r(\cos\theta - 1)/2 - ru^2/(\lambda\sin^2\theta) + \cdots}\,du, \quad u \equiv k - k_{cp}. \tag{8.59}$$

Since the coefficient of the u^2 terms is already negative and real, we choose a path on which u is real. Thus, we write $u = (\lambda\sin^2\theta/r)^{1/2}v$, which results in, after doing the standard integral,

$$\frac{u}{U} \sim \frac{D}{4\mu U}\sqrt{\frac{1}{\lambda\pi r}}\cos\theta\, e^{\lambda r(\cos\theta - 1)/2}. \tag{8.60}$$

Note that the quantity λr is actually a Reynolds number based on distance from the object, and for this asymptotic work to be valid, we actually require that $\lambda r \gg 1$. Note that, since $\cos\theta$ is never larger than 1, this solution always goes to zero exponentially as $r \to \infty$. However, near the a positive real axis, where θ is near zero, the exponential is its largest – that is, in the wake! Near to $\theta = 0$, $\cos\theta \sim 1 - \theta^2/2$ and $r \approx x$, so we have the special case

$$\frac{u}{U} \sim -\frac{D}{4\mu U}\sqrt{\frac{1}{\lambda\pi x}}e^{-\lambda y^2/(4x)}, \quad \text{in the wake}. \tag{8.61}$$

As in the thermal case Section 6.3.7, the wake grows parabolically, with $y \sim \sqrt{x/\lambda}$.

8.7 Method of Stationary Phase; Kelvin's Results

Here we consider integrals of the form

$$f(x) = \int_{\alpha}^{\beta} g(t)e^{ixh(t)}dt,$$

h real, which differs in appearance from (8.28) only in that i is in the exponent. However, a different mechanism is a work; it is sometimes described as "self-cancellation" in a neighborhood of at critical point $t = t_o$, where $h'(t_o) = 0$. The method has an old history dating back to Lord Kelvin and even to Stokes.

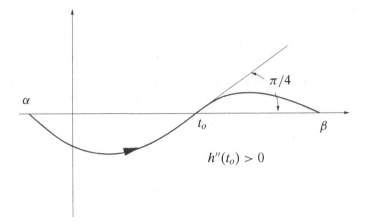

Figure 8.5. Integration path for stationary phase.

Fortunately, the method of steepest descents gives the correct answer if h can be continued analytically near t_o. Thus, we write

$$h(t) = h(t_o) + \frac{1}{2}h''(t_o)(t - t_o)^2 + \frac{1}{6}h'''(t_o)(t - t_o)^3 + \cdots .$$

On a path of steepest descent through t_o,

$$t - t_o = re^{i\gamma}, \quad i = e^{\pi i/2},$$

and

$$ih''(t_o)(t - t_o)^2 = r^2 h''(t_o)e^{2i\gamma + \pi i/2}.$$

If $h''(t_o) > 0$, because we desire $r^2 e^{2i\gamma + \pi i/2} < 0$, take $\gamma = \pi/4$. If $h''(t_o) < 0$, we still desire $r^2 e^{2i\gamma + \pi i/2} < 0$, so take $\gamma = -\pi/4$. (See Figure 8.5.) In either case then,

$$f(x) \sim \sqrt{\frac{2\pi}{x|h''(t_o)|}}g(t_o)e^{ixh(t_o)+i\gamma} + \mathcal{O}(x^{-5/2}) \quad \text{for } x \to \infty,$$

$$\gamma = \operatorname{sgn}(h''(t_o))\pi/4. \tag{8.62}$$

Extensions of the last result to cases of higher order critical points and where $h(t)$ is not analytic are carried out by (Erdélyi).

Example 8.9. **Wave motion** Lord Kelvin derived the expression

$$u(x, t) = \frac{1}{2\pi}\int_0^\infty \cos[k\{x - t\omega(k)\}]dk,$$

which expresses the effect at (x, t) of an impulsive disturbance at $(0, 0)$, when $\omega(k)$ is the velocity of propagation of two-dimensional waves in water corresponding to a wavelength $2\pi/k$.

To employ the method of stationary phase, we see that

$$u(x, t) = \frac{1}{2\pi}\Re\text{eal}\left\{\int_0^\infty e^{ik\{x - t\omega(k)\}}dk\right\}.$$

To make use of the stationary phase results, with $v = x/t$ fixed, consider

$$\int_0^\infty e^{ik\{x - t\omega(k)\}}dk = \int_0^\infty e^{itk\{v - \omega(k)\}}dk,$$

and let $h(k) = k\{v - \omega(k)\}$. We suppose that $h'(k_o) = 0$, and find that $h''(k_o) = -2\omega'(k_o) - k_0\omega''(k_o)$. Hence taking the real part find that

$$u \sim \frac{\cos\{tk_o^2\omega'(k_o) \pm \frac{1}{4}\pi\}}{\left[2\pi t \left|k_o\omega''(k_o) + 2\omega'(k_o)\right|\right]^{1/2}} \quad \text{as } t \to \infty.$$

REFERENCES

M. Abramowitz and I. A. Stegun, eds., *Handbook of Mathematical Functions*, National Bureau of Standards, 1964.

C. M. Bender and S. A. Orszag, *Advanced Mathematical Methods for Scientists and Engineers*, McGraw-Hill, New York, 1978.

N. Bleistein and R. A. Handelsman, *Asymptotic Expansion of Integrals*, Holt, Rhinehart, & Winston, New York, 1975.

G. F. Carrier, M. Krook, and C. E. Pearson, *Functions of a Complex Variable: Theory and Technique*, SIAM Publications, Philadelphia, 2005.

J. D. Cole and J. K. Kevorkian, *Perturbation Methods in Applied Mathematics*, Springer-Verlag Berlin and Heidelberg GmbH & Co., 1981.

A. Erdélyi, *Asymptotic Expansions*, Dover, New York, 1956.

M. L. Glasser, A Function Arising in One-dimensional Percolation, *SIAM Review*, Problem 95-16, **37**, 608 (1995).

M. H. Holmes, *Introduction to Perturbation Methods*, Springer-Verlag, New York, 1995.

H. Jeffreys and B. S. Jeffreys, *Methods of Mathematical Physics*, 3rd edition, Cambridge University Press, Cambridge, UK, 1953.

A. I. Markushevich, *Theory of Functions of a Complex Variable*, Vols. I, II, III, New York : Chelsea Pub. Co., 1977.

F. H. Northover, On the Method of Steepest Descents, *J. Math. Anal. Appl.* **29**, 216–232 (1970).

F. W. J. Olver, Why Steepest Descents? *SIAM Rev.* **12**, 228–247 (1970).

R. B. Paris and A. D. Wood, Stokes Phenomenon Demystified, *Bull. Inst.* Math. Appl. **31**, 21–28 (1995).

L. A. Segel, The Importance of Asymptotic Analysis in Applied Mathematics, *Am. Math. Month.* **73**, 7–14 (1966).

I. N. Sneddon and B. F. Touschek, A Note on the Calculation of Energy Levels in a Heavy Nucleus, *Proc. Camb. Phil. Soc.* **44**, 391–403 (1948).

C. B. Watkins and I. H. Herron, Laminar Shear Layer Due to a Thin Flat Plate Oscillating with Zero Mean Frequency, *J. Fluids Eng.* **100**, 367–373 (1978).

J. E. Wilkins, Jr., Conduction of Heat in an Insulated Spherical Shell with Arbitrary Spherically Symmetric Sources, *SIAM Rev.* **1**, 149–153 (1959).

J. Wojdylo, Computing the coefficients in Laplace's method, *SIAM Rev.* **48**, 76–96 (2006).

EXERCISES

8.1 Use integration by parts to determine the asymptotic behavior of the integral

$$\int_x^\infty \sin(v^3 + v)e^{-x^2 v}dv,$$

for $x \to \infty$. You can also use Watson's lemma or work from first principles. Obtain the *first two terms* in the large-x series.

8.2 The stream function of a certain vortex flow (see Section 2.1.1) is given by

$$\psi(\eta) = \int_0^\eta e^{-x} dx \int_0^x e^y P(y) dy,$$

where P is the originating pressure distribution expressed as

$$P(y) = \int_y^\infty \left(\frac{1 - e^{-z}}{z}\right)^2 dz.$$

Use integration by parts to show that

$$\psi(\eta) \sim \log \eta + A \quad \text{as } \eta \to \infty,$$

and find the value of the constant A.

8.3 The solution of a certain differential equation is known to be

$$y(x) = \int_1^\infty \frac{e^{-xt}}{(t+1)^2} dt.$$

Use integration by parts to develop several terms in the asymptotic expansion of y as $x \to 0^+$. Thus show that $y(x) \to \frac{1}{2}$, while $|y'(x)| \to \infty$ as $x \to 0^+$.

8.4 Show, by direct substitution, that the function $f(x)$ defined by

$$f(x) = \int_0^\infty \frac{e^{-xt}}{1+t^2} dt$$

satisfies the differential equation

$$f'' + f = \frac{1}{x}, \quad x > 0,$$

and that $\lim_{x \downarrow 0} f(x) = \pi/2$. Use Watson's lemma to verify that $\lim_{x \to \infty} f(x) = 0$, while determining its asymptotic expansion as $x \to \infty$.

It was shown in Chapter 2, that $y'' + y = 1/x$ has the particular solution $g(x) = \text{Ci}(x) \sin x - \text{Si}(x) \cos x$, where the functions $\text{Si}(x)$ and $\text{Ci}(x)$ are called the "sine integral function" and "cosine integral function," respectively, and are defined by

$$\text{Si}(x) = \int_0^x \frac{\sin t}{t} dt, \quad \text{Ci}(x) = -\int_x^\infty \frac{\cos t}{t} dt.$$

Draw a connection between $f(x)$ and $g(x)$.

8.5 Use Watson's lemma to find the asymptotic exapnsion of

$$I(x) = \int_0^\infty \frac{t e^{-xt}}{1+t^2} dt$$

as $x \to \infty$.

Find a differential equation satisfied by $I(x)$ and show that

$$I(x) = -\text{Ci}(x) \cos x - \text{Si}(x) \sin x + \frac{\pi}{2} \sin x.$$

8.6 A special function is defined by

$$w(x;a) = A \int_0^\infty \frac{e^{-xt}\, dt}{t^{1-a}(t+1)^a}, \quad \Re\mathrm{eal}(a) > 0.$$

Use Watson's lemma to show that

$$w(x;a) \sim A \left[\frac{\Gamma(a)}{x^a} + \frac{a\Gamma(a+1)}{x^{a+1}} + \cdots \right] \quad \text{as } x \to \infty.$$

Exercises 8.7–8.10 all involve various forms of Bessel functions and modified Bessel functions. The reader is referred to Sections 2.3.2 and 2.4.2 for background information.

8.7 The integral form of the modified Bessel function of order zero of the first kind is

$$I_0(x) = \frac{1}{\pi} \int_{-1}^{1} \frac{e^{xt}}{\sqrt{1-t^2}}\, dt,$$

which was derived in Example 5.8.

Find the asymptotic behavior of $I_0(x)$ as $x \to -\infty$ and as $x \to +\infty$. Notice that these two asymptotic expansions are different. This is an example of Stokes phenomenon, where a function will have different asymptotic expansions in different sectors of the complex plane (Paris and Wood).

8.8 The modified Bessel function of second kind (also known as the Kelvin function) can be written in terms of the following definite integral:

$$K_\nu(z) = \frac{\pi^{1/2} \left(\frac{z}{2}\right)^\nu}{\Gamma\left(\nu + \frac{1}{2}\right)} \int_0^\infty e^{-z\cosh t} \sinh^{2\nu} t\, dt.$$

Work out a two-term correct asymptotic expansion for this function for $z \to \infty$ using the methods given in Section 8.5.

8.9 An integral representation for the Bessel function of first kind is

$$J_n(z) = \frac{1}{\pi} \int_0^\pi \cos(z \sin\theta - n\theta)\, d\theta, \quad \text{for } n = \text{integer}.$$

By writing this as the real part of a complex integral,

$$J_n(z) = \frac{1}{\pi} \Re\mathrm{eal}\left\{ \int_0^\pi e^{i(z\sin\theta - n\theta)}\, d\theta \right\},$$

use the method of stationary phase to show that the large-z behavior of the function is

$$J_n(z) \sim \sqrt{\frac{2}{\pi z}} \cos\left(z - \frac{n\pi}{2} - \frac{\pi}{4}\right), \quad \text{for } z \to \infty.$$

(The critical point is located at $\theta = \pi/2$.)

8.10 An integral representation for the Bessel function of the second kind is

$$Y_n(z) = \frac{1}{\pi} \int_0^\pi \sin(z\sin\theta - n\theta)\, d\theta - \frac{1}{\pi} \int_0^\infty \left(e^{nt} + (-1)^n e^{-nt}\right) e^{-z\sinh t}\, dt, \quad (8.63)$$

for $n =$ integer. Show that

$$Y_n(z) \sim \sqrt{\frac{2}{\pi z}} \sin\left(z - \frac{n\pi}{2} - \frac{\pi}{4}\right), \quad \text{for } z \to \infty.$$

8.11 Consider the following integral

$$f(x) = \int_0^\infty \frac{1}{t^{1/3}} e^{-xt^4 + 2t^2} dt.$$

Give a one-term approximation to the asymptotic behavior for small x. *Hint*: You may want to make a change of variable in the form $t = u/\sqrt{x}$.

8.12 The function $\phi(x, y, t)$ is a linearized water wave potential

$$\phi(x, y, t) = \frac{1}{\pi} \int_0^\infty \int_{-\infty}^\infty \eta(\xi) e^{\omega y} \cos(t\sqrt{g\omega}) \cos(\omega(x - \xi)) d\xi \, d\omega,$$

satisfying

$$\phi_{xx} + \phi_{yy} = 0, \ |x| < \infty, \ y < 0, \ t > 0,$$

together with the boundary condition,

$$g\phi_y(x, 0, t) + \phi_{tt}(x, 0, t) = 0,$$

and the initial conditions,

$$\phi(x, 0, 0) = \eta(x), \ \phi_t(x, 0, 0) = 0.$$

Consider water waves where the pressure is constant except for an impulsive force so that $\eta(x) = -\delta(x)/\rho$ is the Dirac distribution, and ρ is the density. Given that the surface height is derived from the potential by the formula

$$\zeta(x, t) = -\left.\frac{\phi_t}{g}\right|_{y=0},$$

show that

$$\zeta(x, t) = -\frac{1}{\rho\pi\sqrt{g}} \int_0^\infty \sqrt{\omega} \sin(t\sqrt{g\omega}) \cos(\omega x) d\omega.$$

In this expression for ζ, let $v = x/t$, $x > 0$, $t > 0$, and, hence, show that

$$\zeta(x, t) = -\frac{1}{\rho\pi\sqrt{g}} \Im\text{mag} \int_0^\infty \sqrt{\omega} \left(e^{it(\sqrt{g\omega} + v\omega)} + e^{it(\sqrt{g\omega} - v\omega)}\right) d\omega.$$

As $t \to \infty$, we have for fixed $v > 0$, an integral amenable to the *method of stationary phase*. Show that there is only one point of stationary phase, occurring in the second term, giving wave propagation to the right. Obtain the asymptotic form of ζ.

8.13 In a paper by (Sneddon and Touschek) the Laplace inversion integral

$$\rho(Q) = \frac{1}{2\pi i} \int_{\gamma - i\infty}^{\gamma + i\infty} e^{\left(Qz + \frac{\alpha}{z}\right)} dz,$$

γ, α are real and positive. Use the *method of steepest descents* to show that for large values of Q

$$\rho(Q) \sim \frac{\alpha^{1/4}}{2\sqrt{\pi}\,Q^{3/4}} e^{2\sqrt{\alpha Q}} \left(1 - \frac{3}{16}\frac{1}{\sqrt{\alpha Q}} + \cdots\right).$$

8.14 In Section 7.5.1, we encountered the integral solution to the differential equation $y''(x) + xy'(x) - y(x) = 0$, in $0 < x < \infty$, satisfying $y(0) = 1$, $y(\infty) = 0$:

$$y(x) = -\frac{1}{\sqrt{2\pi}} \int_{C_1} \frac{e^{ipx - p^2/2}\,dp}{p^2}.$$

The contour is given in the sketch in Figure 8.6. Use steepest descents to show that,

$$y \sim \frac{1}{x^2}e^{-x^2/2}, \quad \text{for } x \to \infty.$$

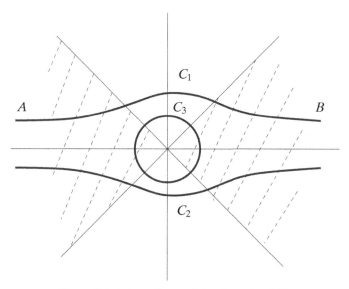

Figure 8.6. Integration path for Exercise 8.14.

8.15 In Chapter 7, a certain differential equation is shown to have the solution

$$y(x) = \int_0^\infty v e^{-x/v - v^2/2}\,dv. \tag{8.64}$$

Obtain the large-x behavior,

$$y \sim \sqrt{\frac{2\pi}{3}}x^{1/3}e^{-3x^{2/3}/2}, \quad x \to \infty.$$

8.16 Consider the function

$$G(t) = \int_0^\infty e^{-t/x} e^{-x^2/2}\,dx,$$

which arises in hopping transport for one-dimensional percolation (Glasser). Determine its asymptotic behavior for small t and as $t \to \infty$.

Hints: For t large, use *Laplace's method*. For small t, note that $G(0)$ is finite, but $G'(0)$ is not. Find a linear third-order ODE satisfied by $G(t)$. Frobenius's method gives a complete answer. You might also be able to integrate by parts in the original integral.

8.17 (a) Use integration by parts to prove the asymptotic equivalence

$$\int_0^\pi t^n \sin t \, dt \sim \frac{\pi^{n+2}}{n^2}, \quad n \to \infty.$$

What is the order of the error?

(b) Derive the same result by a change of variable and Watson's lemma.

8.18 In the paper by (Watkins and Herron) there was encountered the integral:

$$I(y, t) = \int_0^t \frac{(\cos(t - \tau) + \sin(t - \tau))e^{-y^2/4\tau}}{\sqrt{\tau}} d\tau.$$

Use Watson's lemma to show that for $t > 0$ as $y \to \infty$

$$I \sim \frac{4t^{3/2} e^{-y^2/4t}}{y^2}.$$

8.19 Using Watson's lemma, obtain a two term approximation to the following integral for $x \to \infty$

$$I = \int_0^\infty \log(\sqrt{1 + t^p}) \sin(t^n) e^{-xt} dt,$$

where p and n are parameters; $p > 0$, and $p + n > -1$.

8.20 By a simple change of variable and Watson's lemma find the asymptotic expansion of (8.38) in Section 8.6.1 as $x \to \infty$, with s fixed, as is given by (8.40).

8.21 Use the method of generalized Laplace integrals (Section 7.5) to show that

$$xy''' - y = 0$$

has the solution

$$y(x) = \sqrt{\frac{2}{\pi}} \int_0^\infty \sin\left(\frac{x}{v}\right) e^{-\frac{1}{2}v^2} v \, dv,$$

where

$$y(0) = 0, \quad y'(0) = 1, \quad \lim_{x \to \infty} y = 0.$$

Determine the leading asymptotic behavior of y as $x \to \infty$, using the method of steepest descents.

8.22 Use Watson's lemma of Section 8.4 to derive the following variant (Jeffreys and Jeffreys). Consider the integral

$$I = \int_{-a}^b e^{-\frac{1}{2}\xi^2 z^2} f(z) dz,$$

where $a > 0$, $b > 0$ given that $f(z)$ has a Maclaurin's series expansion:

$$f(z) = c_0 + c_1 z + \cdots + c_{n-1} z^{n-1} + R_n(z).$$

Show that

$$I \sim \sqrt{2\pi}\left(\frac{c_0}{\xi} + \frac{c_2}{\xi^3} + 1\cdot 3\frac{c_4}{\xi^5} + \cdots + [1\cdot 3\cdots(2n-1)]\frac{c_{2n}}{\xi^{2n+1}}\right) \quad \text{as } \xi \to \infty. \quad (8.65)$$

Hint: Let $z = t^{1/2}$ for $z > 0$ and $z = -t^{1/2}$ for $z < 0$, obtain two different integrals between which certain terms cancel. You will notice that the solution of Example 8.4 of Section 8.4 has this property.

8.23 In Section 7.6, Exercise 8.8, we derived the solution for θ_s given in terms of the Fourier inversion integral,

$$\theta_s = \frac{e^{-it}}{2\pi}\int_{-\infty}^{\infty} F(k)e^{rh(k)}dk, \quad h(k) = ik\cos\phi - \sqrt{k^2 - i}\sin\phi, \quad (8.66)$$

We seek the behavior of the solution for large r. Using steepest descent on the integral (8.66), obtain the behavior of θ_s for $r \to \infty$. Note that the k-plane has two branch cuts, best chosen as indicated in Figure 8.7.

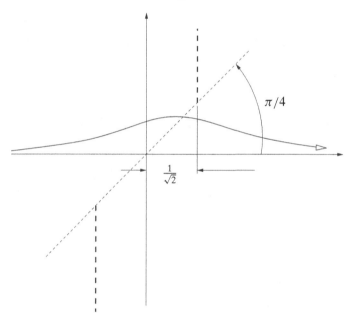

Figure 8.7. Integration path for Exercise 8.23.

(a) Show that there is a critical point at $k_o = \exp(i\pi/4)\cos\phi$. Note that it lies on the dotted line in the figure.

(b) Expand h in the usual Taylor series near k_o, and show that as the steepest path goes through the critical point, it goes through at an angle $-\pi/8$, so the path is like that shown in the sketch on the figure. Finally, show that the asymptotic behavior is

$$\theta_s \sim \frac{F(k_o)}{\sqrt{2\pi r}}e^{-r/\sqrt{2}}\sin\phi\, e^{i(r/\sqrt{2}-t-\pi/8)}, \quad r \to \infty.$$

Obviously, one can take the real part to give the answer for the temperature once F is determined for a given forcing.

8.24 In Exercise 8.5 in Section 7.1, we obtained the expression for that part of the reflected wave field due to a particular bump on the wall. The expression for ϕ'_R given at the end of the problem can be put into the form:

$$\phi'_R = \frac{i\alpha}{\pi} e^{-i\alpha at} \int_C \frac{e^{ikx - y\sqrt{k^2 - \alpha^2}}}{\sqrt{k^2 - \alpha^2}} dk,$$

with the integration path and branch points shown in Figure 8.8. (You do *not* need to show this!)

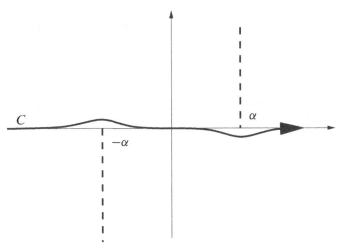

Figure 8.8. Integration path for Exercise 8.24.

(a) Show that the *only* critical point on this path is located at $k = \alpha \cos \psi$, where ψ is a polar coordinate in the x–y plane.

(b) Use the method of steepest descent to show that for large r (where r is a polar coordinate in the x–y plane), ϕ_R is given by

$$\phi_R \sim \sqrt{\frac{\alpha}{\pi r}} \cos \theta \cos \left(\alpha(r - at) - \frac{\pi}{4} \right).$$

(Recall that $\phi_R = \Re\text{eal}(\phi'_R)$.)

8.25 Consider the initial-value problem,

$$\frac{\partial u}{\partial t} + \frac{1}{4} \frac{\partial^4 u}{\partial x^4} = 0, \quad -\infty < x < \infty,$$

$$u = e^{-\beta|x|}, \text{ at } t = 0.$$

(a) Show that the Fourier transform of the solution is given by

$$\mathcal{F}\{u\} = \frac{2\beta}{k^2 + \beta^2} e^{-k^4 t/4}.$$

(b) Use Watson's lemma to determine the behavior of u for long times – but x *not*-large! Obtain the first *two* terms in the asymptotic series. (Actually, x^4/t is small.) Take β to be $O(1)$.

(c) Use the method of steepest descent to determine the behavior of u for large x – actually, it is x^4/t that is large in this case. ONE term is sufficient in this case; again, take β to be $O(1)$. (*Note*: It can be shown that the solution, u, is bounded for all values of x and t; your choice of critical points should give results consistent with that fact.)

8.26 Equation (5.20) gives the forced response of a fluid to an oscillatory motion of a bounding wall. We seek a large-y solution using steepest descent.

(a) Show that writing $s = (y^2/vt^2)\sigma$ gives

$$u_{\text{forced}} = \frac{u_1}{2\pi i} \int_\Gamma \frac{\sigma e^{\eta h(\sigma)}}{\sigma^2 + \frac{\omega^2 t^4 v^2}{y^4}} d\sigma \quad \eta \equiv \frac{y^2}{vt}.$$

What is $h(\sigma)$?

(b) Show that the critical point is located in the complex σ plane at $\sigma = 1/4$. What direction must the path take through the critical point in order that it be a local maximum?

(c) Show that, for $\eta \to \infty$,

$$u_{\text{forced}} \sim \frac{8}{1 + 16\frac{\omega^2 t^4 v^2}{y^4}} \frac{1}{\sqrt{\pi \eta}} e^{-\eta/4}.$$

Note the parameter in the denominator $\omega vt^2/y^2 = (\omega t)/\eta$. Do you believe that this asymptotic result is valid at large η for all time? Discuss what you might do to correct what has been done.

8.27 A plane sound wave propagates toward a circular cylinder that scatters the radiation. It turns out that the complex dimensionless amplitude of the pressure, averaged around the surface, is given by

$$P = \frac{2e^{i\omega t}}{\lambda a} \frac{1}{H_1^{(2)}(\lambda a)},$$

with $\lambda = \omega/c$, where ω is the temporal frequency of the wave, and c is the sound speed. We want to obtain an expression for the response to short waves – that is, for $\lambda \to \infty$. To do that, clearly we need the large-argument asymptotic approximation to the Hankel function, $H_1^{(2)}(z)$. It can be represented by

$$H_1^{(2)}(z) = -\frac{1}{i\pi} \int_{-\infty}^{\infty - i\pi} e^{z \sinh t - t} dt.$$

(This integration path is sketched in Figure 8.9; we discussed this integration path briefly in Chapter 2 as well.)

Use the method of steepest descent to obtain the result

$$H_1^{(2)}(z) \sim \sqrt{\frac{2}{\pi z}} e^{-iz + i\frac{3\pi}{4}}, \quad z \to \infty.$$

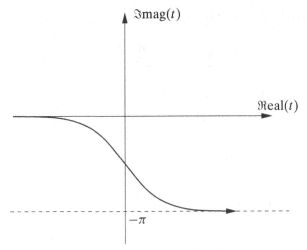

Figure 8.9. Integration path for Exercise 8.27.

8.28 Use steepest descent to obtain the large-x behavior of $G(x)$, where

$$G(x) = \Re\text{eal}\left\{ \int_0^\infty t e^{x(it^2 - 4\sqrt{t})}\, dt \right\}.$$

Show that

$$G \sim \sqrt{\frac{2\pi}{3x}}\, e^{-3\sqrt{3}x/2} \cos\left(\frac{3x}{2} - \frac{\pi}{12} \right) \quad \text{as } x \to \infty.$$

8.29 Given that

$$f(x) = \sqrt{\frac{2}{\pi}} \int_0^\infty \sin \omega x\, e^{-\omega^3/3}\, d\omega.$$

Use the method of steepest descent to show that as $x \to \infty$,

$$f(x) \sim \sqrt{2x}\, e^{-\frac{\sqrt{2}}{3} x^{3/2}} \left[\sin\left(\frac{\sqrt{2}}{3} x^{3/2} - \frac{\pi}{8} \right) - \frac{5}{48 x^{3/2}} \sin\left(\frac{\sqrt{2}}{3} x^{3/2} + \frac{\pi}{8} \right) + \mathcal{O}\left(\frac{1}{x^3} \right) \right].$$

$$(8.67)$$

8.30 Use the ideas of Section 8.4 to derive the following version of Watson's lemma. Suppose that

$$f(x) = \int_0^T e^{-xt} t^\lambda g(t)\, dt.$$

Require $\Re\text{eal}(\lambda) > -1$, to ensure convergence at $t = 0$. Suppose that the integral g has an asymptotic expansion near $t = 0$,

$$g(t) = \sum_{n=0}^N a_n t^n + R_N(t), \quad 0 < t < A,$$

and is exponentially bounded $|g(t)| \le M e^{ct}$ for $A \le t \le T$, then

$$f(x) \sim \sum_{n=0}^{N} \frac{a_n \Gamma(\lambda + n + 1)}{x^{\lambda+n+1}} \quad \text{as } x \to \infty.$$

Note that $T = \infty$ is allowed in this analysis.

8.31 Show by a certain change of variables in the integral given in Exercise 8.8 that for $\Re\text{eal}(v) > -1/2$,

$$K_v(z) = \frac{\pi^{1/2} \left(\frac{z}{2}\right)^v}{\Gamma\left(v + \frac{1}{2}\right)} \int_1^{\infty} e^{-zu}(u^2 - 1)^{v-\frac{1}{2}} du, \quad z \text{ real}.$$

Use the version of Watson's lemma in Exercise 8.30 to show that

$$K_v(z) \sim e^{-z} \sqrt{\frac{\pi}{2z}} \sum_{n=0}^{\infty} \frac{\Gamma\left(v + \frac{1}{2} + n\right)}{\Gamma\left(v - \frac{1}{2} - n\right)} \frac{(2z)^{-n}}{n!} \quad \text{as } z \to \infty.$$

8.32 Consider the integral defined by

$$I(x) = \int_0^{\infty} e^{-xt(t+1)^2} dt, \quad x \text{ real}.$$

Since the exponent is strictly monotonic on $[0, \infty)$, it has a monotonic inverse. Use a suitable change of variable and Watson's lemma to show that

$$I(x) \sim \sum_{n=1}^{\infty} \frac{(-1)^{n-1}(3n - 2)!}{(2n - 1)! x^n} \quad \text{as } x \to \infty.$$

Index

Printed in the United States
By Bookmasters